Introduction to
Heterogeneous Catalysis

Second Edition

Advanced Textbooks in Chemistry

Print ISSN: 2059-7673
Online ISSN: 2059-7681

Published

Introduction to Heterogeneous Catalysis
(Second Edition)
by Roel Prins (Institute for Chemical and Bioengineering, ETH Zurich, Switzerland), Anjie Wang (Dalian University of Technology, China), Xiang Li (Tianjin University of Science and Technology, China) and Foteini Sapountzi (Syngaschem BV, The Netherlands)

Introduction to Heterogeneous Catalysis
by Roel Prins (Institute for Chemical and Bioengineering, ETH Zurich, Switzerland), Anjie Wang (Dalian University of Technology, China), and Xiang Li (Dalian University of Technology, China)

Advanced Textbooks in Chemistry

Introduction to **Heterogeneous Catalysis**

Second Edition

Roel Prins

Institute for Chemical and Bioengineering, ETH Zurich, Switzerland

Anjie Wang

Dalian University of Technology, China

Xiang Li

Tianjin University of Science and Technology, China

Foteini Sapountzi

Syngaschem BV, The Netherlands

NEW JERSEY • LONDON • SINGAPORE • BEIJING • SHANGHAI • HONG KONG • TAIPEI • CHENNAI • TOKYO

Published by

World Scientific Publishing Europe Ltd.
57 Shelton Street, Covent Garden, London WC2H 9HE
Head office: 5 Toh Tuck Link, Singapore 596224
USA office: 27 Warren Street, Suite 401-402, Hackensack, NJ 07601

Library of Congress Cataloging-in-Publication Data
Names: Prins, Roelof, author. | Wang, Anjie (Professor of chemistry), author. | Li, Xiang (Professor of chemistry), author. | Sapountzi, Foteini, author.
Title: Introduction to heterogeneous catalysis / Roel Prins, Institute for Chemical and Bioengineering, ETH Zurich, Switzerland, Anjie Wang, Dalian University of Technology, China, Xiang Li, Tianjin University of Science and Technology, China, Foteini Sapountzi, Syngaschem BV, The Netherlands.
Description: Second edition. | New Jersey : World Scientific, [2022] | Series: Advanced textbooks in chemistry, 2059-7673 | Includes bibliographical references and index.
Identifiers: LCCN 2021053081 | ISBN 9781800611504 (hardcover) | ISBN 9781800611610 (paperback) | ISBN 9781800611511 (ebook for institutions) | ISBN 9781800611528 (ebook for individuals)
Subjects: LCSH: Heterogeneous catalysis--Textbooks. | Catalysis--Textbooks. | Chemistry--Textbooks.
Classification: LCC QD505 .P744 2023 | DDC 541/.395--dc23/eng/20211208
LC record available at https://lccn.loc.gov/2021053081

British Library Cataloguing-in-Publication Data
A catalogue record for this book is available from the British Library.

For any available supplementary material, please visit
https://www.worldscientific.com/worldscibooks/10.1142/Q0340#t=suppl

Desk Editors: Jayanthi Muthuswamy/Michael Beale/Shi Ying Koe

Typeset by Stallion Press
Email: enquiries@stallionpress.com

Preface

Catalysis is essential for the production of energy carriers from oil, gas and coal as well as for the production of many industrially important chemicals. Chemical engineering faculties honour this importance by including a catalysis course, usually in the master's programme. Catalysis builds upon the basic undergraduate courses: Kinetics and thermodynamics are the basis of catalysis, inorganic and physical chemistry provide the basics of the preparation and characterisation of the catalysts and organic chemistry explains the organic reactions in catalysis. This book gives the basics of catalysis (adsorption and its influence on kinetics), the chemistry and process engineering of catalytic reactions and the main types of catalysts. From the beginning, the chemical, materials and engineering aspects of catalysis are illustrated by examples from the refinery, energy and basic chemicals industries that highlight the relevance of catalysis.

Catalysis is a multidisciplinary branch of science and technology, dealing with engineering as well as chemical and materials aspects. Catalytic processes are operated by chemical engineers, the processes and catalysts are developed by chemists and chemical engineers, while chemists and material scientists focus on catalytic materials and supports. Scientists, at all stages of catalysis, from the laboratory to the process, must have the basic knowledge outlined above. To achieve this for all scientists is a challenge. In several countries, chemical engineering students have only a limited knowledge of chemistry, while chemists and physicists have only a limited knowledge of engineering. An introduction to catalysis should enable students from various disciplines to reach the same level.

This book is intended for use in a one-semester course for senior undergraduate, master's and graduate students in chemical engineering, chemistry and materials science, who have no prior knowledge of catalysis but do have a basic knowledge of chemistry, physics and chemical process engineering. Its aim is to enable students and scientists in industry to understand the essence of catalysis and catalysis literature. The book deals with the chemistry of catalytic reactions and catalysts in refineries and in the energy and basic-chemicals industries.

The catalysts dealt with in this book are those used in industry. Metals, solid acids, transition-metal oxides, transition-metal sulfides and bases are used as solid catalysts, usually as small active particles dispersed on the surface of a support. These heterogeneous catalysts find much greater application in industry than transition-metal complexes and liquid acids, which are used in homogeneous catalytic processes, in which the catalyst, the reactants and the products are dissolved in the same liquid. The focus is mainly on heterogeneous catalysis because of its importance in industry. There are examples of homogeneous catalytic processes and explanations of basic principles. Enzymatic catalysis is not dealt with, because it is too broad a topic for the scope of this book.

The selection of reactions to include in an introductory textbook on catalysis depends on the interests of the person teaching the course. Satisfying the interests of many teachers would result in a voluminous textbook covering a large range of topics, more than students can cover in one semester. Instead, this book aims to be user-friendly, focusing on the main topics relevant to catalysis. If the choice of reactions were based on the size of catalytic processes, topics could be limited to refinery processes such as hydrotreating, catalytic cracking and catalytic reforming, with a discussion of only metal and solid acid catalysts. Other energy-related processes in the coal and gas industry, such as coal hydrogenation, the production of synthesis gas from natural gas as well as the use of synthesis gas, would widen the scope but the types of catalysts would be more or less the same. The production of base chemicals is based on transition-metal oxides as catalysts, and the principles of these catalysts differ from those of metal and acid catalysis. By focusing on metal, solid acid and transition-metal oxide catalysts in the energy and base-chemical industry, the main types of industrial catalysts and processes are dealt with, as well as the main principles of catalysis.

To produce and characterise catalysts, a knowledge of material science, crystallography and physical and spectroscopic characterisation techniques is required. Knowledge of chemical bonding theory is required because theoretical calculations play an important role in science and catalysis. Materials science, spectroscopy and chemical bonding are less familiar to students of chemical engineering. In the first edition of our book, we discussed these topics in separate chapters, preceding the chapters on catalysis. Based on our experience in teaching parts of the book in courses over the past years, in this second edition we have integrated materials science, spectroscopy and chemical bonding aspects directly with their applications. Catalyst characterisation methods are now included in Chapter 2, Catalyst Preparation and Characterisation, and zeolite synthesis and characterisation are integrated with its applications in Chapter 7, Catalysis by Solid Acids. For chemists and physicists, who are less familiar with process and kinetic aspects, Chapter 4 includes information on these topics. Because the main catalysts used in industry and academia are metals and acids, more emphasis is put on these two classes of catalysts, but metal sulfides and oxides and their applications are discussed as well.

The book presents the basics of catalysis and illustrates them as much as possible with examples from recent literature. This also serves as an introduction for students to literature. Since the writing of the first edition of this book several interesting new articles have appeared and we have taken the opportunity of the writing of this second edition to include them in the book.

Three new topics have been added to the book. The removal of contaminants like S, N, O and metals from fossil fuels by hydrotreating (Chapter 8), a very important refinery process for the production of clean fuels, offers the opportunity to discuss metal sulfide catalysts. Electrocatalysis (Chapter 10) has the promise to minimise the anthropogenic CO_2 emissions. Solar, wind and hydroelectricity can drive water electrolysis and CO_2 electroreduction and, therefore, excess renewable electricity can be stored in chemicals. Sustainability has not been added as a separate chapter, but is discussed in several chapters. Thus, scenarios for changing the energy source in refineries and in chemical plants by moving away from the classic production of (grey) hydrogen to blue, pink and green forms are discussed in Chapter 1. Platform chemicals and sustainability of chemical feed stocks is discussed in Chapter 9.

Each chapter ends with a list of references and questions. The references were selected for clarity and relevance to catalytic reactions. The questions give the reader the opportunity to test his or her comprehension of the presented material; the answers are given at the end of the book.

We thank the students who attended courses based on parts of this book at the ETH Zurich, Switzerland, the Dalian University of Technology, the Petrochina Petrochemical Research Institute in Beijing, the China University of Petroleum in Beijing, the China University of Petroleum in Qingdao and Tianjin University of Science and Technology, China for helpful comments.

Roel Prins
Anjie Wang
Xiang Li
Foteini Sapountzi
Alkmaar, 6 December 2021

About the Authors

Roel Prins (1941) obtained his Ph.D. at the University of Amsterdam and then worked in the Shell laboratories in Amsterdam and Emeryville, USA. He was professor at the Eindhoven University of Technology, The Netherlands, and the Institute for Chemical and Bioengineering, ETH in Zurich, Switzerland, and guest professor at the Technical University in Delft, the Pierre et Marie Curie University in Paris, the Dalian University of Technology (DUT) and the University of Petroleum in Beijing and Qingdao in China. He was foreign member of the Royal Dutch Academy of Sciences, president of the Federation of European Catalysis Societies, dean of the Department of Chemistry at the ETH and editor-in-chief of the *Journal of Catalysis*. He was Schuit lecturer at the University of Delaware, USA, Ipatieff lecturer at Northwestern University, USA, and received the Distinguished Research Award of the Division of Petroleum Chemistry of the American Chemical Society.

Anjie Wang (1965) obtained his Ph.D. in chemical engineering at Dalian University of Technology (DUT). He is professor in catalysis in the School of Chemical Engineering at DUT and spent a sabbatical year with Professor Toshiaki Kabe at the Tokyo University of Agriculture and Technology, Japan. He worked with Professor R. Prins at the ETH in Zurich, Switzerland, and with Professor Chunshan Song at the Pennsylvania State University, USA, as a visiting professor and with Professors Johannes Lercher and Donald M. Camaioni at the Pacific Northwest National Laboratory, USA. He is a committee member of the Chinese Catalysis Society, director of the Liaoning Key Laboratory of Petrochemical Engineering and Equipment and director of the Centre for Microchannel Reaction Process R&D at DUT. His research interests include hydrotreating catalysis, oxidative desulfurisation, synthesis of novel porous materials, selective hydrogenation, plasma catalysis and microchannel reactors.

Xiang Li (1973) obtained his Ph.D. in industrial catalysis in 2004 at Dalian University of Technology (DUT), China. He spent one year with Professor R. Prins at the ETH in Zurich, Switzerland, and was associate professor at DUT. Now he is a Haihe Distinguished Professor at Tianjin University of Science and Technology and is also a young member of the editorial board of the *Chinese Journal of Catalysis*. His research interests are hydrotreating (hydrodesulfurisation, hydrodenitrogenation and hydrodeoxygenation), hydrogenation and the synthesis of mesoporous and micro-mesoporous materials.

Foteini Sapountzi (1983) obtained her Ph.D. in chemical engineering at the University of Patras, Greece, under the supervision of Prof. C. G. Vayenas. She continued working at the same group and as a visiting lecturer in the Department of Materials Science at University of Patras. She was a postdoctoral researcher with Prof. J. L. Valverde in the University of Castilla La Mancha, Spain, and with Dr. P. Vernoux in the Institute for Research on Catalysis and Environment of Lyon, IRCELYON/France. Since 2015 she has been working as a senior research scientist at Syngaschem BV, The Netherlands, in the group of Prof. J. W. Niemantsverdriet. Since 2021 she has also been working part-time at the Dutch Institute For Fundamental Energy Research, DIFFER/Netherlands with Prof. M. C. M. van de Sanden. Her research is focused on catalytic and electrocatalytic processes, mainly in the framework of electrochemical promotion of catalysis and of fuel cell/water electrolysers.

Contents

Chapter 1

Introduction

1.1 Catalysis and Catalysts

As early as the 18th century it was known that some chemical reactions are accelerated by the addition of compounds and that these compounds can be recovered unchanged after reaction. Some reactions occur only in the presence of such compounds. In 1835, Berzelius introduced the term **catalysis** to indicate the action of a compound (the **catalyst**) that increases the rate of a reaction but which does not change during the reaction. Today we know that a catalyst does change when it catalyses a reaction but then changes back to its original form. Therefore, it is possible to reuse a catalyst over and over again. The following example of the oxidation of SO_2 to SO_3 by a vanadium oxide catalyst illustrates the catalytic cycle.

$$\begin{array}{c} SO_2 + 2\,V^{5+} + O^{2-} \rightarrow SO_3 + 2\,V^{4+} \\ 2\,V^{4+} + {}^1\!/_2\,O_2 \rightarrow 2\,V^{5+} + O^{2-} \\ \hline SO_2 + {}^1\!/_2\,O_2 \rightarrow SO_3 \end{array} \qquad \Delta H = -90\,\mathrm{kJ{\cdot}mol^{-1}}.$$

SO_2 oxidation proceeds by means of a **Mars–van Krevelen mechanism**, by which oxygen does not react directly with the reactant. Instead, the reactant is oxidised by a metal oxide (consuming lattice oxygen) and the reduced metal oxide is re-oxidised by oxygen. Thus, in SO_3 synthesis, SO_2 is first oxidised by V^{5+} to SO_3 and, in a second step, V^{4+} is oxidised back to V^{5+} by oxygen (catalyst regeneration). The first step is an oxidation of SO_2 and a reduction of V^{5+}; in the second step, oxygen is reduced and the catalyst is re-oxidised. Thus, after one full catalytic cycle, it seems that the catalyst has not changed. Schematically this is indicated by a cycle with the catalyst species on the cycle, with reactants entering the cycle

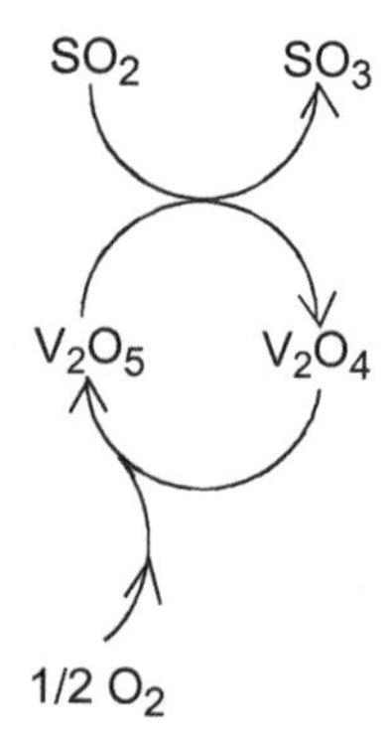

Fig. 1.1 Catalytic cycle for the oxidation of SO_2 to SO_3 over a vanadium oxide catalyst.

and products leaving it (Fig. 1.1), showing that the catalyst can be used an "infinite" number of times.

The oxidation of SO_2 to SO_3 is an important step in the production of sulfuric acid. After burning sulfur to sulfur dioxide and oxidising the SO_2 with oxygen to SO_3, with the help of a vanadium oxide catalyst, the resulting SO_3 is absorbed into 97–98% H_2SO_4 to form oleum ($H_2S_2O_7$). The oleum is then diluted with water to form concentrated sulfuric acid. Sulfuric acid is the number two base chemical with a worldwide production of 260 MT (million tons) in 2018. It is used mainly ($\sim$60%) in the production of ammonium phosphate and ammonium sulfate fertilisers. It is also used to produce chemicals such as hydrochloric acid, nitric acid, sulfate salts, synthetic detergents, dyes, pigments, explosives and pharmaceuticals. This example shows how important catalysis is in chemical reactions. To be able to perform a reaction, or to perform it under milder conditions than usual, is of enormous importance. It should, therefore, be no surprise that about 90% of all chemicals are made by means of catalysed reactions.

Although Berzelius stated that a catalyst does not change during a catalysed reaction and, thus, could be used ad infinitum, in reality things are not that simple. Most catalysts are solids and enable reactions between molecules on their surface where they occur with greater ease than in the gas or liquid phase. Solid catalysts often become progressively covered with coke, with the result that less of the catalyst surface is accessible to the reacting molecules and that the activity of the catalyst decreases. When the deactivation of the catalyst is severe, then it is possible to **regenerate** the catalyst, for instance by burning off the coke in air and re-reducing the catalyst if required. Another reason for the loss of catalytic activity may be **sintering**, i.e. growth of the particles that make up the catalyst. Small particles have a large surface area and, thus, high catalytic activity. Unfortunately, that also means high surface energy, a thermodynamically unstable situation. At elevated reaction temperature particle growth may take place by diffusion and combination of the particles or by scission of atoms from the particles and diffusion to other particles. In either case, there is a decrease in the catalyst activity, not because of a change in the catalyst structure but because of a loss of accessible surface area. The catalytic reactions are still the same, but the reaction rate has decreased. A better description of a catalyst than given by Berzelius was given by Ostwald (Nobel Prize in Chemistry, 1909): a catalyst changes the rate but not the chemical equilibrium of a reaction.

J. J. Berzelius F. W. Ostwald

1.2 Heterogeneous and Homogeneous Catalysis

In **heterogeneous catalysis**, the catalyst is present as a solid phase and the reactants and products are in the gas or liquid phase. Whereas gas–solid heterogeneous catalysis is often used in the refinery and in the base chemicals industry, liquid–solid heterogeneous catalysis is common in the fine chemical industry, where large batch reactors are filled with an organic liquid and a solid catalyst. The V_2O_5 catalyst for SO_2 oxidation (Section 1.1) is a solid catalyst used to convert gaseous SO_2 to SO_3. In heterogeneous catalysis, the molecules can react only on the surface of the solid catalyst; thus, the goal is to maximise the catalyst surface by means of small catalyst particles, because the specific surface area (the surface area per mass unit) is proportional to $1/R$ (surface area is proportional to R^2 and volume to R^3), where R is the radius of the particles. There is a limit to the size of the catalyst particles, because the space between small particles is very small, which hinders the flow of the gas or liquid around the catalyst particles (diffusion problems). Furthermore, the smaller the particles, the greater their tendency to sinter to larger particles. To avoid problems with diffusion and sintering, catalyst particles are often put on the surface of other materials, i.e. **supports**. The particles of the support material are larger than the catalyst particles. This leads to fewer problems with diffusion, and the catalyst particles can be placed far enough apart on the support surface so that they do not come into contact with each other and do not sinter. Therefore, the SO_2 oxidation catalyst V_2O_5/SiO_2 consists of V_2O_5 particles supported on silica. Catalysts and supports will be a topic in Chapter 2.

In **homogeneous catalysis**, catalyst, reactants and products are in the same liquid phase. For instance, acetic acid is produced from methanol and CO in a homogeneous catalytic process that has the largest worldwide production of all homogeneous processes (**14 MT in 2018**). Acetic acid is used in the production of vinyl acetate (and, thus, polyvinyl acetate) and acetic anhydride. It is also used as a solvent in the production of terephthalic acid (to produce polyethylene terephthalic acid, PET). In the homogeneous production of acetic acid, methyl iodide is carbonylated to acetyl iodide ($CH_3I + CO \rightarrow CH_3COI$) with a soluble Rh or Ir carbonyl complex as the catalyst. The catalytic cycle begins with the oxidative addition of methyl iodide to the square planar $RhI_2(CO)_2^-$ complex (with Rh in the Rh(I) oxidation state) to form an octahedral $[Rh(CO)_2(CH_3)I_3]^-$ species (with Rh in the Rh(III) oxidation state):

$$CH_3I + RhI_2(CO)_2^- \rightarrow CH_3RhI_3(CO)_2^-.$$

Methyl migration to a CO ligand forms a five-coordinated square-pyramidal Rh species with an acetyl ligand

$$CH_3RhI_3(CO)_2^- + CO \rightarrow (CH_3CO)RhI_3(CO)_2^-,$$

and reductive elimination of acetyl iodide regenerates the catalytic complex $RhI_2(CO)_2^-$

$$(CH_3CO)RhI_3(CO)_2^- \rightarrow RhI_2(CO)_2^- + CH_3COI.$$

Figure 1.2 shows the catalytic cycle from CH_3I to CH_3COI, with the catalytic complex in the outer cycle and the reactants CH_3I and CO entering the cycle and the product CH_3COI leaving the cycle. The cycle is coupled to the formation of CH_3I from methanol ($CH_3OH + HI \rightleftharpoons CH_3I + H_2O$) and to the hydrolysis of acetyl iodide ($CH_3COI + H_2O \rightleftharpoons CH_3COOH + HI$). In this way, the overall reaction is $CH_3OH + CO \rightarrow$

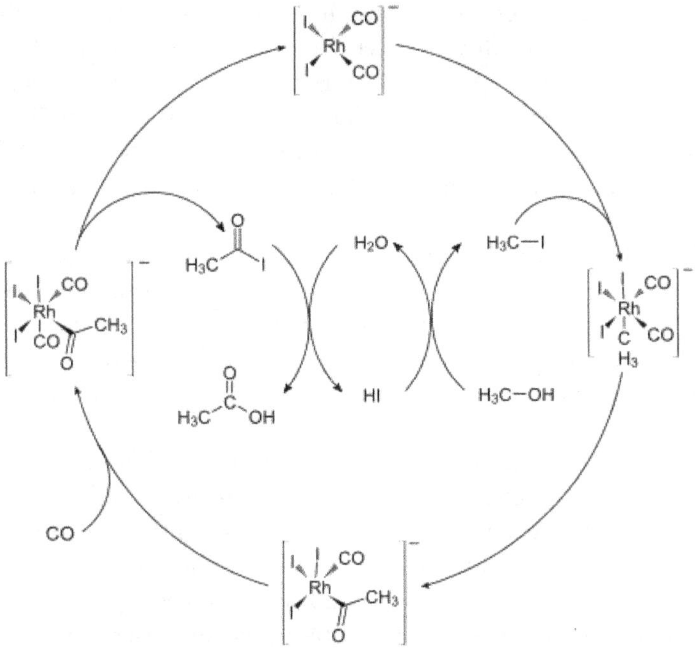

Fig. 1.2 Catalytic cycle for the carbonylation of methanol to acetic acid with a $RhI_2(CO)_2^-$ complex. Reprinted with permission from [1.1]. Copyright (2000) Elsevier.

CH_3COOH. The resulting HI converts, in turn, the reactant methanol to methyl iodide, used in the first step of the catalytic cycle.

Two processes are used industrially for the homogeneous carbonylation of methanol (Fig. 1.2). In the Monsanto process, the $RhI_2(CO)_2^-$ complex functions as a catalyst, while in the Cativa process of BP Chemicals the $IrI_2(CO)_2^-$ complex is used together with $RuI_2(CO)_4$ [1.1]. Lithium iodide is added to increase the solubility of the catalyst at low water concentration. Low water concentration diminishes the loss of CO by the water–gas shift reaction ($CO + H_2O \rightarrow CO_2 + H_2$). The operating conditions are 150–200 °C and 3–6 MPa; the yield (selectivity times conversion) based on methanol consumption is 99% and based on CO 85%.

Examples of the oxidation of SO_2 to SO_3 and the formation of acetic acid from methanol and CO show that the principles of heterogeneous and homogeneous catalysis are the same. Reactants react with the surface of the heterogeneous catalyst or with the metal complex and the resulting product molecules leave the surface or the metal complex. Other large-scale industrial processes apply homogeneous catalysis to synthesise base chemicals: the oxidation of ethene to acetaldehyde in the Wacker reaction ($CH_2{=}CH_2 + 0.5\,O_2 \rightarrow CH_3CHO$) with a Pd^{2+} catalyst in aqueous solution, the hydroformylation of alkenes ($R{-}CH{=}CH_2 + CO + H_2 \rightarrow RCH_2CH_2 - CHO$) catalysed by a $HCo(CO)_4$ or $RhHL_3CO$ complex with L = triphenylphosphine ($P(C_6H_5)_3$) and the Dupont process for the hydrocyanation of butadiene to adiponitrile ($CH_2{=}CH{-}CH{=}CH_2 + 2\,HCN \rightarrow NC{-}CH_2CH_2CH_2H_2{-}CN$), a precursor of 1,6-hexanediamine, a component of nylon-6,6. The catalyst of the hydrocyanation reaction is a $HNiL_2(CN)$ complex, with L = triarylphosphite. In the fine chemicals industry, homogeneous catalysis also plays a role. There is a wealth of published information on homogeneous catalysis reactions and processes, e.g. [1.2].

In heterogeneous catalysis, the atoms at the catalyst surface but not the atoms in the bulk of the catalyst particles are accessible to the reacting molecules and participate in catalysis. The advantage of homogeneous catalysis is that all catalyst complexes and, thus, all metal atoms are accessible. However, a disadvantage of homogeneous catalysis is the difficult separation of the products from the catalyst, because they are both in the same liquid phase. In heterogeneous catalysis, separation is easy, because the products and the catalyst are in different phases. In a heterogeneous continuous process, the catalyst does not leave the reactor and in a heterogeneous batch process the catalyst can be filtered with ease.

The high cost of separation in homogeneous catalysis has led to many attempts to heterogenise the homogeneous catalyst, for instance by immobilisation onto a solid. The Chiyoda Corporation has developed the Acetica process, a heterogeneous methanol carbonylation process for the production of acetic acid [1.3]. The catalyst in the Acetica process is $RhI_2(CO)_2^-$, as in the Monsanto process, and is **immobilised (heterogenised)** on the cationic polymer polyvinyl methylpyridinium iodide $[{-}CH_2{-}CH(C_5H_4N^+CH_3)]I^-$. This polymer is synthesised from polyvinyl pyridine by quaternisation of the N atoms by methylation with CH_3I. The ionic bond between the cationic nitrogen group of pyridine and the anionic Rh complex ensures that the anionic rhodium complex remains inside the polymer. Immobilisation enables a higher catalyst concentration in the reaction mixture without the addition of alkali iodide (for lower corrosion), which gives a higher rate of acetic acid production. The heterogenised catalyst also hinders the formation of by-products as a result of the lower water concentration. This is beneficial because it decreases the loss of CO by means of the water–gas shift reaction. As a consequence, the product yield is claimed to be 99% based on methanol and 92% based on CO consumption.

1.3 Production of Ammonia

To illustrate how important catalysis is for the energy and chemical industries and to show how kinetics (catalysis) and thermodynamics are interwoven, we will take a look at an industrial process in which several catalysts are used to make ammonia, one of the top base chemicals with a worldwide production capacity of 230 MT in 2019. About 80% of the ammonia is used to produce ammonium sulfate ($(NH_4)_2SO_4$) and urea (H_2N–CO–NH_2). These fertilisers release ammonia in the soil, where it is available to the roots of plants. It has been estimated that about one-third of the world's population would starve without the use of these fertilisers in agriculture. In 2014, China produced 32.6% of the world's ammonia, Russia 8.1%, India 7.6% and the USA 6.4% [1.4].

In industrial ammonia synthesis, air is the source of the nitrogen and a carbon-containing material is used to produce synthesis gas (CO and H_2), which is thereafter transformed into hydrogen. Synthesis gas is produced by the reaction of a carbon-containing material with steam or oxygen. When coal is the carbon source, the reaction is

$$CH_x + H_2O \rightarrow CO + (1 + 0.5x)\ H_2, \tag{1.1}$$

and the heat required in the endothermic reaction is provided internally **(autothermally)** by co-feeding air or oxygen. Coal gasification is carried out in large fixed-bed reactors, in which the coal is fed from the top and a gas mixture of steam and oxygen from the bottom. Because of the endothermicity of the reaction, the process operates at elevated temperature (700–1,000 °C) and pressure (3 MPa). Coal impurities end up in slag that exits from the bottom of the reactor. In a fluid-bed type process, pulverised coal particles are fluidised by the gas and gasified at high temperature (1,200–1,600 °C) and elevated pressure (4 MPa). In a fluid-bed process, soft coal can be used, which bakes together in fixed-bed reactors. In the fluid bed process the ashes end up in dust-like particles, which must be separated from the product gas, otherwise they foul the waste heat boiler.

Most industrial processes use natural gas as the carbon source for synthesis gas. In that case, synthesis gas is produced with steam:

$$CH_4 + H_2O \rightarrow CO + 3\ H_2 \qquad \textbf{steam reforming}. \tag{1.2}$$

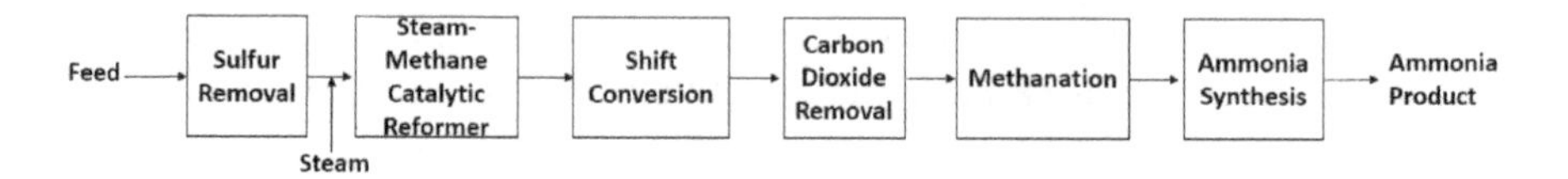

Fig. 1.3 Ammonia-synthesis process.

The production of ammonia with natural gas as carbon feed is a textbook example of industrial catalysis [1.5]. Not only is the ammonia-synthesis reaction ($N_2 + 3\,H_2 \rightarrow 2\,NH_3$) catalysed, but also the synthesis of hydrogen and the purification of natural gas in the feedstock. As Fig. 1.3 shows, several steps lead up to the ammonia-synthesis reaction.

Natural gas not only contains methane, but also sulfur-containing molecules. These molecules must be dealt with, because sulfur deactivates the metal catalysts used in subsequent steps. The removal of sulfur from mercaptans and other sulfur-containing hydrocarbons ($RSH + H_2 \rightarrow RH + H_2S$) is performed with a Co–MoS_2/γ-Al_2O_3 catalyst (Chapter 8). The catalyst contains small particles of molybdenum disulfide, promoted with cobalt atoms at their surface. The small Co–MoS_2 particles are stabilised on the surface of γ-Al_2O_3 to prevent them from growing together (sintering). See Section 2.3.4 for the structure of γ-Al_2O_3. The resulting H_2S is adsorbed on ZnO.

Higher hydrocarbons such as ethane and propane are also present in natural gas and they decompose relatively easily at elevated temperature. The resulting coke (carbon) would deactivate the nickel catalyst used in steam reforming. Therefore, in an adiabatic pre-reformer reactor the higher hydrocarbons are converted over a Ni catalyst at about 450 °C into synthesis gas. At the same time, steam reforming of methane starts and some of the heat load (described in the following sections) is taken off the steam reformers.

After the purification of the natural gas feed, the synthesis of H_2 can begin by means of the steam-reforming reaction, which produces H_2 and CO (1.2). The resulting CO is used to make another molecule of H_2 in the **water–gas shift (WGS) reaction**:

$$CO + H_2O \rightarrow CO_2 + H_2 \qquad \text{WGS.} \tag{1.3}$$

Reactions (1.2) and (1.3) are both performed in two steps to attain full conversion. The process conditions differ and, thus, different catalysts and different reactors are required. As a consequence, not two but four different catalysts are used just for the synthesis of H_2. A fifth catalyst is used to remove small amounts of CO and CO_2 by hydrogenating them to CH_4.

Only then can ammonia be obtained in a sixth catalytic step! Together with the Co–MoS_2/γ-Al_2O_3 catalyst used to purify the natural gas and the Ni/$MgAl_2O_4$ catalyst for the decomposition of higher hydrocarbons, this gives eight catalytic units in one overall process!

1.3.1 *Kinetics and Thermodynamics*

Catalysis has to do with kinetics, with increased rates of reaction, but rates can only be increased if the thermodynamics permit. Two examples illustrate the difference between kinetics and thermodynamics. The equilibrium constant of the reaction $CCl_4 + 2\ H_2O \rightleftharpoons CO_2 + 4\ HCl$ is greater than one at room temperature. This means that CCl_4 should react with water at room temperature. However, CCl_4 does not react, it is unstable (refers to thermodynamics) and inert (kinetics). At room temperature, the equilibrium constant of the reaction $SiCl_4 \rightleftharpoons Si + 2\ Cl_2$ is smaller than one. Thus, $SiCl_4$ is thermodynamically stable and does not decompose, even in the presence of a catalyst.

Hydrogen and nitrogen gas are necessary for the production of ammonia. The synthesis of N_2 is, in principle, easy: it is obtained from air by N_2–O_2 separation or by the consumption of O_2 from air to leave N_2. The latter is the chosen route for the production of ammonia. It is more difficult to produce hydrogen, but an inexpensive source is water and methane. The reaction of methane with steam gives synthesis gas (Eq. (1.2)), but it is highly endothermic (positive enthalpy change, $\Delta H = 206\ \text{kJ}\cdot\text{mol}^{-1}$).

The steam-reforming reaction (1.2) is endothermic ($\Delta H^0 > 0$), but because the entropy change is also positive ($\Delta S^0 > 0$), the change in free enthalpy

$$\Delta G^0 = \Delta H^0 - T\Delta S^0 = -RT \ln K,$$

becomes negative at high temperature. Thus, at high temperature, the equilibrium constant K (ratio of forward rate and backward rate) is higher than 1; the forward reaction rate is higher than the backward reaction rate and conversion becomes favourable. Steam reforming is, therefore, carried out at elevated temperature. It is also carried out at elevated pressure (4 MPa), even though more molecules are produced than are consumed and Le Chatelier's principle predicts that this has a negative effect on conversion. The effect of pressure is predicted by

$$K_x = K_p \cdot P^{-\Delta n} \quad \text{and} \quad \frac{d \ln K_x}{d \ln P} = -\Delta n,$$

where Δn is the change in the number of moles during reaction. A disadvantage of high pressure is that an increase from 0.1 to 3 MPa leads to a higher equilibrium concentration of CH_4 (lower conversion) and lower yields of H_2 and CO (Fig. 1.4). In industry, the application of such high pressure is to reduce the size of the reactors (and thus reduce cost) and to produce H_2 at elevated pressure. This saves compression costs, because high-pressure H_2 is required in the subsequent production of ammonia.

Endothermic steam reforming should be carried out at high temperature, suggesting that a catalyst is not necessary, because, at elevated temperature, kinetics alone may be sufficient (Arrhenius equation, $k = A\exp(-\Delta E/RT)$). However, a catalyst is required in many endothermic reactions, because some materials make it impossible to run reactions at extremely high temperature. For exothermic reactions, it is always advisable to use a catalyst, because with $\Delta H^0 < 0$ the conversion is limited at high temperature. For instance, in the absence of a catalyst, the oxidation of SO_2 to SO_3 (Fig. 1.1) has a low reaction rate; $T > 600\,^\circ\mathrm{C}$ would be necessary to obtain a reasonable rate. Unfortunately, the reaction is exothermic ($\Delta H^0 < 0$) and an increase in temperature has a negative influence on the position of the equilibrium of an exothermic reaction ($\frac{d\ln K}{dT} = \frac{\Delta H^0}{RT^2} < 0$). In such a reaction, thermodynamics requires that the temperature must be low to obtain high conversion, but low temperature also means a low

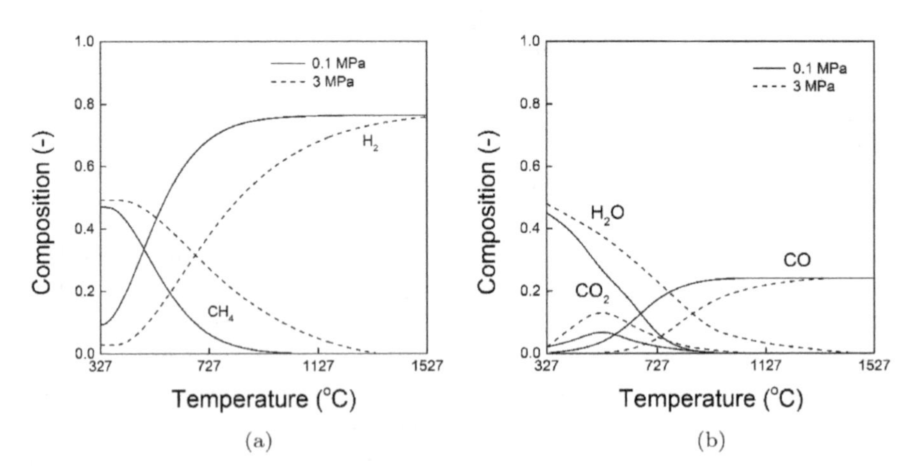

Fig. 1.4 Influence of temperature and pressure on the equilibrium composition of CH_4 and H_2 (a) and CO, CO_2 and H_2O (b) in the steam-reforming gas for $H_2O/CH_4 = 1$. Reprinted with permission from [1.6]. Copyright (2013) John Wiley & Sons, Inc.

rate. The solution to this problem is to use a catalyst and to accelerate the reaction and run it at low temperature.

The exothermicity or endothermicity of a reaction can often (but not always!) be predicted from the reaction stoichiometry, because in many cases the sign of the change in enthalpy ΔH^0 is the same as the sign of the change in the number of molecules Δn during reaction. When the number of product molecules is larger than the number of reactant molecules ($\Delta n > 0$), then, in principle, bonds are broken and enthalpy must be provided (endothermic reaction, $\Delta H^0 > 0$). The reverse often holds true for exothermic reactions ($\Delta n < 0, \Delta H^0 < 0$). Thus, dehydrogenation reactions, such as $C_nH_{2n+2} \rightarrow C_nH_{2n} + H_2$ and $C_6H_{12} \rightarrow C_6H_6 + 3\,H_2$, are endothermic and the reverse hydrogenation reactions are exothermic. Oxidation reactions such as $2\,H_2 + O_2 \rightarrow 2\,H_2O$ ($\Delta H^0 = -484$ kJ·mol^{-1}) and $2\,SO_2 + O_2 \rightarrow 2\,SO_3$ ($\Delta H^0 = -198$ kJ·mol^{-1}) are exothermic; the reverse reduction reactions are endothermic. When $\Delta n \approx 0$, ΔH^0 may be low, as in the WGS reaction $CO + H_2O \rightarrow CO_2 + H_2$ ($\Delta H^0 = -41$ kJ·mol^{-1}). An exception to this is the reaction $N_2 + O_2 \rightarrow 2\,NO$ with $\Delta n = 0$ and $\Delta H^0 = 180$ kJ·mol^{-1}!

This brief diversion into thermodynamics and catalysis explains why, to attain full conversion, the steam reforming of methane to synthesis gas is usually performed in two steps. The feed to the first reforming reactor consists of methane and steam while oxygen and steam are added in the second reactor (Fig. 1.5). The first reactor is operated allothermically (i.e. is heated). Hundreds of tubes (10 m long and 10 cm diameter, filled with Ni/α-Al_2O_3 catalyst), hang in a furnace and are heated by burning a fuel gas around the tubes. The Ni/α-Al_2O_3 catalyst contains nanocrystalline nickel metal particles on the surface of an α-Al_2O_3 support (cf. Section 2.3.3 for the structure of α-Al_2O_3). The Ni/α-Al_2O_3 must withstand high temperatures and, therefore, CaO is added to the catalyst as a structural

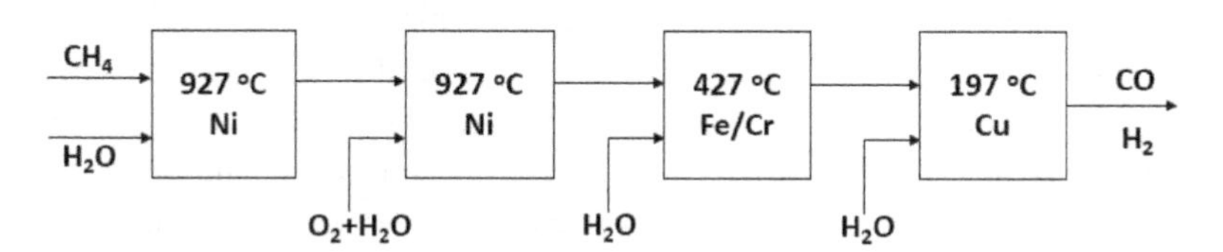

Fig. 1.5 Steam reforming of methane. Synthesis gas is produced in the first allothermic reactor and the second autothermic reactor. In the third and fourth reactors, the CO/H_2 ratio is adjusted by the water–gas shift reaction. Reprinted with permission from [1.7]. Copyright (1996) Elsevier.

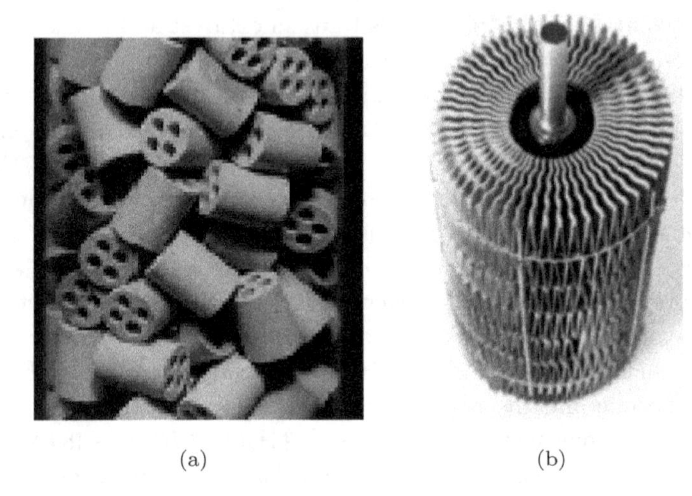

(a) (b)

Fig. 1.6 (a) 4-Hole cylindrical catalyst particles packed in a reformer tube, (b) structured catalyst unit that can be loaded inside a reformer tube.

promoter. The catalyst particles are shaped as cylinders (4 cm long, 2 cm diameter) (Fig. 1.6a). Because stability is more important than activity, the Ni surface area is low, about 0.5 $m^2 \cdot g^{-1}$. Lately, structured catalysts have been introduced to obtain higher hydrogen yields at reduced energy input [1.8]. To improve the gas flow through the reactor tubes, metallic structures are introduced inside the tubes, which are coated with the catalyst (Fig. 1.6b).

1.3.2 *Activity, Selectivity and Stability*

Why are Ni and α-Al_2O_3 used in the steam-reforming catalyst? To answer that question, it is important to consider not only the catalyst activity but also the selectivity and stability of the catalyst. A catalyst should be not only active (accelerate the reaction), but also selective and stable for industrial applications. Activity is often expressed as conversion (fraction or percentage) per unit time and unit catalyst, related to a first-order rate equation $dc_A/dt = -kc_A$, where c_A is the concentration of the reactant and k, the rate constant. Although this information is valuable for industry, detailed scientific information is still missing. To calculate the intrinsic activity, i.e. the activity per catalytic site, the rate constant k should be divided by the number of sites (active atoms at the catalyst surface) in one gram catalyst (see Section 3.3.2 for a determination of the number

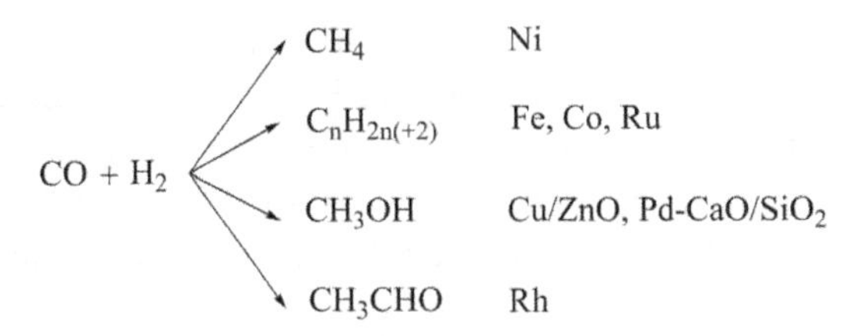

Fig. 1.7 Possible products from synthesis gas and catalysts used to make these products.

of catalytic sites). The intrinsic activity is often referred to as **turnover frequency (TOF)**, the number of molecules that react per unit time and per active site on the catalyst surface.

Selectivity is the fraction or percentage of a reactant that has been converted to a certain product. A high selectivity to the desired product(s) means that fewer reactant molecules are wasted on by-products and that less energy is required for the separation of the desired products from the by-products. An example of the strong effect that a catalyst may have on the selectivity is the transformation of synthesis gas to hydrocarbons and oxo-products (Fig. 1.7).

Metals such as Fe, Co, Ni and Ru lead to hydrocarbons. Ni produces mainly methane and Ru mainly higher hydrocarbons. Fe produces lower alkanes, alkenes and alcohols. Pd, Pt, Ir and Cu are good catalysts for methanol formation in the presence of basic supports or promoters such as ZnO and CaO. Rh is flexible and enables the formation of hydrocarbons, methanol as well as higher oxo-products such as ethanol and acetaldehyde. There is increasing interest in the selectivity of Rh to C_2 and higher oxo-products, especially in countries rich in coal or natural gas, both of which can be transformed into synthesis gas.

Just as Ni is the best metal to produce methane from synthesis gas (Fig. 1.7), Ni is the best catalyst for the reverse reaction, the transformation of methane to synthesis gas by steam reforming. This is because the equilibrium constant K of a reaction does not depend on the catalyst C:

$$\begin{aligned} &A \rightleftharpoons B && \Delta G^0 = -RT\ln K = -RT\ln(k_1/k_{-1}), \\ &A + C \rightleftharpoons B + C && \Delta G'^0 = -RT\ln K = -RT\ln(k'_1/k'_{-1}), \end{aligned}$$

where K is the equilibrium constant of the reactions $A \rightleftharpoons B$ and $A + C \rightleftharpoons B + C$ and k_1 and k_{-1} are the rate constants of the forward and backward reactions, respectively. Because $\Delta G^0 = \Delta G'^0$, $k_1/k_{-1} = k'_1/k'_{-1}$. Thus, the ratio of the forward and backward reactions does not depend on the presence of a catalyst, or, in other words, a catalyst that promotes the forward reaction promotes the backward reaction to the same extent. In the present case, if Ni promotes the methanation reaction, then it also promotes the reverse reaction, i.e. the steam-reforming reaction. See Chapter 6 (Sections 6.4 and 6.6) for information about the high activity and selectivity of Ni.

Industry requires not only active and selective catalysts, but also stable catalysts. High **stability** means that the catalyst can be used for long periods of time with few regenerations or replacements. Regeneration and replacement add to the cost of the catalyst and increase operating costs due to interruptions. Stability is especially important in refineries and the base-chemical industry, where the catalyst must be used for extended periods of time because of the smaller profit margins than in the fine-chemical industry, where higher profits make it feasible to use a catalyst only once. Furthermore, in the refinery and base-chemical industry continuous processes are necessary to produce hundreds or thousands of tons per day; interrupting these processes for the regeneration or replacement of the catalyst results in a substantial decrease in production. Therefore, it is important not only to know the initial activity of a catalyst, but also to know the lifespan of the catalyst, as expressed by the **turnover number** (TON), i.e. the number of molecules that react per catalytic site on the catalyst surface before deactivation of the catalyst occurs. Whereas batch processes, without reusing the catalyst, may afford a TON as low as 10, continuous processes that run for one year may require a TON $> 10^5$. A TON of one means that the reaction is stoichiometric and not catalytic!

The **stability** of a catalyst is often determined by its decline in activity as a function of time. For instance, information such as "The catalyst lost 30% of its activity in the first 24 h and thereafter 10% of the remaining activity over one month" is important for estimating the commercial success of a catalyst. Stability is the reason why α-Al_2O_3 is used as a support for the Ni particles in the steam-reforming catalyst. Steam reforming is an endothermic reaction and, therefore, the reaction is run at high temperature (700–1,000 °C). To prevent the Ni particles from sintering under these conditions, they are separated from each other by depositing them on the surface of the support. The high temperature in steam reforming requires a support with high thermal stability and, as shown in Section 2.3.3, α-Al_2O_3 (with a small surface area of a few $m^2 \cdot g^{-1}$) meets this and other requirements. Other supports, such as γ-Al_2O_3, are unstable at high temperature. They would show a decrease in surface area and, as a result, the Ni particles would come into contact with each other and would sinter. That would lead to a loss of Ni surface area and, thus, of catalytic activity.

1.3.3 H_2 Production

The natural gas feed flows through the tubes of the steam-reforming reactor that are filled with a Ni/α-Al_2O_3 catalyst. The tubes are heated from the outside by burning fuel (Fig. 1.8). Because steel tubes cannot be heated beyond 1,200 °C, the exit gas of the first reformer still contains about 10–14% CH_4 (cf. Fig. 1.4). This CH_4 can reform further to H_2 in a second autothermic reactor, which operates adiabatically and has an inlet temperature of about 800 °C and outlet temperature of 1,000 °C. In the latter reactor, CH_4 reaches almost complete conversion. The energy required for the endothermic transformation is provided *in situ* by the reaction of O_2 or air with CH_4, H_2 and CO in the combustion chamber above the catalyst bed. To prevent damage to the steel walls, the autothermic reactor is lined with ceramic material. The resulting gas is equilibrated in the catalyst bed by the methane-reforming reaction (Eq. (1.2)) and the WGS reaction (Eq. (1.3)). Air is used as feed in the second reformer of an ammonia plant, so that N_2 is automatically present in the mixture. To produce synthesis gas for Fischer–Tropsch and methanol reactions, pure O_2 is used and must first be produced in an air separation plant.

Because steam reforming is an endothermic reaction, it requires a significant input of heat to reach the desired conversion to hydrogen. About 25% extra methane is required to generate the heat needed to compensate for the endothermicity of reaction (Eq. (1.2)). The thermal efficiency of this indirect heating is not more than 50% and the hot flue gas is used to preheat the feed of the steam reformer and to generate high-pressure steam in waste heat boilers (Fig. 1.8).

In modern steam reformers, heat consumption and production from the two reforming reactors is integrated. The first reactor functions as a heat exchanger, in which the hot product gas of the second autothermic reformer is used to heat the tubes of the first reformer (Fig. 1.9). The advantage is that the waste heat energy is used to produce extra hydrogen rather than surplus steam. In this way, thermal efficiency can be increased by as much as 60–70% compared to radiant heating.

Microchannel technology is a developing field of chemical process technology to intensify chemical reactions by reducing the dimensions of the reactor systems. Thus, reactions occur at much faster rates in channels

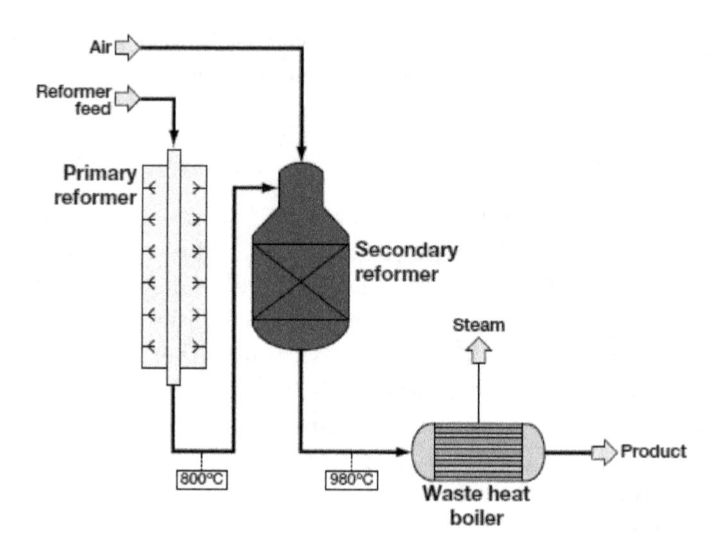

Fig. 1.8 Primary (allothermic) and secondary (autothermal) reformers with waste heat boiler.

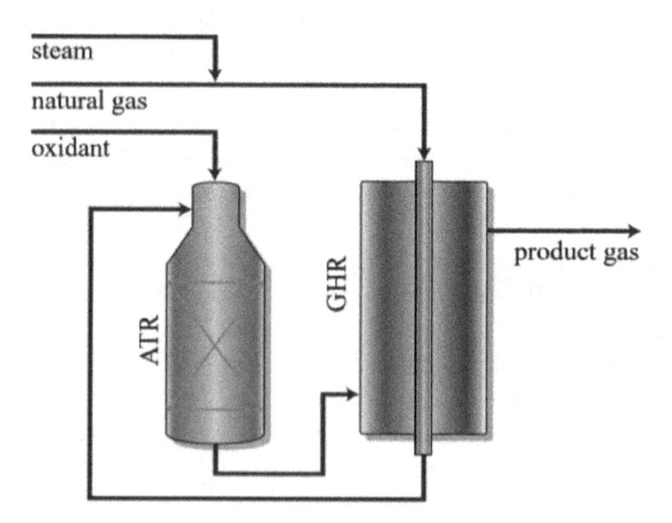

Fig. 1.9 Two-reactor steam reforming with gas–heat exchange. Natural gas and steam enter the right allothermal reactor (GHR) and then go to the left autothermal reactor (ATR). The product gas from the ATR heats the incoming gas from the first reactor in a heat-exchanger configuration.

with a diameter of a few millimeters. The small channels dissipate heat more quickly than conventional reactors with channel diameters of 2–10 cm and more active catalysts can be used. Limitations of mass and heat transfer reduce the efficiency of the large conventional reactors used for Fischer–Tropsch, methane steam reforming and hydroprocessing. The use of microchannels means that plants can be much smaller and can be scaled up or down more flexibly by adding or removing channels. This technology is more efficient than conventional process technology. In a microchannel steam-reforming reactor, methane and steam flow through channels filled with catalyst to produce syngas. Fuel and air flow through adjacent channels, where they combust and generate heat to drive the endothermic steam reforming [1.9].

The product gas from the second reformer contains almost pure synthesis gas. If the goal is to produce H_2, as in the production of ammonia, then CO can be used to produce one more H_2 molecule by means of the WGS reaction (Eq. (1.3)). The WGS reaction is weakly exothermic ($\Delta H = -41$ kJ·mol^{-1}) and should be carried out at low temperature; thus, a highly active catalyst is necessary. Cu is a very active catalyst for this reaction, but Cu reacts strongly even with traces of H_2S. Therefore, the WGS reaction is performed in two reactors in series (Fig. 1.5). In a first high-temperature shift reactor (300–450 °C), an Fe_2O_3 catalyst, stabilised with Cr_2O_3 (**structural promoter**, to prevent sintering), reduces the CO level to 3–5% [1.10]. The Fe_2O_3 catalyst traps the traces of H_2S. The second, low-temperature WGS reactor contains a $Cu/ZnO/Al_2O_3$ catalyst. ZnO functions as a support for the small Cu particles and also as a trap for the last traces of H_2S, while Al_2O_3 functions as a structural promoter that retards sintering of the Cu particles. Because Cu is more active than Fe_2O_3, the reaction can be carried out at lower temperature (200–250 °C) than in the first WGS reactor, so that the CO level goes down to below 0.3% [1.6]. CO conversion should be kept high, as unconverted CO will consume H_2 in the methanation reactor, after the CO_2 absorption unit and before the ammonia-synthesis reactor, and the resulting CH_4 will increase the level of inert gas in the ammonia-synthesis loop. Both sulfur and chlorine poison copper. Thus, both must be removed from the feedstock. The feed purification section usually consists of a unit for catalytic hydrogenation ($RSH + H_2 \rightarrow RH + H_2S$ and $RCl + H_2 \rightarrow RH + HCl$) (with a Co–$MoS_2/\gamma$-$Al_2O_3$ catalyst) and units for H_2S and HCl absorption.

The exit gas from the second WGS reactor consists of H_2, N_2, H_2O and CO_2, as well as a trace of CO. To prepare a mixture of pure H_2 and N_2, as required in the synthesis of ammonia, H_2O, CO_2 and CO must be removed. H_2O is removed by condensation (cooling) and CO_2 by chemical absorption or physical adsorption. Chemical absorption is achieved with a basic liquid such as ethanolamine ($HO–CH_2CH_2–NH_2$), while physical adsorption is performed with zeolites (Section 7.1.2) in a pressure swing adsorption (PSA) unit (Fig. 1.10), where CO, CO_2 and unconverted CH_4 are adsorbed. The H_2 and N_2 pass through the adsorption units with 99.99% purity of H_2. The removed CO, CO_2 and CH_4 are used as fuel gas in the first reformer.

The absorption or adsorption of the gas decreases the CO and CO_2 contents in the $H_2 + N_2$ gas to a few percent, but even this is too high for the Fe catalyst in the ammonia-synthesis reactor. The levels of CO and CO_2 must be reduced to below 5 ppm; even traces of oxygen-containing compounds harm the metallic iron catalysts in the ammonia synthesis. This deep reduction of CO and CO_2 is attained by **methanation reactions,**

Fig. 1.10 Steam-reforming plant for the production of H_2 (left), with two PSA towers for the removal of CO, CO_2 and CH_4 (right).

the reverse of steam reforming.

$$CO + 3\,H_2 \rightarrow CH_4 + H_2O \quad \Delta H = -206 \text{ kJ·mol}^{-1}, \tag{1.4}$$

$$CO_2 + 4\,H_2 \rightarrow CH_4 + 2\,H_2O \quad \Delta H = -165 \text{ kJ·mol}^{-1}. \tag{1.5}$$

The resulting small amount of CH_4 does not harm the Fe catalyst, but the H_2O must be removed by drying and absorption. Ni is the catalyst of choice for the methanation reaction. As discussed above, the equilibrium of a reaction does not depend on the catalyst and a catalyst that functions in the forward reaction also functions in the backward reaction. Therefore, since Ni is suitable for the steam-reforming reaction, it is also suitable for the reverse reaction, the methanation reaction. However, the conditions, under which the forward and backward reactions operate, differ. The methanation reactions (Eq. (1.4)) and (Eq. (1.5)) ($\Delta n < 0$, $\Delta H^0 < 0$) should be performed at low temperature (about 300 °C), while the steam-reforming reaction (Eq. (1.2)) ($\Delta n > 0$, $\Delta H^0 > 0$) should be performed at high temperature (700–1,000 °C). Nevertheless, in both cases the catalyst of choice is Ni. α-Al_2O_3 is used as a support at high temperature because even at 700–1,000 °C it is stable against sintering and, thus, prevents the Ni particles from coming together, sinter and lose surface area. On the other hand, at the lower temperature of 300 °C in the methanation process the lower thermal stability of γ-Al_2O_3 than of α-Al_2O_3 is not a problem. The higher surface area of γ-Al_2O_3 (200–300 $m^2·g^{-1}$) than of α-Al_2O_3 (0.5 $m^2·g^{-1}$) then is a big advantage in stabilising small and, thus, highly active Ni particles.

In this section, we have seen how H_2, one of the reactants in the synthesis of ammonia, is produced in industry today. Many years ago, coal was used as the feed for the production of H_2 (Eq. (1.1)) but later, when cheap natural gas became available, this took over as feed for H_2 (Eqs. (1.2) and (1.3)). However, in recent years the adverse effect of the CO_2, which is produced as by-product in the synthesis of H_2 from CH_4, on global warming has become an issue. Not only 0.25 mol of CO_2 is formed per mol of H_2, an additional 0.17 mol is produced in the combustion needed for driving the endothermic reaction [1.11]. As a result, all over the world research programs have been initiated to produce H_2 by electrolysis of water (Section 10.4.2 describes how this is done). The oxygen by-product of the reaction $2\,H_2O \rightarrow 2\,H_2 + O_2$ can be used in industry in more efficient burning (e.g. in the furnaces of the steel industry, in glass manufacture, in gasification plants) [1.12]. To make the production of H_2 sustainable, the electricity needed for the electrolysis must come from renewable energy sources, such as wind or solar power. Hydrogen that is made this way is called **green hydrogen**. When the electricity is generated by nuclear energy, the phrase **pink hydrogen** is used. Green and pink hydrogen are in contrast to the hydrogen that is made in classic methane steam reforming, which is called **grey hydrogen**.

Industry also looks at intermediate solutions for solving the CO_2 problem. A future way of producing H_2 from natural gas might be to pyrolyse CH_4 at very high temperature into H_2 and solid carbon. The process and the utilisation of the resulting carbon are under study. A solution which needs less research is **capture and storage** (CSS), so that the CO_2 is not dispersed in the atmosphere. The hydrogen that is produced this way in the classic steam reforming of methane is called **blue hydrogen**. But even if the CO_2 that is produced in methane steam reforming would be captured and stored, one also has to deal with the CO_2 that is formed in the heating of the steam-reforming feed in an allothermic reactor (Fig. 1.8). This flue gas is formed in dilute concentration and the CO_2 is not easily captured. Therefore, studies are under way to heat the steam-reformer tubes in the first reactor electrically, instead of heating them from the outside with flames by burning fuel. Electric heating might become an alternative if the electricity price would come down because of increasing interest for a hydrogen and electric economy and if environmental legislation would severely limit the emission of CO_2. Electric heating has the potential to reduce CO_2 emissions and provide flexible and compact heat generation [1.13]. An integrated design could allow a compact reformer unit that is many times smaller than current units. If implemented on a global scale, it could reduce nearly 1% of all CO_2 emissions.

Not all ammonia plants have a CO_2 problem because very large ammonia plants are often combined with urea plants. Urea is a fertiliser and its synthesis ($CO_2 + 2\,NH_3 \rightarrow NH_2{-}CO{-}NH_2 + H_2O$) consumes 40% of the ammonia produced worldwide. Thus, the CO_2 formed in the ammonia process can be used in the synthesis of urea and, at the same time, allows heat integration, because large amounts of heat (steam) are produced in the ammonia process, while large amounts are consumed in the urea process.

1.3.4 *Ammonia Synthesis*

The final step in the ammonia-synthesis process is the reaction

$$N_2 + 3\,H_2 \rightarrow 2\,NH_3 \quad \Delta H = -92.5 \text{ kJ}\cdot\text{mol}^{-1}(N_2). \tag{1.6}$$

The fact that a catalyst that functions in a forward reaction also functions in the reverse reaction was known for 100 years when the search for the optimum catalyst for the synthesis of ammonia began. The forward reaction must be run at very high pressure (>10 MPa) to attain reasonable conversion ($\Delta n < 0$), but the reverse reaction is favoured by low pressure. Experiments are easier to perform at low than at very high pressure; thus, more than 20,000 catalysts were explored in a few years in the laboratory of BASF in Germany under the leadership of Mittasch and Fe, Co, Ni, Ru, Os, Mn, Mo, U and Ce turned out to be good catalysts. Because of availability, ease of preparation and cost, Fe was considered the optimal ammonia-synthesis catalyst. Until 1992, the only catalyst used in industry was unsupported Fe, made by reduction of magnetite (Fe_3O_4) and promoted with Al_2O_3, K_2O and CaO. Since 1992, a more active, but also more expensive, Ru catalyst supported on graphitic carbon and promoted with BaO and Cs_2O has also found industrial application [1.14].

The ammonia-synthesis reaction is exothermic with a low equilibrium constant ($K_{eq} = c^2_{NH3}/c_{N2}c^3_{H2} = 1.6 \times 10^{-4}$ at 400 °C) and, according to thermodynamics, reasonable conversion of the reactants can be reached only at low temperature (Fig. 1.11), but then the rate is low. Therefore, in industrial applications the temperature is relatively high (450 °C) and pressure is increased to about 25 MPa to compensate for the negative effect

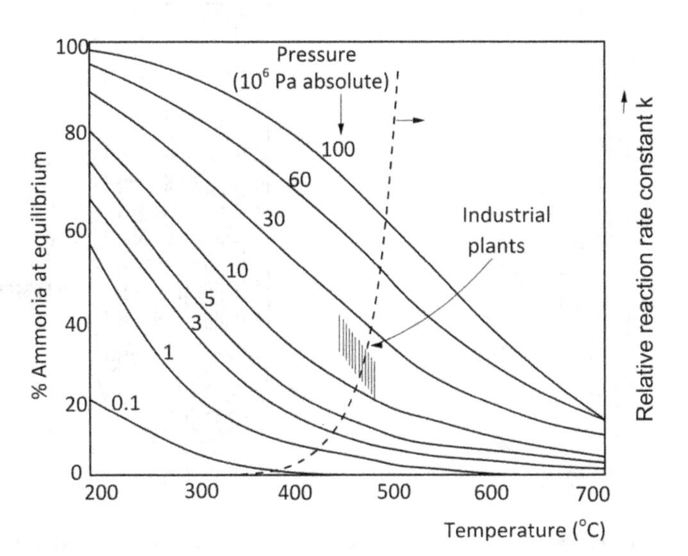

Fig. 1.11 Thermodynamically allowed conversion of $N_2 + 3H_2$ to NH_3. The area of industrial interest is indicated by vertical lines between 10 and 30 × 10^6 Pa (10^5 Pa = 1 bar).

of temperature. Under these conditions the ammonia conversion is about 25%. Higher pressure would give better conversion, but the reactor would be too expensive.

The N_2 and H_2 product gas of the methanation reactor is pressurised, mixed with recycle gas (containing N_2, H_2, NH_3 and some CH_4 and Ar), compressed a second time by the reaction pressure and fed to the ammonia reactor. To protect the outer wall of the high-pressure NH_3 reactor (25 MPa!), the cold feed gas flows first between the catalyst beds and the outer wall of the reactor (Fig. 1.12) and is then heated in a heat exchanger before entering the first catalyst bed. To prevent the temperature in the NH_3 reactor from becoming too high during the exothermic ammonia-synthesis reaction, cold feed gas **(cold shot)** is injected between the catalyst beds and heat is removed in a heat exchanger. This heat control is essential; otherwise temperature would increase and conversion would decrease.

The ammonia-synthesis reaction never reaches full conversion and the remaining gases ($N_2 + H_2 + NH_3$) must be separated from the product and

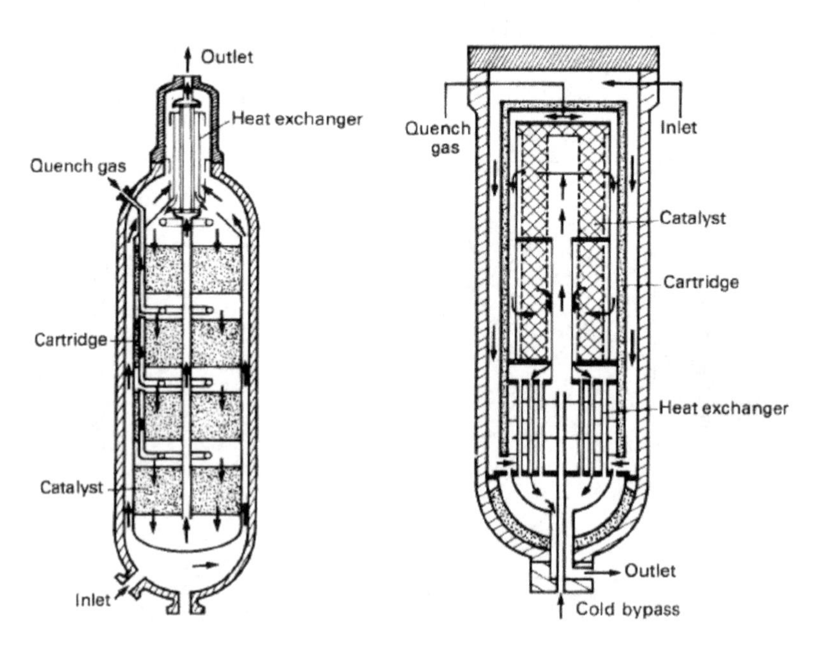

Fig. 1.12 NH_3 synthesis reactors with axial flow (left, Kellog) and radial flow (right, Topsøe).

recycled to the reactor. Separation is done by cooling the exit gas of the reactor to 30 °C to liquefy and remove the ammonia. NH_3 is decompressed to 2.4 MPa so that impurities such as CH_4 and H_2 become gases. The heat released by the reaction is removed and used to heat the incoming gas mixture. A purge is necessary, because the purge gas is relatively rich in argon, which comes from the air that is used in the second, autothermic steam-reforming reactor (air contains **0.9% argon!**).

Figure 1.13 shows the complete ammonia process, with gas pre-treatment, production of synthesis gas, WGS, CO_2 removal, methanation and final ammonia synthesis. Table 1.1 gives all the catalysts. Catalysts are used in the gas pre-treatment (two catalysts), steam-reforming (two catalysts), WGS (two catalysts), methanation and ammonia synthesis. Only the removal of CO_2 (by absorption or adsorption) is not catalytic, indicating that eight of nine process steps are catalytic!

The above example of a process with seven integrated catalytic steps represents all the aspects of catalysis. Thermodynamics determine the process conditions (T and P), while kinetics (activity, selectivity and stability) determine the choice of the catalyst. Many types of active materials (metals, metal oxides and metal sulfides) and supports are

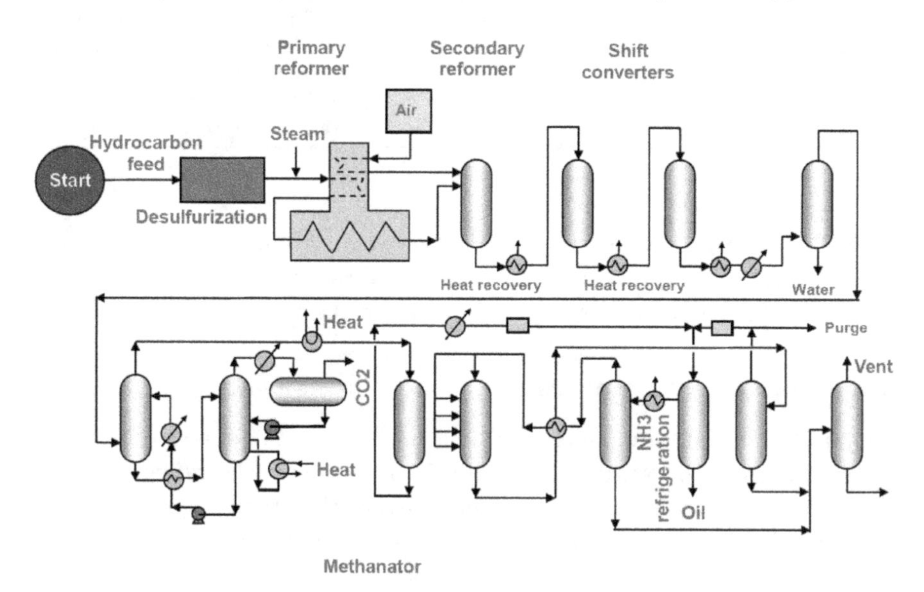

Fig. 1.13 Overall process for the production of ammonia from natural gas. Reprinted with permission from Serenix (2015).

Table 1.1 Process steps and catalysts used in the production of ammonia.

Process step	Catalyst	T(°C)	P(MPa)
Gas cleaning	$Co–MoS_2/\gamma$-Al_2O_3	300	3
Allothermic reforming	Ni/α-Al_2O_3	700	3
Autothermic reforming	Ni/α-Al_2O_3	1,000	3
High-temperature shift	Fe_2O_3	350	3
Low-temperature shift	$Cu/ZnO/Al_2O_3$	220	3
Methanation	Ni/γ-Al_2O_3	300	3
Ammonia synthesis	Fe	450	25

Table 1.2 Global production capacity of some industrial catalytic processes (2018).

Process	Product	Catalyst	MT/y
Hydrotreatment of oil fractions	Clean fuel	$Co–MoS_2/\gamma$-Al_2O_3	600
Catalytic cracking of gasoil	Gasoline	Zeolite	300
Catalytic reforming	Gasoline	$Pt–Sn/\gamma$-Al_2O_3	250
Ammonia synthesis	NH_3	Fe	230
SO_2 oxidation	H_2SO_4	V_2O_5	260
CO hydrogenation	Methanol	Cu/ZnO	110
Polymerisation	$(C_2H_4)_n$	Ti/MgO	55
Ethylene oxide	Ethylene oxide	Ag/α-Al_2O_3	26
Alkylation	Cumene	Zeolite	15
Methanol carbonylation	CH_3COOH	Rh complex	12

possible. In ammonia synthesis, only two supports are used, Al_2O_3 and ZnO, but there are two types of Al_2O_3. Why were these metals and supports selected? These questions will be addressed in the following chapters.

1.4 Relevance of Catalysis

The last focus in this chapter is on the scope of industrial catalysis. Table 1.2 gives 10 examples of catalytic processes in the refinery and in the base-chemical industry. They demonstrate that refinery processes are on a much larger scale than most processes for the production of base chemicals. Hydrotreatment is the largest catalytic process worldwide in terms of the quantity of feed reacted per year, because sulfur and nitrogen must be removed from almost all refinery streams and because we all like to drive a car! It is the third largest process in terms of catalyst consumption. Only catalytic cracking and the cleaning of car exhaust gas consume more catalysts per year. Inorganic base chemicals, e.g. NH_3 and H_2SO_4, are produced yearly on a large scale, but the production of organic base chemicals is about an order of magnitude smaller than in the refinery and NH_3 and H_2SO_4 processes.

References

1.1 G. J. Sunley, D. J. Watson, High productivity methanol carbonylation catalysis using iridium. The Cativa^{TM} process for the manufacture of acetic acid, *Catal. Today* 58, 293–307, 2000; J. H. Jones, The Cativa^{TM} process for the manufacture of acetic acid, *Platinum Metals Rev.* 44, 94–105, 2000.
1.2 P. W. N. M. van Leeuwen, *Homogeneous Catalysis*, Kluwer, Dordrecht, 2004.
1.3 N. Yoneda, T. Minami, J. Weiszmann, B. Spehlmann, The Chiyoda/UOP acetica^{TM} process: A novel acetic acid technology, *Stud. Surf. Sci. Catal.* 121, 93–98, 1999.
1.4 V. Pattabathula, J. Richardson, *AIChE* Sept. 2016, pp. 69–75.
1.5 G. Ertl, H. Knözinger, F. Schüth, J. Weitkamp, Eds., in *Handbook of Heterogeneous Catalysis*. Wiley-VCH, Weinheim, 2008, Vols. 4 and 6.
1.6 J. A. Moulijn, M. Makkee, A. van Diepen, *Chemical Process Technology*, 2nd Edition, Wiley, 2013.
1.7 M. A. Peña, J. P. Gómez, J. L. G. Fierro, New catalytic routes for syngas and hydrogen production, *Appl. Catal. A* 144, 7–57, 1996.
1.8 W. Whittenberger, P. Farnell, Structured catalysts for steam reformers, 2018, www.digitalrefining.com/article/1001067 and https://matthey.com/news/2018.
1.9 A. Y. Tonkovich, S. Perry, Y. Wang, D. Qiu, T. LaPlante, W. A. Rogers, Microchannel process technology for compact methane steam reforming, *Chem. Eng. Sci.* 59, 4819–4824, 2004.
1.10 C. Ratnasamy, J. P. Wagner, Water gas shift catalysis, *Catal. Rev. Sci. Eng.* 51, 325–440, 2009.
1.11 Estimating The Carbon Footprint of Hydrogen Production (forbes.com).
1.12 T. Kato, M. Kubota, N. Kobayashi, Y. Suzuoki, Effective utilisation of by-product oxygen from electrolysis hydrogen production, *Energy* 30, 2580–2595, 2005.
1.13 S. T. Wismann, J. S. Engbæk, S. B. Vendelbo, F. B. Bendixen, W. L. Eriksen, K. Aasberg-Petersen, C. Frandsen, I. Chorkendorff, P. M. Mortensen, Electrified methane reforming: A compact approach to greener industrial hydrogen production, *Science* 364, 756–759, 2019.
1.14 D. E. Brown, T. Edmonds, R. W. Joyner, J. J. McCarroll, S. R. Tennison, The genesis and development of the commercial BP doubly promoted catalyst for ammonia synthesis, *Catal. Lett.* 144, 545–552, 2014.

Questions

1.1 Should a Mars–van Krevelen-type catalytic reaction be done in one or in two reactors?

1.2 How can one confirm the Mars–van Krevelen mechanism for the oxidation of SO_2 over V_2O_5 with isotope labelling?

1.3 Which side reactions in the process for the carbonylation of methanol can lead to a less efficient use of methanol? How can water have a negative effect on the carbonylation of methanol?

1.4 Do you expect ΔH^0 of the reaction $2\,NCl_3 \rightleftharpoons N_2 + 3\,Cl_2$ to be positive or negative? The equilibrium of this reaction is on the right-hand side at room temperature. Is ΔH^0 large or small? If shock is applied to NCl_3 it explodes. Is this due to an increase in temperature or pressure?

1.5 Why is the same catalyst (Ni/Al_2O_3) used in both the methanation reaction and the steam-reforming reaction? Is there really no difference between the two catalysts?

1.6 The hydrogenations of N_2 to ammonia and of CO to methanol are both exothermic. Why?

1.7 Zinc oxide traps H_2S and HCl, which may be present as impurities in synthesis gas and thus in the feed of the Cu–ZnO–Al_2O_3 low-temperature WGS catalyst. Give the chemical reactions that explain why zinc oxide traps H_2S and HCl.

1.8 Why can the CO level become as low as 0.3% in the second WGS reactor, with the Cu/ZnO/Al_2O_3 catalyst, although it operates at lower temperature than the Fe_2O_3 catalyst in the first WGS reactor, which achieves a CO level of 4%?

1.9 Why are low levels of CO and CO_2 in the feed gas of the ammonia-synthesis reactor still too high for the Fe catalyst?

1.10 The methanation of CO ($CO + 3\,H_2 \rightarrow CH_4 + H_2O$) and the synthesis of NH_3 ($N_2 + 3\,H_2 \rightarrow 2\,NH_3$) are both exothermic and both have the same decrease in the number of molecules ($\Delta n = -2$). Why does ammonia synthesis require a much higher pressure than the CO methanation reaction?

Chapter 2

Catalyst Preparation and Characterisation

2.1 Supported Catalysts

Reactions that are catalysed by solid catalysts take place on the surface of the catalyst particles. To increase the surface area of the catalytic solid, it is necessary to decrease the size of the particles, because the specific surface area is proportional to $1/R$ (surface area is proportional to R^2 and volume to R^3), where R is the radius of the particles. For instance, cubic Pt particles, 1 mm long, have a specific surface area of $6 \times 10^{-6}\,\text{m}^2/21.45 \times 10^{-3}\,\text{g} = 2.8 \times 10^{-4}\,\text{m}^2{\cdot}\text{g}^{-1}$, because the density of Pt is $21.45\,\text{g}{\cdot}\text{cm}^{-3}$. If the Pt cubes were 1 nm, then the specific surface area would be $280\,\text{m}^2{\cdot}\text{g}^{-1}$ and the activity of the catalyst per unit weight would be one million times larger!

The very large surface area of small particles makes them unstable, because their surface energy is high. The particles tend to grow together (sinter) and, as a consequence, their surface area and surface energy decrease. If a reactor were filled with nano-sized Pt particles and a catalytic reaction run at several hundred degrees, then sintering of the particles would occur. Neighbouring particles make contact through corners, edges and planes and at elevated temperature the surface atoms are mobile and grow together to give large particles with a very much smaller surface area. To circumvent sintering, catalyst particles are placed on a material with a large surface area. The material acts as a support for the catalytic particles. When, during preparation, the catalyst particles are well spread over the surface of the support they do not contact each other. The support must, therefore, have a high surface area, and must interact with the catalyst particles to avoid diffusion of the catalyst particles over the support surface and, thus, avoid sintering of the catalyst particles. Catalysts in refineries and in base-chemical processes are usually used for several years and often at elevated temperature. That means that the support must also be stable during long periods of operation at high temperature and, thus, must have a high thermal and chemical stability.

Common supports in refineries and in the base-chemical industry are alumina (Al_2O_3), silica (SiO_2) and amorphous silica–alumina ($SiO_2{\cdot}Al_2O_3$, abbreviated as ASA). The specific surface area of these supports varies from 100 to 800 $m^2{\cdot}g^{-1}$. Other supports, such as carbon, barium sulfate ($BaSO_4$), magnesia (MgO), zinc oxide (ZnO) and titania (TiO_2) are used in the fine chemicals industry. Table 2.1 gives ten examples of catalysts and supports in refinery and base-chemical processes. The reactions are varied and include the removal of sulfur from oil fractions, the increase in the octane number of gasoline, the hydrocracking of heavy oil fractions to light fractions, the steam reforming of methane to synthesis gas, the water–gas-shift (WGS) reaction of CO to H_2, the Fischer–Tropsch reaction of synthesis gas to hydrocarbons, the alkylation of benzene, the reaction of ethylene to ethylene oxide, the reaction of propene to acrylonitrile and the oxidation of SO_2 to SO_3 (to produce H_2SO_4).

Catalysts have several shapes. In a laboratory reactor with a volume of about 1 cm^3 powdered catalysts ($< 100\,\mu m$) are often employed. The particles consist of small, **primary particles** (10–50 nm), which agglomerate randomly to **secondary particles** up to several hundred nm. These secondary particles can agglomerate further to **tertiary structures** up to a few hundred μm, the macroscopic powder catalyst particles. Random packing of particles leaves spaces between the particles, which are several times smaller than the particles themselves. Although these interparticle spaces are not cylindrical, they are referred to as pores and often represented in diagrams as cylinders. Nevertheless, based on this

Table 2.1 Examples of heterogeneous catalysts used in industry.

$Co–MoS_2/\gamma-Al_2O_3$	Sulfur removal from oil	Co/SiO_2	Fischer–Tropsch
$Pt/\gamma-Al_2O_3$	Octane number enhancement	H_3PO_4/SiO_2	Alkylation benzene
$Ni–MoS_2/SiO_2{\cdot}Al_2O_3$	Hydrocracking of oil	$Ag/\alpha-Al_2O_3$	Ethylene oxide
$Ni/\alpha-Al_2O_3$	Steam reforming	Bi_2MoO_6/SiO_2	Acrylonitrile
Fe_3O_4, $Cu/ZnO–Al_2O_3$	Water–gas-shift	V_2O_5/SiO_2	Sulfuric acid

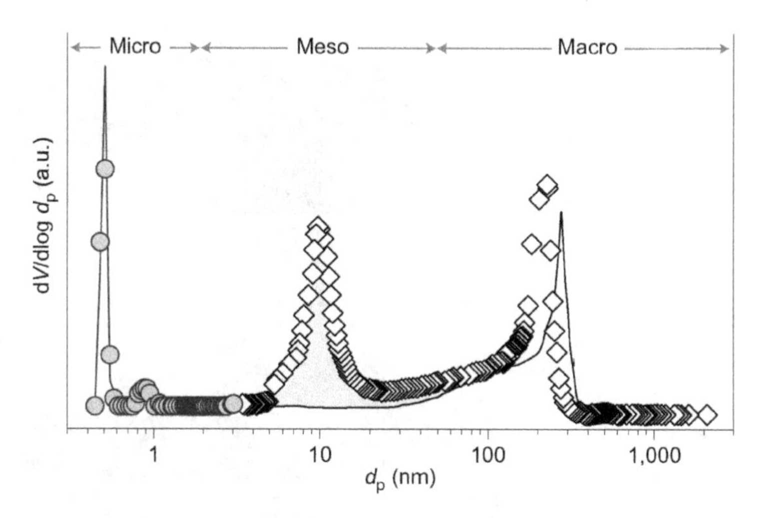

Fig. 2.1 Micro-, meso- and macroporosity of extrudates of ZSM-5 (drawn line) and mesoporous zeolite ZSM-5 (○ and ◇). The mesopores were created by base leaching. Reprinted with permission from [2.1]. Copyright (2012) Macmillan Publishing Ltd.

representation it is possible to compare the pore size with the size of a molecule and to determine whether the molecule can diffuse through the pore. The International Union of Pure and Applied Chemistry (IUPAC) has defined three pore sizes (Fig. 2.1):

- **micropores** ($< 2\,\text{nm}$),
- **mesopores** (2–50 nm),
- **macropores** ($> 50\,\text{nm}$).

This may indicate that, in powdered supports, the space between the primary particles forms the microporous space and that the space between the secondary particles determines the mesopores.

The macropore space between particles with a tertiary structure up to a few hundred μm is small. If such particles were used in large industrial fixed-bed reactors with a volume above $1\,\text{m}^3$, then it would require high pressure to force the gas or liquid feed through the reactor. This would lower the energy efficiency of the process. The solution is to use larger catalyst particles and, thus, create a macroporous interparticle space to enable fast diffusion of the molecules through the catalyst bed to all the catalyst particles. Therefore, the powdered particles are "glued" together with an organic or inorganic binder (boehmite is often used as inorganic binder, see Section 2.3.1 for its formula and structure). A paste of powder,

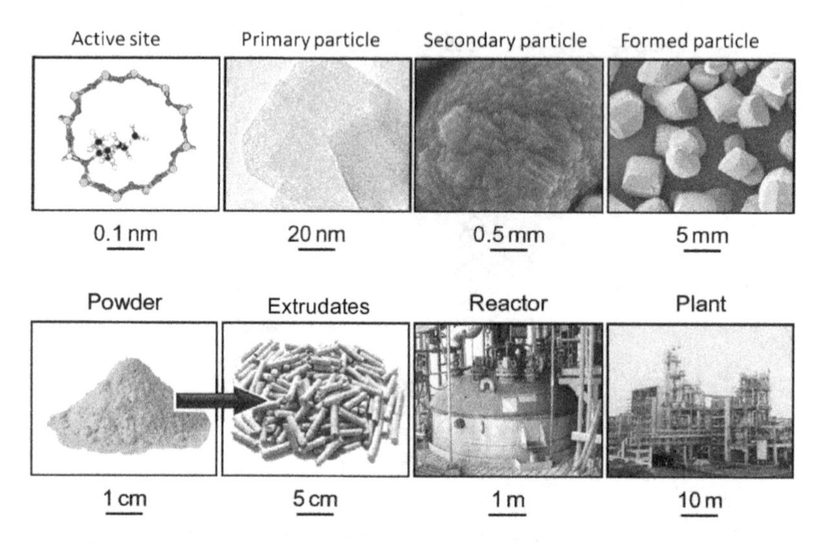

Fig. 2.2 Change in the catalyst from molecular (sub nm) to macroscopic (m) dimension. Reprinted with permission from [2.1]. Copyright (2012) Macmillan Publishing Ltd.

binder, and water is pressed through an extruder, which consists of a screw drive that forces the paste through holes (circles, triangles, trilobed or quadrilobed circles) in a plate. The resulting ribbon (with a cylindrical or lobed cylindrical section) of paste is baked in a furnace and cut into pieces, the so-called extrudates (Fig. 2.2), which may have a diameter of several mm and a length of several cm. The empty space in a reactor between the extrudates is called the bed porosity ($\sim$30–40%). Spherical particles (beads) are prepared by letting droplets of paste (powder + binder + water) descend in a vessel filled with hot oil. The water evaporates, the particles bake together and the viscous flow forces the particles into a spherical shape. The beads are collected at the bottom of the vessel. Beads give a bed porosity of about 30%. Figure 2.2 demonstrates the change in scale when going from a catalyst material to a catalytic plant.

Many materials are used to support catalytic particles, but alumina and silica have found widespread use, because they are the most versatile materials in terms of acid/base resistance, thermal stability, adjustment of chemical/physical properties and relative ease of forming. In addition, sources for these raw materials are plentiful. Alumina can be made in many structures that are typically differentiated by their Greek precursors. The worldwide annual production of alumina is about 130 MT (mainly used in the aluminium industry) and of catalyst grade aluminas is about 0.6 MT. γ-Alumina (γ-Al_2O_3) is used widely as a catalyst support in refineries for catalytic reforming (increasing the octane number of naphtha, Section 7.5.3), hydrotreating (removal of S and N atoms from fuel fractions, Sections 8.4.1 and 8.4.2), hydroconversion of residues to clean fuels (Section 8.4.3), as well as in the synthesis of petrochemicals and fine chemicals [2.2]. Furthermore, γ-Al_2O_3 and θ-Al_2O_3 are used as supports for automotive-exhaust catalysts (Section 9.1.2). Because aluminas can be both acidic and basic (Section 2.3), they are also suitable for use as a catalyst. Aluminas are used as Claus catalysts in the conversion of H_2S into sulfur ($2\,H_2S + SO_2 \rightarrow 2\,H_2O + 3\,S$) and in the synthesis of methyl chloride ($CH_3OH + HCl \rightarrow CH_3Cl + H_2O$), which is used to make methyl chlorosilane, an intermediate in the synthesis of silicones. In Section 2.3, we describe the synthesis and structures of aluminas and in Section 2.4 the synthesis and structure of silica but, in order to understand the structures of the aluminas, we will first take a look at the structures of metal oxides in general and how to determine if the correct material has been prepared.

2.2 Crystal Structures

2.2.1 *Crystal Lattices*

To maximise the interatomic forces and, thus, lower the energy, the atoms in a solid are packed as closely as possible. In metals, this leads to the most densely packed face-centred cubic (fcc) and hexagonal close-packed (hcp) structures, in which layers of atoms are arranged in a hexagonal pattern (Fig. 2.3). This arrangement gives the maximum number of atoms

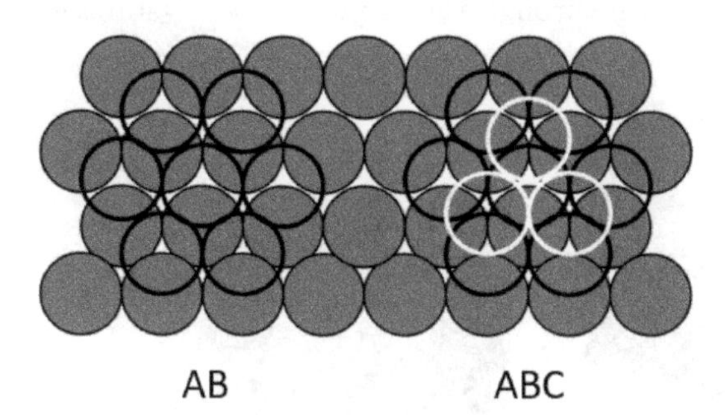

Fig. 2.3 The atoms in each close-packed layer fill the threefold hollow sites between the atoms in the layer beneath it. The atoms in the third layer can either be located right above the atoms in the first layer (hcp structure with AB sequence) or above the threefold hollow sites between the atoms in the first layer (fcc structure with ABC sequence).

in plane A. A second close-packed layer B can be positioned on top of layer A by placing the B atoms in the middle of three A atoms, in the threefold hollow positions of the A layer. The atoms of a third close-packed layer C can be put in **threefold hollow** positions on top of the B layer in two ways:

(1) placing them right above the atoms in the A layer gives the **hcp structure** with **AB sequence**, in which the A and B planes alternate (Fig. 2.3, left);
(2) placing the atoms of the third layer in new hollow positions (Fig. 2.3, right), giving the **fcc structure**, which has the **ABC sequence**.

Not only are metals built up of close-packed layers, but so are many metal halides, oxides and sulfides. For instance, the lattice of **rock salt** (NaCl) can be considered to be made up of close-packed layers of Cl^- ions, packed in ABC sequence (fcc structure). The smaller Na^+ cations are located between the Cl^- layers and each Na^+ is surrounded by six Cl^- ions. The six Cl^- anions are considered to be at the corners of an octahedron, with eight triangular side planes (Fig. 2.4c). Therefore, the Na^+ cations are in an octahedral position. One can also consider the NaCl structure to consist of close-packed layers of Na^+ cations, with Cl^- anions in the octahedral positions between the Na^+ layers.

A crystal is considered to be built up by the repetition of a unit cell in three dimensions. Directions in the crystal are indicated by vectors [uvw]. The indices u, v and w indicate that the direction of the vector is between the point 0, 0, 0 (the origin of the axis system) and the point u, v, w along the x, y and z axes, respectively, indicated by a, b and c in Fig. 2.5. Planes can be drawn through the atoms in the crystal and the orientation

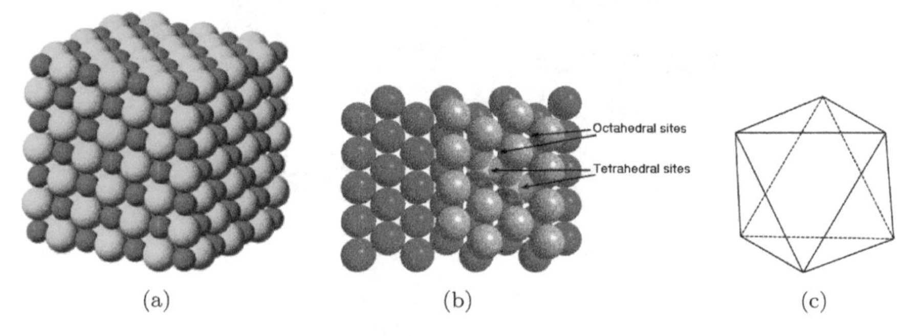

Fig. 2.4 (a) Rock salt structure of NaCl and MgO, (b) layers of anions with octahedral and tetrahedral holes for cations, (c) octahedron.

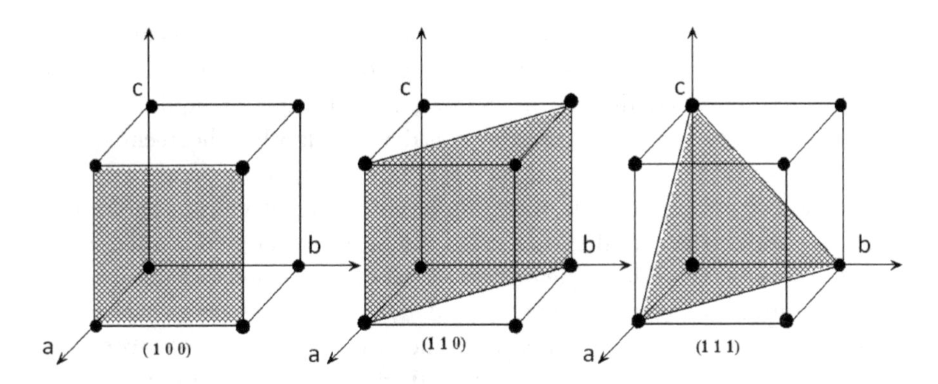

Fig. 2.5 Most important crystal planes with low Miller indices.

of the planes is indicated by **Miller indices (*hkl*)**, the reciprocals of the fractional intercepts of the plane with the crystallographic axes (Fig. 2.5).

When one of the Miller indices of a plane is zero, as in $(hk0)$, then the axis intercept is $1/0 = \infty$ and the plane $(hk0)$ is parallel to the c axis. Figure 2.5 shows that the (100) plane is parallel to the b and c axes, while the (110) plane is parallel to the c axis. Planes and directions with the same Miller indices are perpendicular to each other: $(hkl) \perp [hkl]$. Thus, the [100] direction (the a axis) is perpendicular to the (100) plane. The [110] direction is the diagonal in the ab plane and the [111] direction is the diagonal from one corner of the cube ($a = b = c = 0$) to the opposite corner ($a = b = c = 1$).

Most crystals have surface planes with low Miller indices, because these planes have a high atom density. The most important surfaces are, therefore, the (100), (110) and (111) planes. Rock salt structures, such as NaCl and MgO, usually have (100) and (110) surfaces, because they have an equal number of cations and anions in the surface and are, thus, electrically neutral. The (111) surface of rock salt contains only anions or only cations in a close-packed layer (Fig. 2.4b). Because of the strong electric field perpendicular to the surface, the surface will reconstruct to a surface that is a mixture of cations and anions, so that the total charge and the electric field at the surface are reduced (see also Section 2.3.5.3).

2.2.2 *X-ray Diffraction*

The structure of crystalline compounds can be determined with X-ray diffraction (XRD). This technique can also be used to tell which crystalline phases are present in a material and this makes it a standard technique for checking if the preparation of a catalyst has been successful. XRD can be used under *in situ* conditions, in the presence of a gas or liquid, because X-rays penetrate the material to a considerable depth. The technique is based on the principle that an X-ray is scattered by atoms and molecules when the X-ray wavelength is of the dimension of the atoms and molecules. An XRD apparatus contains an X-ray tube, in which electrons are created and accelerated in a high electric field. Upon impingement on a metal surface (often copper) holes are created in the inner shells of the metal atoms (e.g. a Cu 1s hole). The holes are filled by transfer of an electron from a higher shell to the hole (Cu 2p $\rightarrow$ 1s) and the transition energy is released as monochromatic X-rays. The XRD apparatus furthermore contains an X-ray detector. The X-ray source and detector move synchronously on a so-called Rowland circle or only the detector rotates along the Rowland circle. The **Bragg equation**

$$n\lambda = 2d \sin\theta, \tag{2.1}$$

relates the angle θ between the incoming X-ray beam and crystal plane to the distance between subsequent planes in a crystal lattice (Fig. 2.6). Because the angle between the incoming and the reflected X-ray beam is experimentally measured, the reflected X-ray intensity is usually plotted as a function of 2θ (Fig. 2.7). When measuring a single crystal, it is possible to rotate the crystal and measure the reflections coming from the planes of the crystal lattice. This enables the determination of the structure of the crystal, as well as the positions of the atoms in the crystal. The distance between crystal planes can be expressed as $\frac{1}{d_{hkl}^2} = \frac{h^2}{a^2} + \frac{k^2}{b^2} + \frac{l^2}{c^2}$, where h, k and l are the Miller indices and a, b and c are the unit distances along the three main crystallographic axes (cf. Fig. 2.5).

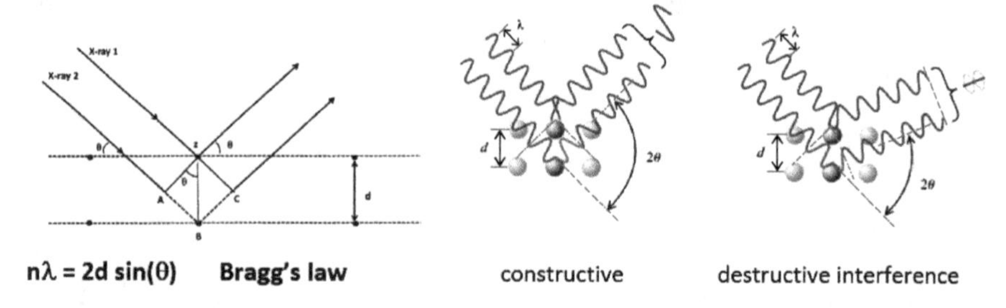

Fig. 2.6 Interference between X-ray waves reflecting from two subsequent crystal planes.

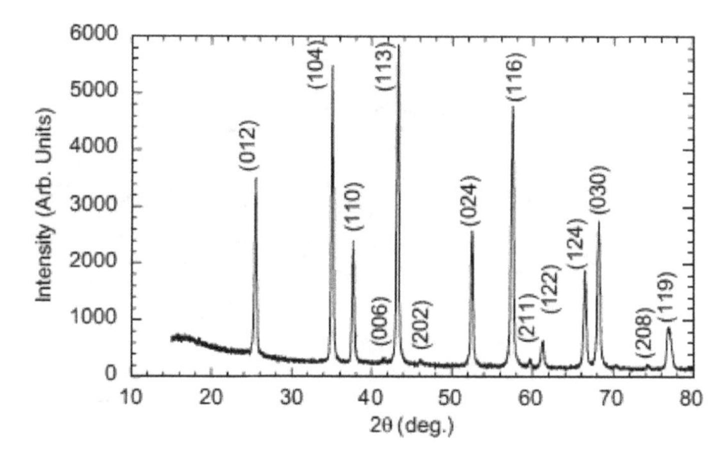

Fig. 2.7 Powder X-ray diffraction pattern of α-Al_2O_3. The Miller indices (*hkl*) indicate which reflecting planes are responsible for the peaks. Reprinted with permission from [2.3]. Copyright (2009) Elsevier.

The measurement of the positions and intensities of the X-ray reflections is carried out automatically. The determination of the crystal structure is performed with the help of computer programs. The greatest challenge is the preparation of a sufficiently large and pure single crystal. When such a crystal is not available, it is possible to measure the XRD pattern of its powder; in many cases the XRD pattern of the powder enables the determination of the crystal structure. For instance, the structure of α-Al_2O_3 (Section 2.3.3) can be determined from its powder XRD pattern (Fig. 2.7). This is especially important for heterogeneous catalysts, which contain small crystals of a catalytic material supported on a support. The support is microcrystalline and therefore the catalyst particles are oriented in a random orientation relative to the incoming X-ray beam.

With XRD one can even determine the size of the crystallites from the width of the XRD peaks with the **Scherrer equation**

$$D_{hkl} = K \cdot \lambda / [\cos \theta_{hkl} \cdot (\delta\theta)_{hkl}], \tag{2.2}$$

where D_{hkl} is the diameter of the particle perpendicular to the crystal plane (*hkl*), K is a constant (usually one takes $K = 0.9$), θ is the angle at which the X-ray is reflected from the crystal plane (*hkl*) (Fig. 2.6) and $\delta\theta$ is the width at half height of the XRD peak. A derivation of the Scherrer equation is given in [2.4]. The Scherrer equation explains why very small particles ($< 3\,\text{nm}$) cannot be measured by XRD, because the line width

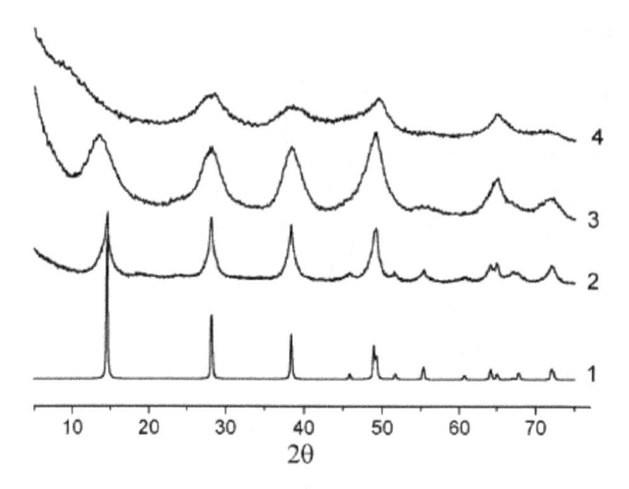

Fig. 2.8 XRD patterns of boehmites: (1) theoretical pattern, (2) and (3) crystalline boehmite with different crystallite size, (4) pseudoboehmite. Reprinted with permission from [2.5]. Copyright (2010) Springer Science + Business Media.

would become too large to observe an XRD peak. Very large particles can be measured by XRD but their particle size cannot be determined because the contribution of particle size to line width becomes so small that other effects predominate. Examples of line broadening caused by the small size of the crystallites are presented in Fig. 2.8 [2.5]. Boehmite is an aluminium oxyhydroxide (AlO(OH), cf. Section 2.3.1) and its crystallites are often small and have broad XRD lines (patterns 2 and 3 in Fig. 2.8). Pseudoboehmite has very broad lines and the (020) peak, which is at about 14° (2θ) in large crystallites, is very broad and shifted to lower angle (pattern 4). This is because pseudoboehmite consists of disordered plate-like crystallites with a thickness of only a few atomic layers. These primary particles are agglomerated to particles larger than 100 nm. Pseudoboehmite is considered as the initial stage of crystallisation of boehmite, in which two-dimensional platelets with the structure of boehmite have already formed, but the three-dimensional crystal structure is not yet fully developed.

2.3 Aluminas

2.3.1 *Aluminium Hydroxides and Oxyhydroxides*

All aluminium hydroxides, oxyhydroxides and oxides are made from bauxite. Bauxite is formed by weathering of silicate rock and consists of aluminium hydroxides, iron hydroxides (hence its reddish-brown colour), kaolinite (China clay, $Al_2(OH)_4Si_2O_5$, the name kaoline comes from the Chinese word Gao Ling, high hill) and titanium hydroxide. In the Bayer process, the aluminium hydroxides are dissolved in hot NaOH to $NaAl(OH)_4$. After filtration and cooling, aluminium hydroxide (gibbsite, $Al(OH)_3$) is precipitated by seeding the $NaAl(OH)_4$ solution with gibbsite.

In the laboratory, aluminum hydroxides can be prepared by acidification of aluminate, neutralisation of aluminum salts and hydrolysis of aluminium alkoxides:

$$NaAlO_2 + HCl + H_2O \rightarrow NaCl + Al(OH)_3, \tag{2.3}$$

$$Al_2(SO_4)_3 + 6\,NaOH \rightarrow 2\,Al(OH)_3 + 3\,Na_2SO_4, \tag{2.4}$$

$$Al(OR)_3 + 3\,H_2O \rightarrow Al(OH)_3 + 3\,HOR. \tag{2.5}$$

Four $Al(OH)_3$ hydroxides are known: bayerite, gibbsite, nordstrandite and doyleite, all of which have a HO–Al–OH sandwich-like structure with a layer of Al^{3+} cations in the octahedral positions between two OH^- anion layers. Due to the difference in the charge of one Al^{3+} cation and two OH^- anions, one third of the octahedral positions are not filled, leading to a honeycomb distribution of the Al^{3+} cations (Fig. 2.9).

The $Al(OH)_3$ hydroxides differ in the orientation of the OH groups at the interface between the sandwich sheets. Gibbsite has an AB BA AB stacking order of the OH layers, with the OH groups in one layer directly above the OH groups in the next layer (Fig. 2.10). Bayerite has an AB AB AB stacking order, with the OH groups from adjacent layers in a close-packed array. In both structures, there are inter-layer and intra-layer hydrogen bonds between OH groups. Gibbsite is a major constituent of bauxite and laterite, which are formed by weathering of aluminium silicate rocks in

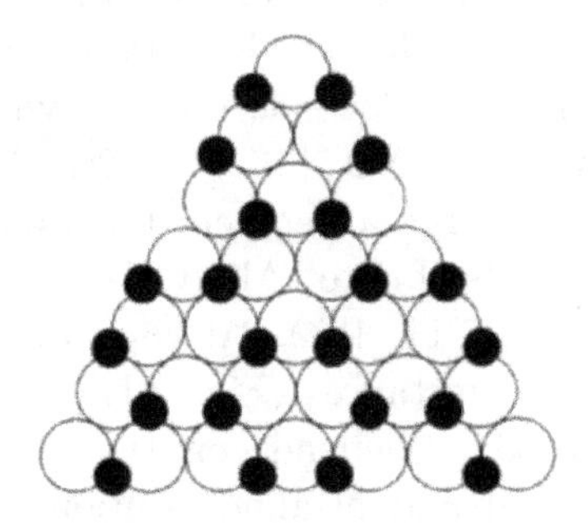

Fig. 2.9 Location of the Al^{3+} cations (black balls) between the OH^- layers (white balls) of gibbsite and bayerite.

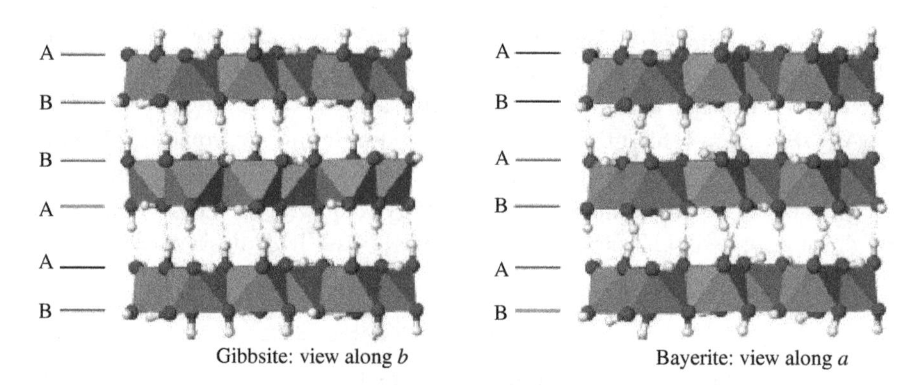

Fig. 2.10 Layer structures of gibbsite and bayerite. Reprinted with permission from [2.6]. Copyright (2008) Elsevier.

tropical and subtropical regions. Bayerite forms when bauxite is dissolved in caustic solution.

Not only aluminium hydroxides but also the aluminium oxyhydroxide boehmite (AlO(OH)) may be formed in reactions (Eqs. (2.3)–(2.5)) and subsequent dehydration (Eq. (2.6)):

$$Al(OH)_3 \rightarrow AlO(OH) + H_2O. \tag{2.6}$$

The resulting boehmite is dehydrated to aluminium oxide by heating (Eq. (2.7)):

$$2\,AlO(OH) \rightarrow Al_2O_3 + H_2O. \tag{2.7}$$

Two AlO(OH) oxyhydroxides are known, boehmite and diaspore, both of which are found in nature. Boehmite is a constituent of bauxite, while diaspore is found in rocks and is considered to be a gemstone (zultanite). Whether hydroxide or oxyhydroxide is obtained after hydrolysis of Al compounds (Eqs. (2.3)–(2.5)) depends on the synthesis conditions (pH, temperature and anions (Cl^-, NO_3^-, SO_4^{2-})) [2.7]. For instance, the XRD patterns indicate that, at room temperature, amorphous $Al_2O_3{\cdot}xH_2O$ is obtained at pH < 6, boehmite AlO(OH) at pH ~ 7 and bayerite $Al(OH)_3$ at pH > 9 (Fig. 2.11, left). At higher temperature, boehmite is obtained under all circumstances (Fig. 2.11, right). The formation of the different (oxy)hydroxides depends on the aluminium complex that forms in the acid, neutral and alkaline solution. In an acidic medium, $[Al_{13}O_4(OH)_{24}(H_2O)_{12}]^{7+}$ polycations are present and the addition of a base leads to the formation of an amorphous gel between 25 °C and 80 °C,

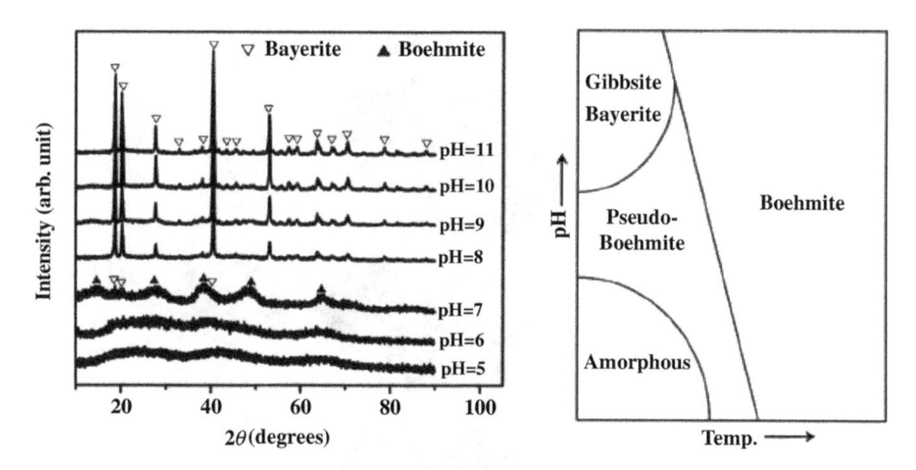

Fig. 2.11 Left: XRD patterns of Al (oxy)hydroxides precipitated at different pH from a solution of $Al(NO_3)_3$. Reprinted with permission from [2.7]. Copyright (2009) Elsevier. Right: pH-T dependence of the stability of gibbsite, bayerite and boehmite.

and boehmite above 80 °C [2.7]. At neutral pH and higher temperature, trinuclear Al_3 clusters evolve to $Al_4O(OH)_{10}(H_2O)_5$ tetramers, which condense to form boehmite. At high pH, anionic Al clusters form and condense to bayerite.

The oxyhydroxide boehmite consists of two layers of Al^{3+} cations and O^{2-} anions, sandwiched between OH^- anion layers (Fig. 2.12). The Al^{3+} and O^{2-} layers form sheets of AlO_6 octahedra, with O^{2-} anions in the corners and Al^{3+} cations in the centre. The octahedra are fused by edge sharing. The hydroxyl ions hold the sandwich sheets together through hydrogen bonding. Boehmite, thus, consists of HO–AlOOAl–OH sandwich layers. The transitions from bayerite and gibbsite to boehmite ($Al(OH)_3 \rightarrow AlO(OH) + H_2O$) are reversible. Bayerite can change to boehmite by heating an aqueous slurry, while a boehmite slurry converts to bayerite upon ageing at 40 °C. This reversibility has to do with the similarity of the crystal structures. All three compounds have a layered structure and can, therefore, interconvert easily.

2.3.2 *Transition Aluminas*

Many aluminium oxides exist (α, γ, δ, η, θ, κ, ρ, χ). They are prepared by dehydration of bayerite, gibbsite or boehmite and the resulting products depend on the precursor (Fig. 2.13). Bayerite, gibbsite and boehmite transform to η-Al_2O_3, χ-Al_2O_3 and γ-Al_2O_3, respectively, and on to

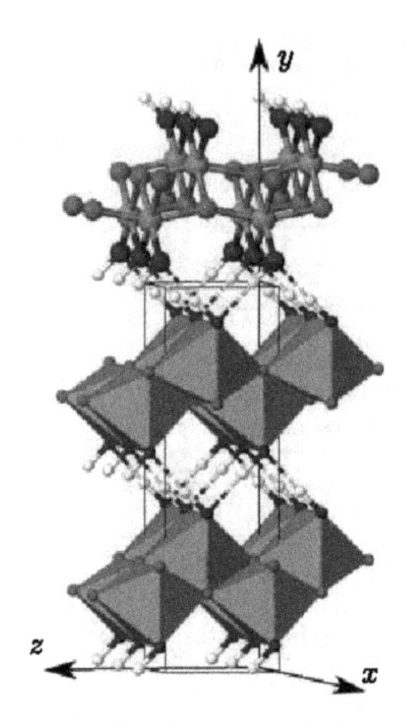

Fig. 2.12 Double layer structure of boehmite, with hydrogen bonds between the double sheets of octahedra. Reprinted with permission from [2.8]. Copyright (2009) Springer Science + Business Media.

other aluminas. The transformations do not occur at one temperature but over a range of temperatures. The ranges depend amongst others on the particle size and on impurities. The sequence of transformation is not reversible. Neither α-Al_2O_3 nor any of the high-temperature aluminas can be converted to one of the lower-temperature aluminas. In the presence of water (steam), the aluminas convert to the hydroxides (bayerite or gibbsite) or oxyhydroxide (boehmite). Figure 2.13 is, thus, not a phase diagram; it only indicates the temperature ranges in which the aluminas occur. Therefore, these aluminas are called transition aluminas. They are thermodynamically unstable, but reasonably reproducible states of structural reordering, and the structure of the starting material determines the type and sequence of transition forms during thermal decomposition.

Figure 2.14 shows the XRD patterns of the aluminas that are formed during the heating of bayerite, boehmite and gibbsite and the temperatures required for these transformations to occur. Upon heating to 1,000 °C, the hydroxide bayerite transforms first to η-Al_2O_3, then to θ-Al_2O_3 and finally to α-Al_2O_3. The hydroxide gibbsite transforms first to χ-Al_2O_3, then to κ-Al_2O_3 and, with time, to α-Al_2O_3. The oxyhydroxide boehmite first

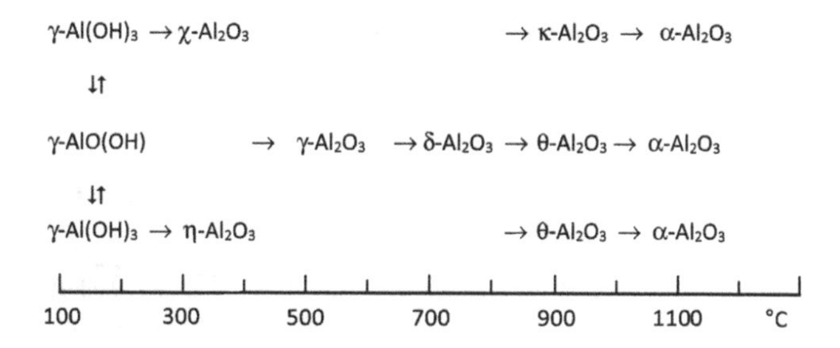

Fig. 2.13 Transitions of Al (oxy)hydroxides to aluminas.

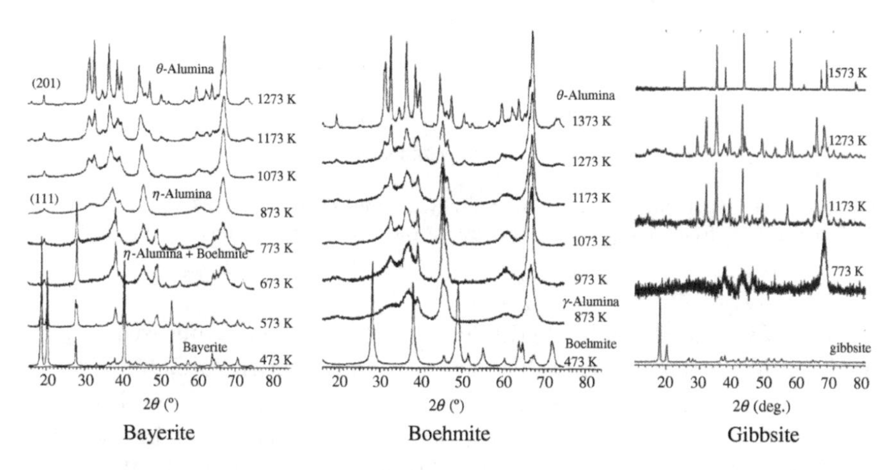

Fig. 2.14 XRD powder patterns of bayerite, boehmite and gibbsite and of the aluminas that form upon heating. Left and middle figure reprinted with permission from [2.9]. Copyright (1991) the International Union of Crystallography (http://dx.doi.org/10.1107/S0108768191002719). Right figure reprinted with permission from [2.10]. Copyright (2010) Elsevier.

forms γ-Al_2O_3, then δ-Al_2O_3, θ-Al_2O_3 and finally α-Al_2O_3. Boehmite is the only aluminium compound to produce γ-Al_2O_3 upon heating. Therefore, to prepare γ-Al_2O_3, the most widely used catalyst support, the pH and temperature must be controlled during the hydrolysis of the aluminium salt, because otherwise aluminium hydroxides (Fig. 2.11) and thus other aluminas are obtained.

The stability diagram (Fig. 2.15) shows the free enthalpies ΔG^0 of the reactions α-$Al_2O_3 + nH_2O \rightarrow Al_2O_3{\cdot}nH_2O$, with $n = 0.2$ for tohdite ($Al_5O_7(OH)$, also referred to as akdalaite), $n = 1$ for $AlO(OH)$ and $n = 3$ for $Al(OH)_3$, as calculated by density functional theory (DFT) [2.11].

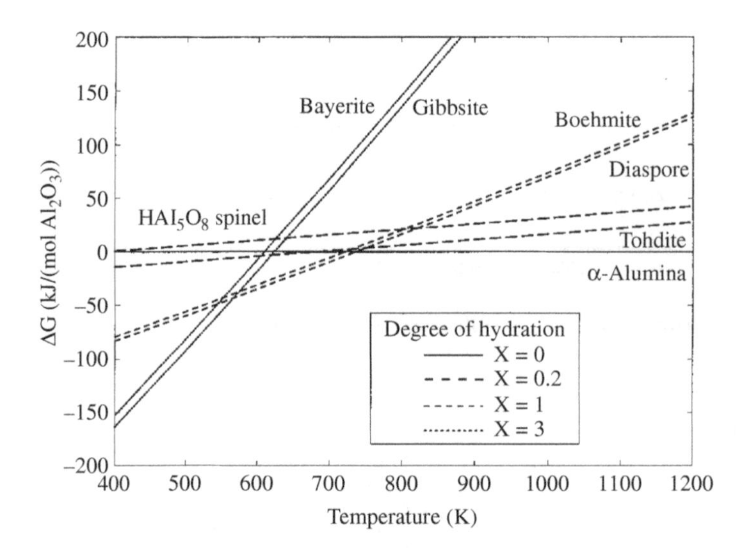

Fig. 2.15 Free enthalpies of aluminium oxyhydroxides compared to α-Al_2O_3. Reprinted with permission from [2.11]. Copyright (2002) the American Chemical Society.

This shows that the fully hydroxylated bayerite and gibbsite are thermodynamically the most stable forms at low temperature, but that at higher temperature dehydrated structures become more stable. Boehmite (γ-AlO(OH)) and diaspore (α-AlO(OH)) are more stable at intermediate temperature and fully dehydrated α-Al_2O_3 is the most stable alumina at high temperature.

The transitions from boehmite to γ-Al_2O_3, bayerite to η-Al_2O_3 and gibbsite to χ-Al_2O_3 can be understood by comparing the crystal structures of the (oxy)hydroxides and the resulting aluminas. Many aluminas have a structure consisting of alternating layers of O^{2-} anions and Al^{3+} cations. As described in the foregoing, boehmite, bayerite and gibbsite also have a layered structure, be it a slightly different one. The reactions of boehmite to γ-Al_2O_3, bayerite to η-Al_2O_3 and gibbsite to χ-Al_2O_3 take place by reaction of neighbouring hydroxyl anions ($2\,OH^- \rightarrow H_2O + O^{2-}$). The resulting water molecules diffuse from the crystal along the plane of the original hydroxyls and a hydroxyl vacancy and an O^{2-} ion form in a hydroxyl lattice position. A shift parallel to the layers brings together the chains of the octahedrally-coordinated Al^{3+} ions of the original boehmite. Therefore, some of the Al^{3+} ions move to different positions, including tetrahedral positions [2.12, 2.13]. At the relatively low temperature of the transformation, the thermal energy is insufficient to allow rearrangement to a perfectly ordered material. This explains the high degree of disorder in γ-Al_2O_3.

2.3.3 α-Al_2O_3

When transition aluminas are heated to high temperature they transform into α-Al_2O_3 (corundum), the most stable form of alumina (Fig. 2.13). Because of the high temperature, rather large α-Al_2O_3 particles form and the peaks in the XRD pattern of α-Al_2O_3 are narrow (Fig. 2.7). α-Al_2O_3 contains close-packed layers of oxygen ions and cations in octahedral holes between the oxygen layers. The oxygen anion layers have the hcp structure, with AB stacking (Fig. 2.3). Because of the different charge of the Al^{3+} cations and O^{2-} anions, the Al^{3+} cations in α-Al_2O_3 have the same honeycomb-type ordering between the layers of O^{2-} anions as the Al^{3+} cations between the OH^- layers in bayerite and gibbsite (Fig. 2.9). If Al^{3+} cations in single crystals of corundum are substituted by traces of other cations, gemstones and semi-precious gemstones are formed: sapphire (blue, Fe^{3+}, Ti^{3+}), ruby (red, Cr^{3+}), topaz (yellow, Fe^{3+}, Fe^{2+}), amethyst (purple, Fe^{3+}, Mn^{3+}, Ti^{3+}) and emerald (green, Fe^{2+}), see Fig. 2.16. The Logan sapphire (Smithsonian Museum of Natural History, Washington, D.C., USA) is one of the largest sapphires ever found (423 carat, 85 g).

Corundum is a hard material that is used as an abrasive in finger files and to polish lenses and spectacle glasses. It is also used as a support for catalysts used in high-temperature applications, as in the Ni/α-Al_2O_3 catalyst, which must withstand high temperatures in steam reforming (Section 1.3.1). Another catalytic application of α-Al_2O_3 is as a support for

Fig. 2.16 The Logan sapphire (left) and a ruby (right), as examples of corundum doped with Fe^{3+} and Ti^{3+} cations and with Cr^{3+} cations respectively.

silver particles in the catalysed synthesis of ethylene oxide from ethylene and oxygen. This reaction must be monitored carefully; otherwise the ethylene oxide product would oxidise further to CO_2 (Section 9.3.1). The support must be inert, have wide pores (so that the intermediate ethylene oxide can diffuse quickly out of the pores) and be stable at high temperature. A more recent application of α-Al_2O_3 is as support for Pd particles. This catalyst combines CO molecules oxidatively to dialkyl oxalate [2.14].

$$2\,CO + 2\,C_2H_5ONO \rightarrow 2\,NO + C_2H_5OOC - COOC_2H_5,$$

$$2\,C_2H_5OH + 0.5\,O_2 + 2\,NO \rightarrow 2\,C_2H_5ONO + H_2O,$$

$$2\,CO + 2\,C_2H_5OH + 0.5\,O_2 \rightarrow C_2H_5OOC - COOC_2H_5 + H_2O.$$

Because dialkyl oxalate can be hydrogenated to ethylene glycol, this allows the production of glycol from coal.

$$C_2H_5OOC - COOC_2H_5 + 4\,H_2 \rightarrow 2\,C_2H_5OH + CH_2OHCH_2OH.$$

2.3.4 γ-Al_2O_3

γ-Al_2O_3 is the most widely used catalyst support, because it can be produced with a high surface area (200–300 $m^2 \cdot g^{-1}$) and is thermally and chemically stable. Its large surface area allows it to be loaded with a large amount of active material (metal, metal oxide or metal sulfide). Even though γ-Al_2O_3 is widely used, some aspects of its structure are still unknown. It is generally agreed that the O^{2-} anions in γ-Al_2O_3 form fcc close-packed layers with ABC stacking, because the (222) XRD peak is sharp (Fig. 2.17), its intensity being due mainly to X-ray scattering of O^{2-} anions [2.9]. Also the sharp (440) and (400) peaks point to an fcc structure, because in this structure the intensity of the (440) peak is determined by contributions from the oxygen anion planes as well as from the metal cation planes, both with the tetrahedral and the octahedral positions, and the (400) XRD peak receives intensity contributions from the oxygen anion planes and from the metal cation planes with octahedral positions. It is also undisputed that the cations occupy octahedral as well as tetrahedral positions between the O^{2-} anion layers.

The XRD pattern of γ-Al_2O_3 is similar to that of spinels, with a small peak at about 20° (2θ), several peaks at around 40° (2θ) and two peaks at around 65°(2θ). Therefore, it is assumed that γ-Al_2O_3 has a **spinel structure**. Spinels are ternary metal oxides with the stoichiometry AB_2O_4. Medium-sized A^{2+} B^{3+} and A^{4+} B^{2+} cation pairs may have a spinel

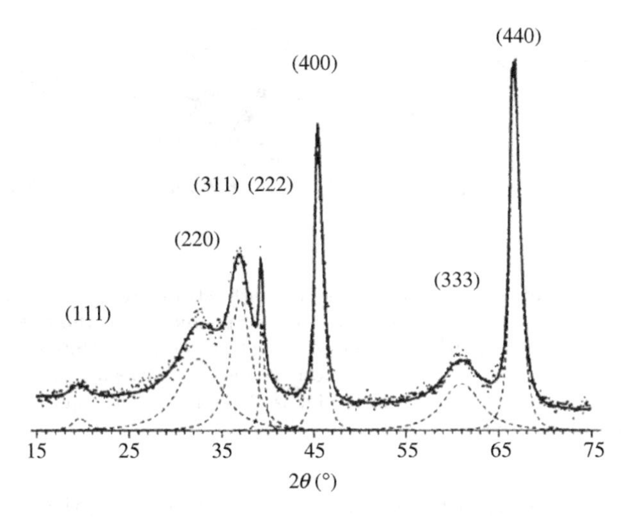

Fig. 2.17 XRD pattern of γ-Al_2O_3, with the diffracting (*hkl*) planes indicated above the XRD peaks. Reprinted with permission from [2.9]. Copyright (1991) the International Union of Crystallography (http://dx.doi.org/10.1107/S0108768191002719).

structure. Spinels can be considered to be built up of close-packed layers of oxygen ions, but the cations occupy not only octahedral positions between the oxygen layers (as in MgO and α-Al_2O_3), but also tetrahedral positions, in which the cation is surrounded by four anions, which form a tetrahedron.

There are two different cation layers sandwiched between close-packed oxygen layers in the [111] direction of a $[A^{2+}][B^{3+}]_2[O^{2-}]_4$ spinel (Fig. 2.18). One cation layer contains only octahedrally coordinated B^{3+} cations, while the other layer contains both octahedrally coordinated B^{3+} cations and tetrahedrally coordinated A^{2+} cations. Such spinels are referred to as **normal spinels**; examples are $MgAl_2O_4$, $CoAl_2O_4$, $FeCr_2O_4$ and $ZnFe_2O_4$. $MgFe_2O_4$ and Fe_3O_4 are **inverse spinels**, because in their $[B^{3+}][A^{2+}, B^{3+}][O^{2-}]_4$ structures one B^{3+} cation occupies a tetrahedral position and the A^{2+} cation and second B^{3+} cation occupy octahedral positions. To locally compensate the positive charge of a cation, higher valent cations should be surrounded by a larger number of anions and will, thus, prefer an octahedral rather than a tetrahedral position. This explains why compounds of Ti^{4+}, such as $TiMg_2O_4$ and $TiZn_2O_4$ (usually written as Mg_2TiO_4 and Zn_2TiO_4), are inverse spinels, $[M^{2+}][Ti^{4+}, M^{2+}][O^{2-}]_4$, and why in $MgAl_2O_4$ both Al^{3+} cations are octahedrally coordinated, even though the Mg^{2+} cation is "too" big for a tetrahedral position (which is smaller than an octahedral position).

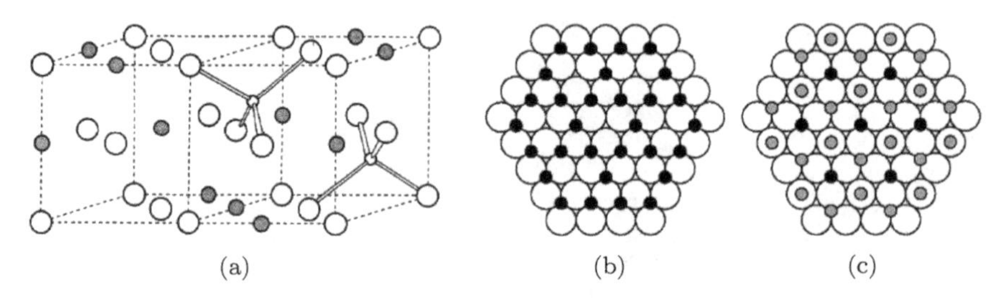

Fig. 2.18 Structure of an AB_2O_4 normal spinel with tetrahedrally coordinated A^{2+} and octahedrally coordinated B^{3+} cations. (a) view perpendicular to the (100) plane. O^{2-} anions are indicated by large white spheres, A^{2+} cations by small white spheres and B^{3+} cations by grey spheres. (b) and (c) view perpendicular to two subsequent (111) planes. O^{2-} anions are indicated by large white spheres, A^{2+} cations by small grey spheres and B^{3+} cations by black spheres.

Spinels are present in the earth crust as aluminates (MAl_2O_4), ferrites (MFe_2O_4) and chromates (MCr_2O_4) and in the earth mantle as M_2SiO_4 ortho-silicate spinels. M is a transition metal in these compounds. In industry, ferrites are used as antiferromagnetic or ferrimagnetic materials and $LiMn_2O_4$ is used as the cathode in lithium ion batteries. Spinels also play a role in jewellery. When $MgAl_2O_4$ contains a trace of Cr^{3+} it is red-violet and is referred to as red spinel. Red spinels are found in nature (Myanmar) and look exactly the same as rubies, because both contain Cr^{3+} cations in an octahedral position. The Black Prince's Ruby in the Imperial State Crown of the United Kingdom is not a ruby but a 170-carat spinel. The largest spinel gemstone is the Samarian spinel (500 carats, 100 g); it is part of the Iranian crown jewels.

In an actual AB_2O_4 spinel the metal-to-oxygen ratio is 3:4, but in aluminas it is 2:3. Thus, if γ-Al_2O_3 were to have a spinel structure, then it would lack metal cations and there would be cation vacancies in the lattice. Therefore, aluminas are referred to as **defect spinels**. The structure of γ-Al_2O_3 is still under debate, because, thus far, it has not been possible to prepare macrocrystalline γ-Al_2O_3, thus hindering the determination of the crystal structure. It is generally accepted that the oxygen anions are located in cubic close-packed layers, and that most of the Al^{3+} cations are at tetrahedral and octahedral spinel positions, but there is still disagreement with regard to the positions of some of the Al^{3+} cations and of the location of the lacking Al^{3+} cations, the vacancies. Neutron diffraction, transmission electron microscopy (TEM) and nuclear magnetic resonance (NMR) studies showed that the structure of γ-Al_2O_3 is more complicated than that of an ideal spinel with cation vacancies. An introduction to TEM is presented in [2.15] and an introduction to NMR in [2.16].

In the past, ^{27}Al NMR measurements had indicated that 25% [2.17] or 31% [2.18] of the Al^{3+} cations in γ-Al_2O_3 are on tetrahedral positions and that, as a consequence, the cation vacancies are mainly on tetrahedral positions. Based on these results, DFT calculations had predicted that γ-Al_2O_3 does not only contain Al^{3+} cations in spinel positions, but also in non-spinel positions and that the vacancies are mainly in tetrahedral positions [2.17–2.19]. However, ^{17}O NMR measurements and more recent ^{27}Al NMR measurements showed that there are 37.5% tetrahedrally coordinated Al^{3+} cations and, thus, that all vacancies are in octahedral positions, as in the isostructural γ-Fe_2O_3 and in accordance with DFT predictions [2.20]. A different model proposed that γ-Al_2O_3 consists of nanocrystals with all the Al^{3+} cations occupying spinel positions [2.21]. The irregularly stacking of the primary crystallites provides distortions in the cation sublattice while maintaining order in the oxygen sublattice. The planar defects in the cation sublattice are equivalent to filling non-spinel cation sites in other models of the γ-Al_2O_3 structure. This model explains the experimental XRD pattern of γ-Al_2O_3 without having to assume the presence of non-spinel cation sites in the bulk γ-Al_2O_3 structure [2.22].

2.3.5 *Surface of γ-Al_2O_3*

2.3.5.1 *Lewis acid sites*

Knowledge about the surface of γ-Al_2O_3 is important for many applications but especially for catalysis, because γ-Al_2O_3 is not only the main support for catalytic metal, metal oxide or metal sulfide particles, but also an acid catalyst. The morphology of γ-Al_2O_3 particles is inherited from the morphology of boehmite particles, because boehmite transforms topotactically to γ-Al_2O_3 (meaning that the crystal shape is preserved). Boehmite nanoparticles often have a rhombohedral shape (Fig. 2.19), with a predominant (010) basal surface and (100), (001) and (101) edge surfaces. These four boehmite surfaces transform to the (110), (110), (100) and (111) surfaces of γ-Al_2O_3, respectively [2.12].

Most Al^{3+} cations at the surface are **coordinatively unsaturated sites (CUS)**, coordinated to fewer oxygen anions than in the bulk. Their distribution on the surface and coordination states (5-, 4- or 3-coordinated) are important for the catalytic activity of γ-Al_2O_3, because the CUS Al^{3+}

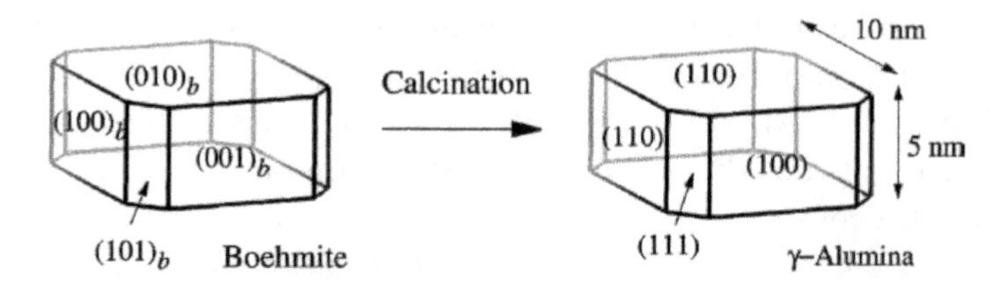

Fig. 2.19 Transformation of boehmite to γ-Al_2O_3 nanoparticles with the corresponding surface orientations. Reprinted with permission from [2.17]. Copyright (2004) Elsevier.

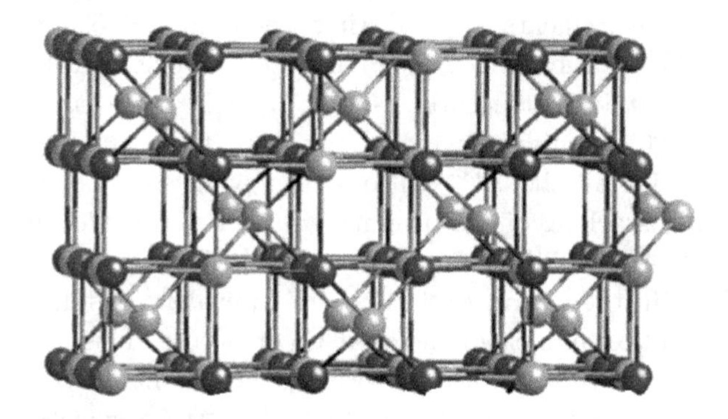

Fig. 2.20 Spinel structure with octahedrally and tetrahedrally coordinated Al^{3+} cations indicated as grey spheres and O^{2-} anions as dark spheres. The top layer is the high-density (100) surface and the layer below the top layer would form the low-density (100) surface.

cations act as **Lewis acid sites** that bind basic molecules

$$M^{n+} + B \rightarrow [M - B]^{n+}. \tag{2.8}$$

When the surface does not reconstruct, its structure is determined by the bulk structure. For γ-Al_2O_3 with a spinel structure, there are two types of planes in the [100] direction, one high-density plane contains O^{2-} anions and octahedrally coordinated Al^{3+} (Al_O) cations (Fig. 2.20, top layer) and a low-density plane that contains only tetrahedrally coordinated Al^{3+} (Al_T) cations (Fig. 2.20, second layer from the top) [2.23]. Also in the [110] direction (Fig. 2.20, in the diagonal direction) there are two different alternating types of planes; both contain Al^{3+} cations as well as O^{2-} anions. In one surface there is an equal number of Al_O and Al_T cations, while the other surface only contains Al_O cations. As discussed in Section 2.3.4, in the [111] direction γ-Al_2O_3 contains planes that alternatingly contain Al^{3+} cations or O^{2-} anions. There are two different Al planes, one contains only

Al_O cations, while the other contains both Al_O cations and Al_T cations (Fig. 2.18).

Many techniques can provide information about the bulk structure of a powdered solid, for instance XRD, but few can give information about its surface structure. Infrared (IR) spectroscopy is one of the techniques that can do that. IR radiation excites vibrations of molecules or solids and measuring the IR absorption allows distinguishing the molecules or solids. Catalyst samples can be deposited on an IR-transparent KBr disk and transmission spectra can be measured, or reflection spectra can be measured by irradiating powder samples and detecting the reflected IR radiation. The transmitted or reflected IR radiation is measured as a function of the wavelength; the spectrum reveals how much radiation is absorbed at each wavelength. Analysis of the position, shape and intensity of peaks in the IR spectrum reveals details about the structure of the sample. IR spectroscopy is a very versatile tool, because IR spectra can be measured in the presence of gaseous or liquid reactants, and it is a sensitive technique that allows measurement of small amounts of material.

An important application of IR spectroscopy in catalysis is the measurement of the catalyst surface. Although the M^{n+} cations and O^{2-} anions at the surface of a metal oxide cannot be measured directly by IR, because M–O vibrations at the surface show up as IR bands in the same low wavenumber range (300–800 cm^{-1}) as the bands of M–O bulk vibrations, they can be measured indirectly, by making use of the missing oxygen anions to chemisorb basic probe molecules on the CUS cations (Eq. (2.8)). The CUS cations act as Lewis acid sites that bind basic molecules B, such as pyridine, amines, alcohols or CO. The cation bonded to B effects the vibrations of molecule B and changes its IR frequencies, thus providing information on the **Lewis acid site**, the metal cation.

Pyridine is often used in studies of catalysts or support surfaces. Gaseous pyridine has IR bands at 1,436, 1,481, 1,577 and 1,583 cm^{-1} but after adsorption on γ-Al_2O_3 at room temperature there are six bands at 1,450, 1,490, 1,575, 1,595, 1,613 and 1,623 cm^{-1}, which disappear upon heating (Fig. 2.21) [2.24]. The increase in the number of bands is due to the fact that pyridine adsorbs on three Lewis acid sites, giving rise to IR bands at 1,595 cm^{-1} (weak sites), 1,613 cm^{-1} (medium strong sites) and 1,623 cm^{-1} (strong sites). The stronger the acidity of the site, the higher the temperature at which pyridine desorbs. The weak Lewis sites may consist of penta-coordinated Al^{3+} cations (μ_5-Al^{3+}), which originate from bulk octahedral Al, i.e. an Al_O cation that misses one O^{2-} anion and

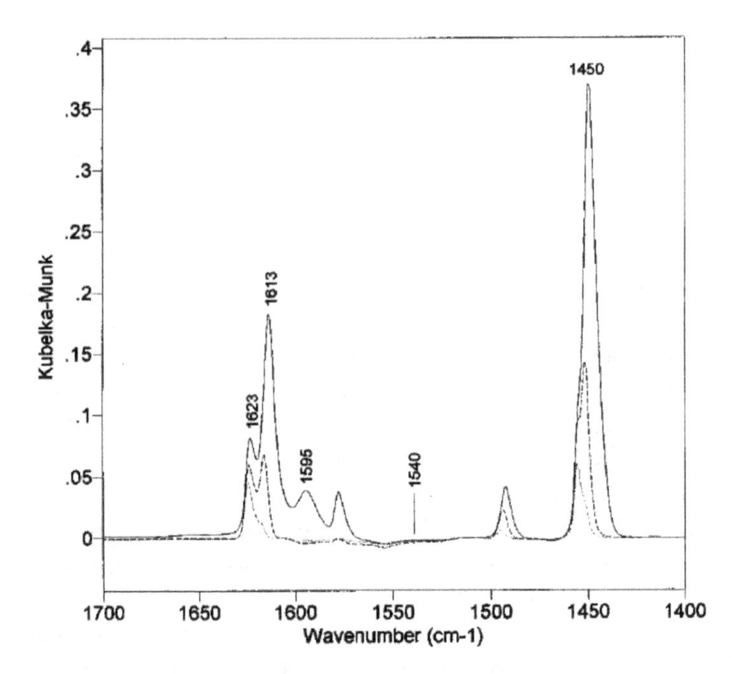

Fig. 2.21 FTIR spectra of γ-Al_2O_3 after pyridine adsorption at room temperature (upper curve), followed by heating at 180 °C for 1 h (middle curve) and followed by heating at 400 °C (lower curve) minus the FTIR spectrum of γ-Al_2O_3 before pyridine adsorption. The γ-Al_2O_3 was calcined at 450 °C under flowing N_2 before adsorption. Reprinted with permission from [2.24]. Copyright (1997) the American Chemical Society.

hence has one free coordination site. The medium strong sites may consist of neighbouring μ_5-Al^{3+} and μ_3-Al^{3+} cations (one free coordination site, formed by removing an oxygen atom that is bonded to a bulk Al_O cation as well as to a Al_T cation) and the strong sites may consist of μ_3-Al^{3+} cations, which originate from bulk Al_T cations, with one free coordination site [2.25].

Penta-coordinated μ_5-Al^{3+} cations have been observed in the ^{27}Al NMR spectra of γ-Al_2O_3. These Lewis acid sites are located at the (100) surface of γ-Al_2O_3 and act as anchors for BaO and La_2O_3 particles, thus preventing the phase transition of γ-Al_2O_3 to θ-Al_2O_3 [2.26]. They also act as anchors for PtO particles and inhibit sintering of supported Pt metal particles [2.27]. μ_5-Al^{3+} sites play a key role in the reaction of ethanol to ethylene on the (100) surfaces of γ-Al_2O_3 and θ-Al_2O_3 [2.28]. DFT calculations of the reaction of isopropanol to propene and diisopropylether on γ-Al_2O_3 indicated that μ_5-Al^{3+} neighbouring Lewis acid sites and basic species are required to catalyse these reactions. The (100) surface is more suited for these reactions than the (110) surface, because the latter remains partially hydrated during reaction and, thus, lacks basic enough sites [2.29]. Three-coordinated μ_3-Al^{3+} cations have not been observed, but these strong Lewis acid sites are held responsible for the activation of H_2, CH_4 and N_2 at the (110) surface of γ-Al_2O_3 [2.30]. In combination with specific O^{2-} anions, they form extremely reactive Al^{3+}, O^{2-} Lewis acid–base pairs that trigger the low-temperature heterolytic splitting of CH_4 and H_2 to yield Al–CH_3 and Al–H species, respectively.

2.3.5.2 *Brønsted acid sites*

The surface of metal oxides not only contains M^{n+} cations and O^{2-} anions (which may act as Lewis acids and bases respectively) but also OH^- anions, which are formed when the metal oxide surface chemisorbs water molecules

$$M^{n+} + O^{2-} + H_2O \rightarrow [M - OH_2]^{n+} + O^{2-} \rightarrow [M - OH]^{(n-1)+} + OH^-. \tag{2.9}$$

Because metal oxides always come into contact with air, the cations on their surfaces are covered by OH^- anions. These OH^- anions are responsible for the bands that are visible in the IR spectrum of γ-Al_2O_3 between 3,800 and 3,000 cm^{-1} (Fig. 2.22). The O–H vibrations have much higher wavenumbers than the Al–O vibrations (300–800 cm^{-1}) because the wavenumber σ is proportional to the square root of the force constant k of the bond between atoms divided by the reduced mass m, $\sigma = 1/\lambda = \nu/c = \frac{\sqrt{k/m}}{2\pi c}$, with $1/\mathrm{m} = 1/\mathrm{m}_1 + 1/\mathrm{m}_2$. Therefore X–H vibrations have a high wavenumber, $1/\mathrm{m} \cong 1/\mathrm{m}_\mathrm{H} = 1$, because the light atom moves more readily than the heavy atom. H–O vibrations therefore have a higher wavenumber than Al–O vibrations.

Knözinger and Ratnasamy explained the positions of the IR bands and the acidity or basicity of the OH^- anions by the bonding of the OH^- anions to one, two or three μ_5-Al^{3+} or μ_3-Al^{3+} cations [2.23]. Using the rules of Pauling (Nobel Prize in Chemistry in 1954 and Nobel Prize for Peace in 1962) for ionic solids, they calculated the formal charges on the OH^- anions of γ-Al_2O_3. For instance, when the +3 charge of the Al^{3+} cations is divided equally among the neighbouring O^{2-} and OH^- anions, the charge of a μ_1-OH^- anion bonded to one tetrahedrally coordinated Al^{3+} cation (Fig. 2.23, left) is $-1 + 3/4 = -0.25$. For a μ_2-OH^- anion bonded to two Al^{3+} cations, one with octahedral and the other with tetrahedral

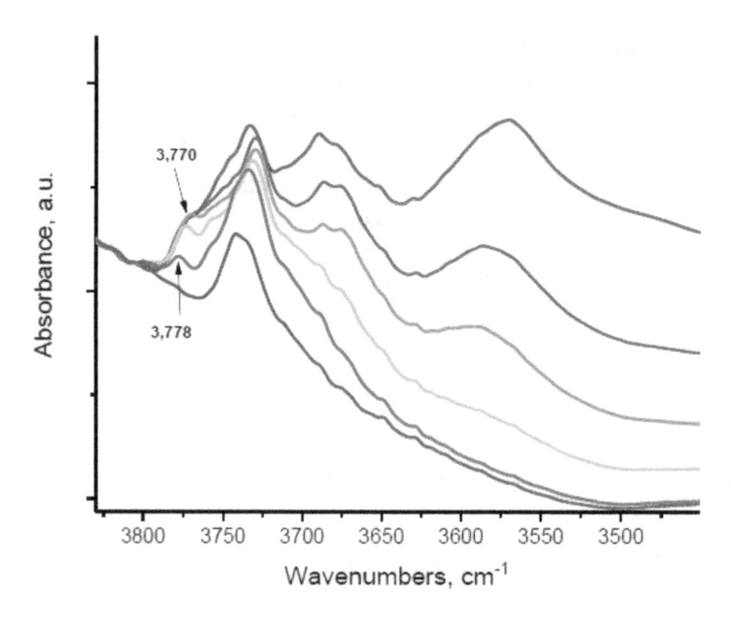

Fig. 2.22 *In situ* FTIR spectra of γ-Al_2O_3 samples (measured under vacuum at 25 °C) that were heated from 150 °C (top spectrum) to 700 °C (lowest spectrum). Water and CO_2 molecules desorb at lower temperature and OH groups dehydrate at higher temperature. Reprinted with permission from [2.31]. Copyright (2021) John Wiley & Sons, Inc.

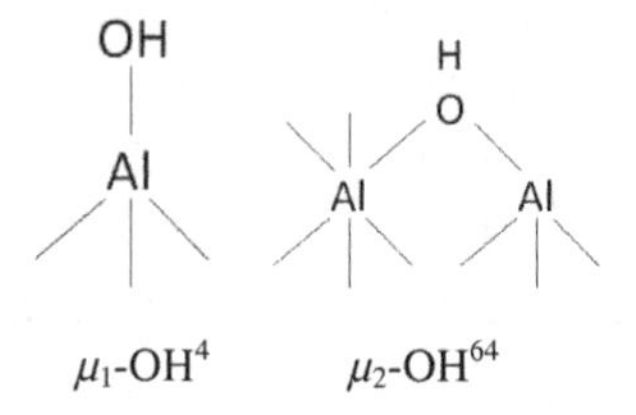

Fig. 2.23 μ_1-OH^4 is an OH group bonded to one tetrahedrally coordinated Al^{3+} cation and μ_2-OH^{64} is an OH group bonded to two Al^{3+} cations, one octahedrally and the other tetrahedrally coordinated.

coordination (Fig. 2.23, right), the charge on the OH^- anion is $-1+3/6+3/4=+0.25$.

The higher the charge, the stronger the repulsion of the proton by the positive charge, the greater the ease with which the proton leaves the OH^- anion and the higher the acidity of the OH group. At the same time, the OH bond will be weaker and the frequency of the OH vibration lower. The more Al^{3+} cations are bonded to an OH^- anion, the lower the OH vibration frequency: $\nu(\mu_1\text{-OH}) > \nu(\mu_2\text{-OH}) > \nu(\mu_3\text{-OH})$. Five different OH^- anions are predicted with this simple model, depending on the number (1, 2 or 3) and type (tetrahedral or octahedral) of the Al^{3+} cations that bind with the OH^- anion [2.23]. Other scientists came to a slightly different assignment of the OH bands [2.32]. Fivefold coordinated Al^{3+} cations at the surface of alumina have been considered as well, but the conclusion that singly bonded OH groups are more basic and have a higher OH vibration frequency remained valid [2.33].

Combined measurement of the OH and pyridine IR regions of the same γ-Al_2O_3 samples has shown that the pyridine molecules adsorbed on the Lewis acid sites (Fig. 2.21) diminish the intensity of certain OH vibrations. It indicates that the pyridine molecules on the Lewis sites hydrogen-bond with adjacent OH groups and that different Lewis acid sites have different adjacent OH groups [2.24]. This allows to study the proximity of Lewis acid sites and OH groups on the surface.

Although IR is a very valuable technique, it does not give unambiguous proof of the actual nature of the various surface OH groups, neither in terms of terminal versus bridging character, nor regarding the coordination of the attached Al sites. Double resonance NMR techniques should in principle be able to provide such information. The terminal or bridging character of the OH groups of γ-Al_2O_3 observed in the ^{1}H MAS–NMR spectrum could be determined by their difference in the rate of relaxation in ^{1}H–^{27}Al resonance-echo saturation-pulse double-resonance experiments. Assignment of the OH groups to μ_4-Al, μ_5-Al or μ_6-Al neighbours was made by $\{^1H\}$-^{27}Al dipolar heteronuclear multiple quantum correlation [2.34]. With increasing chemical shift (which is equivalent to decreasing frequency in the IR spectrum) the surface hydroxyls were found to have the following structures: HO-μ^1-Al^4, HO-μ^2-Al^6, Al^n, HO-μ^2-Al^5, Al^n, HO-μ^2-Al^4, Al^n and HO-μ^3-Al^n, Al^n, Al^n. These results constitute the first direct proof that the highest frequency IR OH band is from an OH group bonded to a single tetrahedrally coordinated Al^{3+} cation rather than an octahedrally coordinated Al^{3+} cation. The low sensitivity of solid-state ^{27}Al NMR can be greatly enhanced by combining low-temperature magic-angle spinning (MAS) with dynamic nuclear polarisation (DNP), a technique in which the polarisation of an electron spin is transferred to a nuclear spin. With DNP-enhanced $^1H \rightarrow {}^{27}Al$ cross-polarisation, the spin polarisation of an electron can be transferred to hydroxyl groups and on to Al^{3+} ions at the surface of γ-Al_2O_3. In this way, μ_5-Al^{3+} ions could be clearly observed [2.35]. An experiment that eliminated surface signals showed that the ratio of Al_O to Al_T cations in a fully hydrated γ-Al_2O_3 sample is much greater in the surface than in the bulk and that the surface reconstructs substantially. The results show that the combination of DNP with MAS–NMR is a very promising technique for studying solid surfaces.

The γ-Al_2O_3 surface contains several OH groups ranging from basic to acidic (Fig. 2.22), but if pyridine is adsorbed on γ-Al_2O_3 the band of protonated pyridine around 1,540 cm^{-1} is absent in the IR spectrum (Fig. 2.21). Apparently, pyridine adsorbs only on Lewis-acid Al^{3+} sites; the OH groups of γ-Al_2O_3 are not strong enough to protonate pyridine. 2,6-Dimethylpyridine is more basic than pyridine and can be protonated by weakly acidic OH groups. Unfortunately, IR measurements of 2,6-dimethylpyridine adsorbed on γ-Al_2O_3 gave conflicting results. One group observed a weak band of the 2,6-dimethylpyridinium cation [2.36] but another group did not observe such a band [2.37]. On the other hand, the IR spectra of chlorided γ-Al_2O_3 revealed the 1,540 cm^{-1} band of the pyridinium cation and the disappearance of the high wave number OH bands. Apparently, during chlorination with chlorine-containing molecules (which decompose to HCl) basic μ_1-OH^- groups are replaced by Cl^- anions. The remaining, more isolated, acidic OH groups have less opportunity for hydrogen-bonding, and this increases their acidity [2.38]. Pt on chlorided γ-Al_2O_3 is used as catalyst for the improvement of the octane number of hydrocarbon molecules. The Brønsted acidity of Cl-γ-Al_2O_3 is needed for the protonation of alkenes and subsequent isomerisation of the carbenium ions (Section 7.5.3).

2.3.5.3 *Surface reconstruction*

In the previous section it is assumed that the surface of γ-Al_2O_3 does not reconstruct, keeps the same structure as in the bulk. However, only ionic crystal surfaces that have no dipole moment perpendicular to the surface are relatively stable [2.39]. They may undergo relaxation (an inward shift of the atoms in the outer surface layer or layers) but no reconstruction. This is the case when the surface contains an equal number of equally charged cations and anions and, thus, is nonpolar, as in the checker-board arrangement of cations and anions in the (100) surfaces of NaCl and MgO. The surface should also be stable if the anions and cations are arranged in a symmetrical way in a repeat unit, as in successive Cl–Cd–Cl layers in $CdCl_2$. On the other hand, surfaces with a dipole moment, such as the (111) surface of MgO, which consists of a surface layer of O^{2-} anions with a subsurface layer of Mg^{2+} cations, or vice versa, should undergo reconstruction [2.39].

For γ-Al_2O_3, this means that only the high-density (100) surface is stable, but that the low-density (100) surface, both (110) surfaces and the (111) and $(11\bar{1})$ surfaces are expected to reconstruct.

DFT calculations of γ-Al_2O_3 surfaces, which were allowed to relax and reconstruct, confirmed that reconstruction of the high-density (100) surface with μ_5-Al^{3+} cations does not take place, but that extensive reconstruction takes place of the low-density (100) plane, because this surface contains strongly undercoordinated μ_2-Al^{3+} cations originating from Al_T cations [2.40, 2.41]. These μ_2-Al^{3+} cations reconstruct to μ_5-Al^{3+} cations by sinking into the surface. Both (110)-terminated surfaces undergo massive reconstruction into a sawtooth surface with (111) and $(11\bar{1})$ microfacets [2.40]. μ_5-Al^{3+} cations are present in the (111) facets and μ_3-Al^{3+} and μ_4-Al^{3+} cations in the $(11\bar{1})$ facets. TEM experiments confirmed that the (110) surface of γ-Al_2O_3 is not atomically flat but significantly reconstructed, forming nanoscale (111) facets [2.42].

In contrast to the DFT calculations based on the spinel structure, Digne *et al.* performed DFT calculations of bare and hydrated (100), (110) and (111) surfaces with a non-spinel model of γ-Al_2O_3 [2.13, 2.17]. Also these DFT calculations showed that the bare relaxed (100) surface of γ-Al_2O_3 contains only μ_5-Al^{3+} cations. The (110) surface contains μ_4-Al^{3+} cations (originating from bulk Al_O cations) and μ_3-Al^{3+} cations (originating from bulk Al_T cations) (Fig. 2.24). The surface relaxation induces large geometric modifications. The μ_4-Al^{3+} cations relax inward to a pseudoregular tetrahedral configuration and all oxygen surface anions relax upward. The μ_3-Al^{3+} cations and their surrounding oxygen anions relax to a planar AlO_3 surface species.

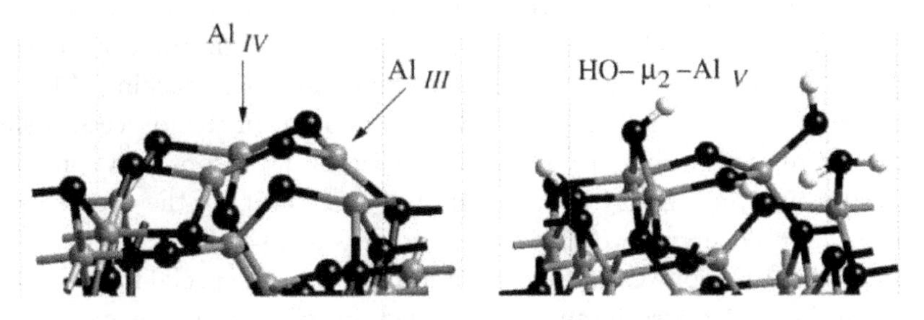

Fig. 2.24 Relaxed configurations of the γ-Al_2O_3 (110) surface before (left) and after (right) adsorption of H_2O. Adsorption of H_2O on two Al_{IV} sites gives a μ_2-OH^{55} group and on an Al_{II} site a μ_1-OH^4 group. O black, Al grey, H white. Reprinted with permission from [2.17]. Copyright (2004) Elsevier.

The surface of γ-Al_2O_3 relaxes when exposed to water vapour [2.17]. The μ_5-Al cations in the water-free (100) plane react to μ_1-OH^6 and μ_3-OH^6 upon adsorption of water molecules and the μ_3-Al^{3+} and μ_4-Al^{3+} cations in the water-free (110) plane react to μ_1-OH^4 and μ_2-OH^{55} upon water adsorption (Fig. 2.24) [2.17]. The DFT calculations showed that different crystal surfaces had different OH frequencies and confirmed that the IR frequency of an OH group is lower when it binds to a greater number of Al^{3+} cations, the OH group becomes more acidic.

2.4 Silica

Silica (SiO_2) is prepared by hydrolysis of a silicon compound, usually sodium silicate ($Na_4SiO_4 + 4\,HCl \rightarrow Si(OH)_4 + 4\,NaCl$). The resulting silica sol is an aqueous slurry of SiO_2 particles (10–20 nm) that are stabilised by the presence of base ($Si(OH)_4 + OH^- \rightarrow Si(OH)_3O^- + H_2O$). When the pH is lowered to 6–7, the silica OH groups are no longer charged ($Si(OH)_3O^- + H^+ \rightarrow Si(OH)_4$), and the particles make contact to form a gel, a hydrogen-bonded network. During drying, condensation of OH groups to water and the formation of siloxane bridges occur ($2\,Si(OH)_4 \rightarrow (HO)_3Si{-}O{-}Si(OH)_3 + H_2O$); an amorphous structure of small primary SiO_2 particles, bonded together by siloxane bridges, forms. The resulting silica has a broad distribution of mesopores and a high surface area (300–$400\,m^2{\cdot}g^{-1}$). It has many applications: as desiccant (drying agent), as stationary phase in chromatography and as catalyst support. The silica surface contains some OH groups, which make it suitable for grafting. These OH groups are weakly acidic. Silica is used in the chemical industry as a support for oxidation catalysts such as V_2O_5 (SO_2 to SO_3, *o*-xylene to phthalic anhydride) and $(VO)_2P_2O_7$ (butane to maleic anhydride) (Section 9.3.3.2). To produce silica with a specific particle size distribution (particle diameter 100 to 200 μm), as required in fluid catalytic cracking (FCC) in a riser reactor (Section 7.5.1), spray drying is used. In this process, a silica slurry is pressed through a nozzle and disperses into drops of a controlled size (10 to 500 μm). Spray drying (rapid drying in air) of the drops results in a dry, free-flowing powder with very small silica particles.

MCM-41 (Mobile Crystalline Material nr. 41) is a special form of silica. It is synthesised hydrothermally from sodium silicate in the presence of the surfactant cetyltrimethylammonium bromide, a **structure-directing agent (template)**. Surfactants exist as micelles in aqueous solution with the aliphatic chains in a lipophilic environment and the polar heads on the outside, in contact with the water (Fig. 2.25, left).

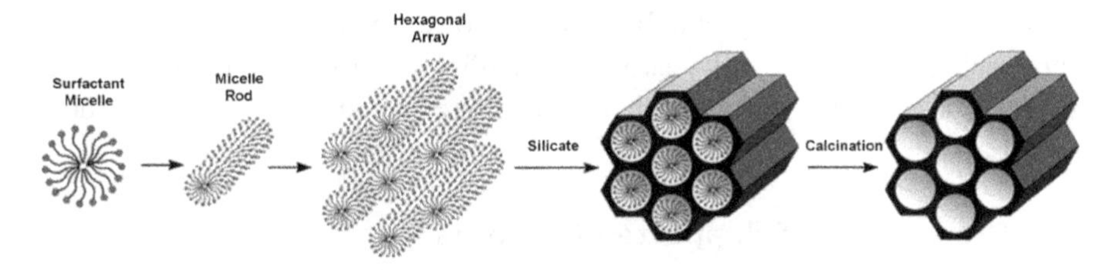

Fig. 2.25 Formation of MCM-41 by adsorption of silicate ions around a surfactant micelle and hexagonal stacking of the micellar rods. Reprinted with permission from [2.43]. Copyright (2006) John Wiley & Sons, Inc.

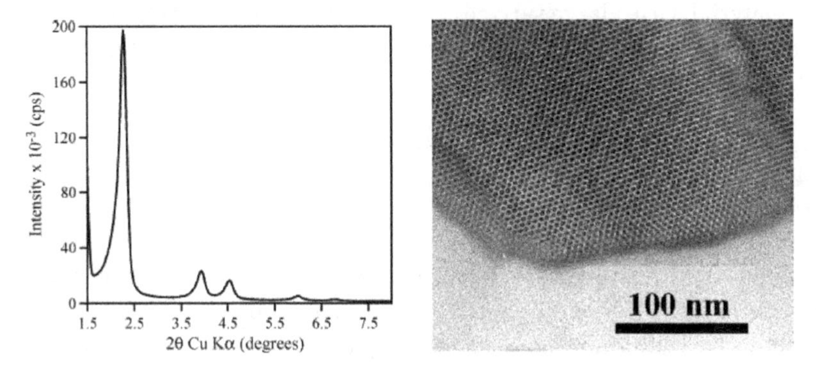

Fig. 2.26 XRD and TEM micrograph of MCM-41. Reprinted with permission from [2.44]. Copyright (2006) Elsevier.

When the surfactant concentration is high enough, the micelles assemble into rods, forming a hexagonal array. Silicate anions adsorb on the positively charged outer surface of the micellar rods. During hydrothermal synthesis (elevated temperature and pressure) the silicate anions polymerise to a silicate layer around the rods. After removal of the template by calcination, a hexagonal array of hollow silicate cylinders is left (Figs. 2.25, right). Because of the surfactant, MCM-41 has parallel uniform mesopores (Fig. 2.26) with amorphous silica walls and a large surface area of about $1000\,m^2{\cdot}g^{-1}$. The pore diameter is controlled (1.5 to 20 nm) by using surfactants of different chain lengths and swelling agents.

SBA-15 is a similar material. It is synthesised with Pluronic 123 $EO_kPO_lEO_m$ triblock copolymers as the template (EO is ethylene oxide, PO is propylene oxide) and tetraethyl orthosilicate as the silica source. These mesoporous materials have found application in several areas. They are not yet employed in industrial catalysts, because they are expensive (the template can be removed only by burning) and, due to their thin walls, they are unstable at high temperature and high water vapour pressure. Furthermore, it is difficult to introduce acidity into these materials.

2.5 Preparation of Supported Catalysts

As explained in Section 2.1, most catalysts consist of catalytically active particles deposited on the surface of a support. Otherwise the particles would be in contact, would grow together (sinter) and their surface area would decrease. To circumvent sintering the catalyst particles are deposited on a material with a large surface area; they spread out and do not come into contact with each other. The three most common methods for depositing material on a support are:

- **Impregnation**,
- **Adsorption**,
- **Precipitation**.

In **pore-volume impregnation** (also referred to as **incipient wetness impregnation**) the required amount of a solution containing a precursor of the active material (e.g. a metal salt) is added to a specific amount of support so as to just fill the pores of the support. The support can be a powder or pre-shaped (extrudates, beads). During the dropwise addition of the solution, the powder is stirred to achieve homogeneous impregnation. When the volume of the added solution is equal to the pore volume, the resulting impregnated support will appear to be dry, because all the liquid is taken up by the pores. After impregnation, the material is dried at room temperature and then at 100 °C in air, inert gas or under vacuum.

The pore volume impregnation method is easy to perform and requires a minimal quantity of solvent. In industry, water is the solvent of choice, because it is cheaper to recover or discard than organic solvents. Highly soluble salts (metal nitrates and chlorides) are preferred when high metal loadings must be deposited. However, the anion should not influence the catalyst or catalysis. For example, to prepare a metallic Ni/Al_2O_3 catalyst, $NiSO_4$ is unsuitable, because the sulfate anions will reduce to H_2S, which adsorbs on the resulting Ni metal surface and poisons it. Halide salts should not be used to prepare a catalyst supported on Al_2O_3 when acidity is undesirable, because adsorbed halide ions lead to an acidic Al_2O_3 surface. Thus, it is important to consider how the anion influences the final steps of catalyst preparation or the use of the resulting catalyst. Organic anions such as acetate, formate and oxalate do not influence the resulting catalyst, because they decompose to H_2O, CO_2 and/or CH_4 during calcination, but the disadvantage is that their metal salts are not particularly soluble in water.

The **impregnation method** may lead to an inhomogeneous distribution of the active component over the support particles. If the adsorption of the active component on the surface of the support particles is faster than the diffusion into the interior of the support particles, then most material will be adsorbed near the pore mouths. A poor distribution may also occur during evaporation of the water from the impregnated pores, caused by the capillary forces that are stronger in narrow pores than in wide pores. During evaporation, water is removed quickly from the wide pores, but the narrow pores tend to retain some of the liquid, leading to crystallisation of the metal salt near the centre of the particles and to a so-called egg-yolk distribution. The opposite type of maldistribution may occur when drying occurs too quickly; because of a temperature gradient, evaporation starts deep in the pores and drives the metal salt to the pore mouths (so-called egg-shell distribution). In industrial impregnation of extrudates of several mm or even cm, this may be a severe problem or it can be used to minimise pore diffusion in the finished catalyst, which can be useful for catalytic reactions of large molecules. Especially with precious metals, the depth of surface penetration is critical, as the metal that is located in the deep interior pores is essentially lost, or worse, can cause unwanted side reactions. One can also use maldistribution intentionally, to deposit one catalyst component near the outside of the extrudates and another component in the centre. For example, egg-shell catalysts give higher selectivities for consecutive reactions where the intermediate is the desired product, since side reactions are suppressed. If the reactions contain poisons, an egg-yolk distribution may be favourable so that the poison can be captured at the edge of the catalyst particle, where few active sites are present. In three-way automotive catalysts (Section 9.1.2) the aim is to protect the more expensive Rh component and deposit it in the middle of the supported catalyst particles, while Pt is deposited on the outside of the particles. This is done by selecting a Pt salt that adsorbs stronger than a Rh salt, or by means of competitive adsorption with an organic acid such as citric acid, which adsorbs stronger than the metal salts and occupies the sites on the support surface near the pore mouths. The organic acid is removed by calcination (heating) in air.

The **adsorption method** brings the support into contact with an excess amount of solution and, with time, adsorption equilibrium of the metal salt is established between the solution and the support surface. The cations adsorb on the support by exchange with protons:

$$\text{Si–OH} + \text{M(H}_2\text{O)}_6^{n+} \rightarrow [\text{Si–O–M(H}_2\text{O)}_5]^{(n-1)+} + \text{H}_3\text{O}^+. \tag{2.10}$$

Because the OH groups of silica are only weakly acidic, the adsorption equilibrium must be shifted to the right by working at high pH, so that the protons are removed by reaction with OH^-. However, many metal cations form insoluble metal hydroxides at high pH and, therefore, ammonia rather than NaOH or KOH is used to increase the pH. The ammonia forms soluble $M(NH_3)_m^{n+}$ amine complexes with the metal cations, with m usually equal to 4 or 6. The adsorption process is then:

$$\text{Si–OH} + \text{M(NH}_3)_m^{n+} \rightarrow [\text{Si–O–M(NH}_3)_{m-1}]^{(n-1)+} + \text{NH}_4^+. \tag{2.11}$$

The pH during adsorption depends on the support. For Al_2O_3, the zero-potential point (where protons and hydroxyl anions adsorb equally) is about pH $=$ 8, while for SiO_2 it is about 2. Thus, below pH $=$ 8 alumina is protonated, becomes positively charged and preferentially adsorbs anions (Fig. 2.27), while silica is deprotonated above pH $=$ 2 and, thus, can adsorb cations. H_2PtCl_6 (with $PtCl_6^{2-}$ anions) is usually used to impregnate Pt on alumina, while $Pt(NH_3)_4Cl_2$ (with $Pt(NH_3)_4^{2+}$ cations) is used for silica.

The adsorption method leads to a homogeneous distribution of the cations on the support surface. A disadvantage is that no more than the maximum adsorption capacity can be loaded onto the support. In general,

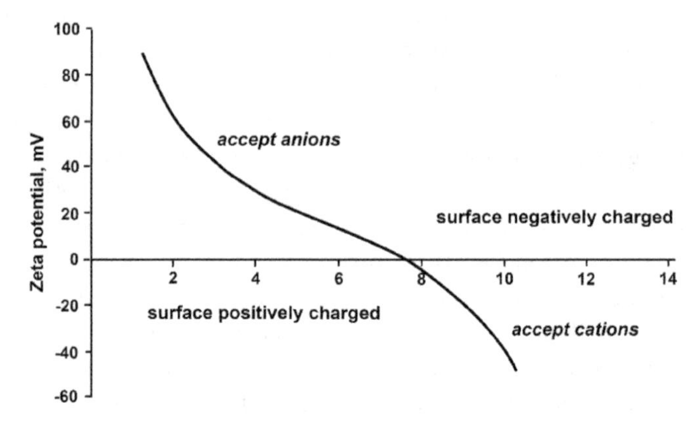

Fig. 2.27 Adsorption of cations and anions on γ-Al_2O_3 as a function of pH.

this is relatively small ($< 3\%$) but is not a problem when low loadings are desired, as for expensive noble metals. Other disadvantages are the large quantities of water that must be used (a problem in industry) and the long equilibration time.

In the **co-precipitation method** the catalyst and support are prepared in one step. For example, to prepare a Ni/Al_2O_3 catalyst with a very high Ni loading, Ni and Al hydroxides are precipitated together by increasing the pH of a solution that contains salts of both metals. Heating the mixed hydroxides to high temperature gives a NiO/α-Al_2O_3 material, in which most of the Ni has segregated to the surface as nickel oxide and some nickel cations are present in the bulk of the formed support. This Ni/α-Al_2O_3 catalyst is used in the steam reforming of methane at high temperature (Section 1.3). Another example of an industrial catalyst made by co-precipitation is the $Cu/ZnO/Al_2O_3$ catalyst, used in the synthesis of methanol from synthesis gas. It is usually made from a solution of Cu^{2+}, Zn^{2+} and Al^{3+} salts with Na_2CO_3 as the precipitating agent (Section 6.4.2) [2.45]. Again, the final catalyst is obtained by calcining (heating) the precipitate in air. In the **deposition precipitation method**, a support is added to a solution of the required metal salt and base (ammonia) is added slowly, increasing the pH. The metal hydroxide precipitates onto the walls of the pores of the support. A controlled homogeneous increase in pH is achieved by hydrolysing urea, $CO(NH_2)_2$:

$$M^{2+} + CO(NH_2)_2 + 3\,H_2O \rightarrow M(OH)_2 + CO_2 + 2\,NH_4^+. \tag{2.12}$$

After preparing a supported catalyst by impregnation, adsorption or precipitation, it is usually not in the active form yet. To achieve a metallic, metal oxide or metal sulfide form, the catalyst precursor must be **activated**. This can be done *in situ* in the reactor in which the catalysis is carried out, or *ex situ* in a separate reactor. If a metal salt on a support must be transformed into a metal-on-support catalyst, then H_2 gas is usually used to reduce the metal salt. For example, a $Ni(NO_3)_2/SiO_2$ precursor reduces directly in H_2 to Ni/SiO_2, and NO_x and NH_3 are removed as gas. Alternatively, it is possible to first remove the anion by calcination (heating) in air to NO_x and then reduce the resulting NiO/SiO_2 with H_2 to Ni/SiO_2. If the aim is a NiO/SiO_2 catalyst, then calcination of the $Ni(NO_3)_2/SiO_2$ precursor suffices. To obtain a Ni_3S_2/SiO_2 catalyst, it is possible to calcine the nitrate precursor to NiO/SiO_2 followed by sulfidation in a mixture of H_2S and H_2.

In the industry as well as in the laboratory, catalysts are normally prepared *ex situ* to obtain a sufficiently large batch. The catalyst is removed from the preparation reactor and stored for later use. Most catalyst surfaces are very sensitive to air and will (partially) oxidise when exposed to air during transfer from the preparation reactor to a storage vessel. Uncontrolled oxidation of the catalyst, e.g. by too fast exposure to air, may lead to local oxidation and severe heating of the catalyst, which causes surface reconstruction and complete oxidation of the catalyst particles. To avoid this, controlled limited oxidation (**passivation**) of the catalyst is applied, which means that increasing concentrations of oxygen or air are very slowly added to the catalyst. For example, a Ni/SiO_2 catalyst is first exposed to N_2 gas and then to increasing amounts of O_2. This should lead to adsorption of O atoms on the top layer of the Ni surface only. After passivation, the reactor is opened and the passivated catalyst can be transported safely in air to a storage vessel. Because of the passivation, the catalyst must later be reformed by removing the oxygen ad-layer. With a passivated metal-on-support catalyst this can be done by reduction in H_2 at moderate temperature (lower than required for reduction of the pure metal oxide). With a passivated metal sulfide it can be achieved with a mixture of H_2S and H_2 at moderate temperature.

References

2.1 S. Mitchell, N.-L. Michels, K. Kunze, J. Pérez-Ramírez, Visualisation of hierarchically structured zeolite bodies from macro to nano length scales, *Nature Chem.* 4, 825–831, 2012.

2.2 P. Euzen, P. Raybaud, X. Krokidis, H. Toulhoat, J. L. Le Loarer, J. P. Jolivet, C. Froidefond, in *Handbook of Porous Solids*, Eds. F. Schüth, K. S. W. Sing, J. Weitkamp. Wiley, Chichester, 2002, pp. 1591–1677.

2.3 A. Boumaza, L. Favaro, J. Lédion, G. Sattonnay, J. B. Brubach, P. Berthet, A. M. Huntz, P. Roy, R. Tétot, Transition alumina phases induced by heat treatment of boehmite: An X-ray diffraction and infrared spectroscopy study, *J. Solid State Chem.* 182, 1171–1176, 2009.

2.4 The Scherrer equation can be semi-quantitatively derived as follows. Suppose the reflection of the X-ray beam of a crystal plane k would deviate slightly from the angle θ, needed for obtaining a coherent interference between incoming and reflected beam. If the difference in path length between this reflection and the reflection from a plane $k + n$ would be equal to $\lambda/2$, these two reflections would be in anti-phase and annihilate each other. If the crystal is very large, then for every plane there will be another plane from which the two reflections will annihilate each other if the reflections deviate by $\delta\theta$ from the ideal Bragg angle θ. In that case the

XRD lines are infinitely narrow; they are lines, not peaks. However, when the crystals are limited in size, then reflections from some planes in the direction $\theta + \delta\theta$ may not have a counterpart that is in anti-phase. In that case there will be some intensity in the direction $\theta + \delta\theta$ and the XRD line will broaden to a peak. Total annihilation of all reflections will occur when the reflections from planes in the upper half of the crystal are annihilated by planes in the lower half. In that case the difference in path length of the reflected beams from the upper and lowest crystal planes should be one X-ray wavelength λ (the phase difference should be 2π).

$$2\,\mathrm{Nd}\ \sin(\theta + \delta\theta) = \mathrm{n\,N}\,\lambda + \lambda$$
$$2\,\mathrm{Nd}\ [\sin\theta \cdot \cos\delta\theta + \cos\theta \cdot \sin\delta\theta] = \mathrm{n\,N}\,\lambda + \lambda.$$

Because $\cos\delta\theta \cong 1$ and $\sin\delta\theta \cong \delta\theta$ for $\delta\theta$ small, when $\delta\theta$ is measured in radians, we obtain

$$2\,\mathrm{Nd}\ [\sin\theta + \cos\theta \cdot \delta\theta] = \mathrm{n\,N}\,\lambda + \lambda.$$

With $\mathrm{N} \cdot 2\mathrm{d}\sin\theta = \mathrm{nN}\lambda$ this gives $2\,\mathrm{Nd} \cdot \cos\theta \cdot \delta\theta = \lambda$ and $\mathrm{D} = \mathrm{Nd} = \lambda/2\cos\theta \cdot \delta\theta$. A more rigorous treatment gives $\mathbf{D} = \mathbf{K} \cdot \boldsymbol{\lambda}/\mathbf{cos}\boldsymbol{\theta} \cdot \boldsymbol{\delta\theta}$, the **Scherrer equation**. For Cu Kα X-rays ($\lambda = 0.154\,\mathrm{nm}$), $\delta\theta$ in degrees and $\mathrm{K} = 0.9$ the Scherrer equation becomes $\mathrm{D} = 7.9/\cos\theta \cdot \delta\theta$.

N is the number of planes and $\mathrm{D} = \mathrm{Nd}$ the diameter of the particle. The intensity would decrease to zero for the angles $\theta + \delta\theta$ and $\theta - \delta\theta$. If we assume a triangular intensity profile, then the width of the XRD peak at half height will be $\delta\theta$. To determine the size of a crystal in a certain direction, we thus must look at the corresponding XRD peak and measure its reflection direction θ and width $\delta\theta$. If we want to know the crystal thickness in the x and y directions, then we must determine θ and $\delta\theta$ of the (n00) and (0m0) reflections, respectively.

2.5 K. I. Shefer, D. A. Yatsenko, S. V. Tsybulya, E. M. Moroz, E. Yu. Gerasimov, Structural features of finely dispersed pseudoboehmite obtained by a sol-gel method, *J. Struct. Chem.* 51, 322–326, 2010.

2.6 R. Demichelis, B. Civalleri, Y. Noel, A. Meyer, R. Dovesi, Structure and stability of aluminium trihydroxides bayerite and gibbsite: A quantum mechanical *ab initio* study with the Crystal06 code, *Chem. Phys. Lett.* 465, 220–225, 2008.

2.7 X. Du, Y. Wang, X. Su, J. Li, Influences of pH value on the microstructure and phase transformation of aluminum hydroxide, *Powder Technol.* 192, 40–46, 2009.

2.8 Y. Noel, R. Demichelis, F. Pascale, P. Ugliengo, R. Orlando, R. Dovesi, *Ab initio* quantum mechanical study of γ-AlOOH boehmite: Structure and vibrational spectrum, *Phys. Chem. Miner.* 36, 47–59, 2009.

2.9 R.-S. Zhou, R. L. Snyder, Structures and transformation mechanisms of the η, γ and θ transition aluminas, *Acta. Cryst. B* 47, 617–630, 1991.

2.10 L. Favaro, A. Boumaza, P. Roy, J. Lédion, G. Sattonnay, J. B. Brubach, A. M. Huntz, R. Tétot, Experimental and *ab initio* infrared study of χ-, κ- and α-aluminas formed from gibbsite, *J. Solid State Chem.* 183, 901–908, 2010.

2.11 M. Digne, P. Sautet, P. Raybaud, H. Toulhoat, E. Artacho, Structure and stability of aluminum hydroxides: A theoretical study, *J. Phys. Chem. B* 106, 5155–5162, 2002.

2.12 P. Nortier, P. Fourre, A. B. Mohammed Saad, O. Saur, J. C. Lavalley, Effects of crystallinity and morphology on the surface properties of alumina, *Appl. Catal.* 61, 141–160, 1990.

2.13 X. Krokidis, P. Raybaud, A.-E. Gobichon, B. Rebours, P. Euzen, H. Toulhoat, Theoretical study of the dehydration process of boehmite to γ-alumina, *J. Phys. Chem. B* 105, 5121–5130, 2001.

2.14 X. C. Gao, Y. J. Zhao, S. P. Wang, Y. L. Yin, B. W. Wang, X. B. Ma, A Pd-Fe/α-Al_2O_3/cordierite monolithic catalyst for CO coupling to oxalate, *Chem. Eng. Sci.* 66, 3513–3522, 2011.

2.15 With an **electron microscope (EM)** it is possible to observe very small objects by focusing a beam of electrons rather than ordinary light on the objects. De Broglie (Nobel Prize in Physics 1929) predicted that particles can also be regarded as waves: $\lambda = h/mv = \mathrm{h}/\sqrt{2mE(kin)}$, where λ is the wavelength of the particle, h is Planck's constant and m, v and E(kin) are the mass, velocity and kinetic energy of the particle, respectively. An electron with E(kin) = 80 keV is equivalent to a wavelength of 0.14 nm, which means that highly accelerated electrons can diffract on objects of atomic dimensions.

The electron microscope has electrostatic and electromagnetic "lenses" to control and focus the electron beam so that an image is formed. The lenses have the same function as the glass lenses of an optical microscope, which produces a magnified image by focusing light on or through the specimen. In a **transmission electron microscope (TEM)**, electrons are emitted by a tungsten filament cathode. The electron beam is accelerated by an anode (40–400 kV difference with the cathode), focused by electrostatic and electromagnetic lenses, and transmitted through the specimen. The electron beam that emerges from the specimen carries information about the structure of the specimen that is magnified by the objective lens system of the microscope. The resulting image may be viewed by coupling a high-resolution phosphor to the sensor of a charge-coupled device camera. The resulting image is displayed on a monitor or computer.

Hardware correction of spherical aberrations in the **high-resolution transmission electron microscope (HRTEM)** results in images with a resolution below 0.5 Å (0.05 nm, 50 pm) and magnifications above 50 million (Fig. 2.28). HRTEM can determine the position of atoms in materials, making it an important tool in nanotechnological research and development. A disadvantage of TEM is the need for extremely thin sections of the specimens, typically about 100 nm.

In a **scanning electron microscope (SEM)**, the electron beam (E = 0.2–40 keV) is focused by condenser lenses to a spot about 0.5 to 5 nm in diameter. The beam passes through pairs of scanning coils or deflector plates into the electron column and the final lens deflects the beam in the x and y axes so that it scans raster-like a rectangular area

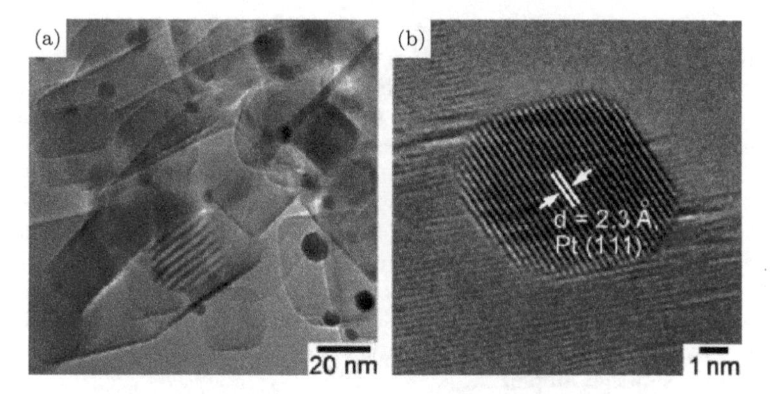

Fig. 2.28 TEM (a) and HRTEM (b) images of Pt particles supported on MgO. Reprinted with permission from [2.46]. Copyright (2012) Elsevier.

Fig. 2.29 SEM image of ZSM-5 crystals.

of the sample surface. When the electron beam interacts with the sample, elastic as well as inelastic scattering may occur. Elastical scattering gives reflected electrons. Inelastic scattering occurs when the electrons in the beam exchange energy with the sample, giving rise to emission of secondary electrons and electromagnetic radiation. Imaging of the secondary electrons produces very high-resolution images of a sample surface, revealing details smaller than 1 nm in size. Due to the very narrow electron beam, SEM micrographs have a large depth of field, yielding a characteristic three-dimensional image, which aids the understanding of the surface structure of a sample. This is exemplified by the micrograph of zeolite ZSM-5 crystals (Fig. 2.29). Because SEM makes use of reflection, it does not need thin samples. Usually, the sample is coated with a thin metal layer, to make the surface more conductive and prevent charge build up.

Not only an image of the sample surface can be obtained with SEM, but also its elemental composition can be measured by using the electrons that are elastically back-scattered from the sample. Heavier elements scatter more strongly than light ones and thus appear brighter in a back-scattered electron image. Thus, back-scattered electrons provide information about the distribution of the elements in the sample. Characteristic X-rays are created in a SEM instrument by the bombardment of the atoms in the sample by the beam electrons, just as in an X-ray tube. The bombarding electrons remove an electron from inner shells of the atoms and the resulting holes are filled by electrons from energetically higher shells. The resulting X-ray fluorescence (XRF) is characteristic for the element and is used to identify the elemental composition and measure the abundance of the elements in the sample.

2.16 **Nuclear magnetic resonance** (NMR) spectroscopy uses the nuclear spin of atomic nuclei to determine physical and chemical properties of the molecules that contain the nuclei. A spinning charge generates a magnetic moment proportional to the spin. In the presence of an external magnetic field, two spin states α and β exist for a nucleus with spin $1/2$ (e.g. ^{1}H and ^{13}C): One spin aligns parallel to the magnetic field and the other antiparallel to it. The energy difference ΔE between the two spin states increases with the strength of the field. Irradiation of the sample with energy corresponding to the exact separation of the spin states of a specific group of nuclei excites the nuclei from the lower to the higher energy state. The frequency required to excite ^{1}H nuclei is about 400 MHz when the strength of the magnetic field is 10 T. Other atomic nuclei in a molecule absorb at different radio frequencies for the same magnetic field strength. This means that the NMR frequencies of nuclei present in a molecule provide chemical and structural information about the molecule. Because the energy difference ΔE is very small, the difference in the Boltzmann population of the two states is small and the NMR signal is weak. Fourier-transform NMR allows to measure spectra much more quickly and improve the signal-to-noise ratio.

The frequency values of the resonant signals depend not only on the nucleus, but also on the electrons surrounding the nucleus. Upon application of an external magnetic field, the electrons generate local magnetic fields that oppose the much stronger applied external field. The local field "shields" the nucleus from the applied magnetic field, which must be increased to achieve absorption of radio frequency energy. Such increments are very small, usually parts per million (ppm). NMR signals are given relative to a reference signal and on a relative scale by dividing the frequencies of the NMR peak positions by the spectrometer frequency. The resulting **chemical shifts** are given in ppm and provide information about the structure of the molecule. For instance, the ^{1}H-NMR spectrum of ethanol (CH_3CH_2OH) has three signals with different chemical shifts for the H atoms in the CH_3, CH_2 and OH groups. Because of the molecular motion in solution at room temperature, the three methyl protons average out during the NMR experiment and only one NMR peak is observed for the protons of the CH_3 group.

Because hydrogen and carbon nuclei can be measured, chemists and biochemists rely on NMR to investigate the properties of organic molecules, from small molecules to large proteins or nucleic acids. NMR spectra can be measured of solutions and solids and many nuclei can be measured that are important in catalysis: ^{1}H, ^{13}C, ^{19}F, ^{27}Al, ^{29}Si and ^{31}P. Apart from ^{27}Al, these nuclei have a spin $1/2$, which makes it easier to carry out NMR spectroscopy. ^{27}Al and ^{17}O have a spin 3/2 and have a quadrupole moment. Only the transition from the spin state $1/2$ to $-1/2$ is useful.

The shielding effect of the electrons on the nucleus depends on the orientation of the molecule with respect to the external field. As a consequence, the NMR spectrum of a solid sample is usually very broad because the position of the NMR line depends on the orientation of the molecule relative to the magnetic field. Figure 2.30 shows the liquid and solid state ^{13}C-NMR spectra of an organic molecule. Five lines are visible in the liquid phase spectrum, whereas in the solid-state spectrum each of these lines broadens to a profile that has a maximum in the middle (for asymmetric molecules with $x \neq y \neq z$) or a maximum on the left or right side (when $x = y \neq z$). Molecules in solution rotate fast and the so-called chemical shift anisotropy of the NMR spectrum averages out during the course of the NMR experiment (which typically requires a few ms) to give a single peak. To obtain narrow lines for solid samples, one must "rotate" the molecules, as it were, and average the orientation dependence in order to obtain values close to the average chemical shifts. Most of the chemical

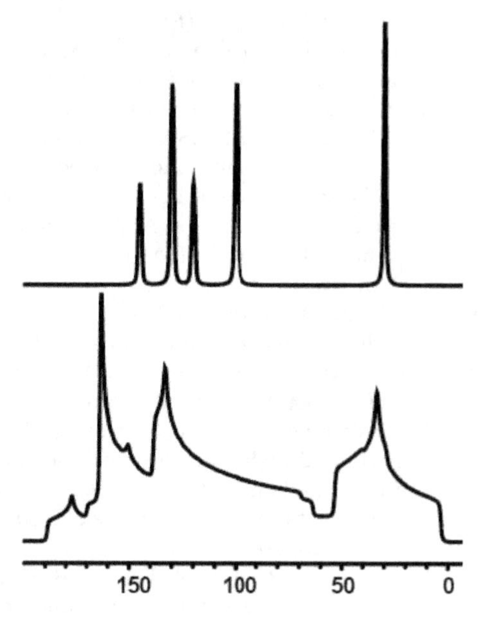

Fig. 2.30 ^{13}C NMR spectra of an organic molecule in the liquid (upper spectrum) and solid state (lower spectrum) [2.47].

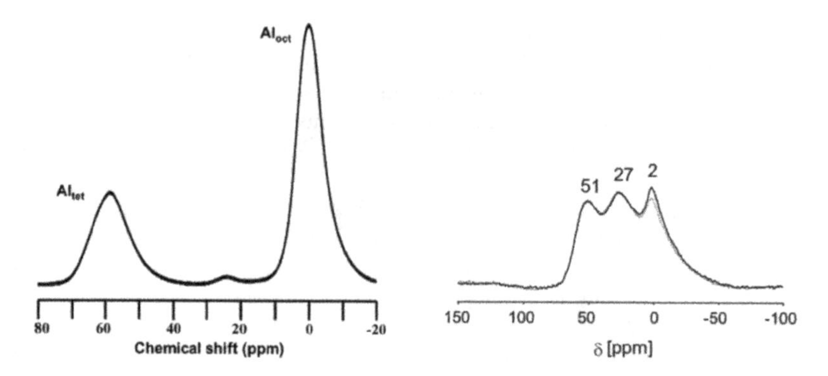

Fig. 2.31 ^{27}Al NMR spectra of left γ-Al_2O_3 and right of bimetallic Pt–Pd/ASA (dashed line) and monometallic Pt/ASA on ASA with SiO_2:Al_2O_3 = 80:20. Reprinted with permission from [2.48] and [2.49]. Copyright (2012) Elsevier.

shift anisotropy is averaged out if the sample is rotated fast, at an angle to the magnetic field for which $3\cos^2\theta = 1$. When spinning is carried out at this so-called magic angle (MAS), an almost liquid-phase-like spectrum is obtained.

Examples of ^{27}Al MAS NMR spectra are given in Fig. 2.31. In catalysis, ^{27}Al MAS NMR is used to study solid acids such as aluminas, zeolites (Section 7.1.1) and amorphous silica–alumina (Section 7.1.2). Aluminas show two peaks at 59 and 0 ppm (Fig. 2.31, left), corresponding to tetrahedrally and octahedrally coordinated Al^{3+} cations, respectively [2.48]. Amorphous silica–alumina (ASA) often shows a third peak due to pentagonally coordinated Al^{3+} cations, as shown by the three signals at 51, 27 and 2 ppm in Fig. 2.31 (right) of Pt/ASA and Pt–Pd/ASA. These signals are due to tetrahedrally, pentagonally and octahedrally coordinated Al^{3+} cations, respectively. The shift of the peak of tetragonally coordinated Al^{3+} from 59 ppm for γ-Al_2O_3 to 51 ppm for ASA is due to the fact that the Al^{3+} cations in ASA have Si instead of Al next-nearest neighbour atoms and disordered Al–O–Si angles.

The resolution and sensitivity of NMR is better in a higher magnetic field and, thus, a higher frequency of the absorbed radiation (which is proportional to the strength of the magnetic field). Nowadays spectrometers with frequencies of 400 or 500 MHz are standard and several laboratories even have 900 MHz instruments (21.1 T field strength). Superconducting magnets provide the required high magnetic field strengths.

2.17 M. Digne, P. Sautet, P. Raybaud, P. Euzen, H. Toulhoat, Use of DFT to achieve a rational understanding of acid–basic properties of γ-alumina surfaces, *J. Catal.* 226, 54–68, 2004.

2.18 G. Paglia, C. E. Buckley, A. L. Rohl, B. A. Hunter, R. D. Hart, J. V. Hanna, L. T. Byrne, Tetragonal structure model for boehmite-derived γ-alumina, *Phys. Rev. B* 68, 144110, 2003.

2.19 G. Paglia, A. L. Rohl, C. E. Buckley, J. D. Gale, Determination of the structure of γ-alumina from interatomic potential and first-principles calculations: The requirement of significant numbers of nonspinel positions to achieve an accurate structural model, *Phys. Rev. B* 71, 224115, 2005.
2.20 R. Prins, On the structure of γ-Al_2O_3, *J. Catal.* 392, 336–346, 2020.
2.21 S. V. Tsybulya, G. N. Kryukova, Nanocrystalline transition aluminas: Nanostructure and features of X-ray powder diffraction patterns of low-temperature Al_2O_3 polymorphs, *Phys. Rev. B* 77, 024112, 2008.
2.22 V. P. Pakharukova, D. A. Yatsenko, E. Yu. Gerasimov, A. S. Shalygin, O. N. Martyanov, S. V. Tsybulya, Coherent 3D nanostructure of γ-Al_2O_3: Simulation of whole X-ray powder diffraction pattern, *J. Solid State Chem.* 246, 284–292, 2017.
2.23 H. Knözinger, P. Ratnasamy, Catalytic aluminas: Surface models and characterisation of surface sites, *Catal. Rev. Sci. Eng.* 17, 31–70, 1978.
2.24 X. Liu, R. E. Truitt, DRFT-IR Studies of the surface of γ-alumina, *J. Am. Chem. Soc.* 119, 9856–9860, 1997.
2.25 C. Morterra, G. Magnacca, A case study: Surface chemistry and surface structure of catalytic aluminas, as studied by vibrational spectroscopy of adsorbed species, *Catal. Today* 27, 497–532, 1996.
2.26 J. H. Kwak, J. Z. Hu, A. Lukaski, D. H. Kim, J. Szanyi, C. H. F. Peden, Role of pentacoordinated Al^{3+} ions in the high temperature phase transformation of γ-Al_2O_3, *J. Phys. Chem. C* 112, 9486–9492, 2008.
2.27 D. Mei, J. H. Kwak, J. Z. Hu, S. J. Cho, J. Szanyi, L. F. Allard, C. H. F. Peden, Unique role of anchoring penta-coordinated Al^{3+} sites in the sintering of γ-Al_2O_3-supported Pt catalysts, *J. Phys. Chem. Lett.* 1, 2688–2691, 2010.
2.28 J. Lee, E. J. Jang, H. Y. Jeong, J. Kwak, Critical role of (100) facets on γ-Al_2O_3 for ethanol dehydration: Combined efforts of morphology-controlled synthesis and TEM study, *Appl. Catal. A* 556, 121–128, 2018.
2.29 K. Larmier, C. Chizallet, N. Cadran, S. Maury, J. Abboud, A.-F. Lamic-Humblot, E. Marceau, H. Lauron-Pernot, Mechanistic investigation of isopropanol conversion on alumina catalysts: Location of active sites for alkene/ether production, *ACS Catal.* 5, 4423–4437, 2015.
2.30 R. Wischert, P. Laurent, C. Copéret, F. Delbecq, P. Sautet, γ-Alumina: The essential and unexpected role of water for the structure, stability, and reactivity of "defect" sites, *J. Am. Chem. Soc.* 134, 14430–14449, 2012.
2.31 K. Khivantsev, N. R. Jaegers, J.-H. Kwak, J. Szanyi, L. Kovarik, Precise identification and characterisation of catalytically active sites on the surface of γ-alumina, *Angew. Chem. Int. Ed.* 60, 17522–17530, 2021.
2.32 G. Busca, G. Lorenzelli, G. Ramis, R. J. Willey, Surface sites on spinel-type and corundum-type metal oxide powders, *Langmuir* 9, 1492–1499, 1993.
2.33 A. A. Tsyganenko, P. P. Mardilovich, Structure of alumina surfaces, *J. Chem. Soc. Farad. Trans.* 92, 4843–4852, 1996.
2.34 M. Taoufik, K. C. Szeto, N. Merle, I. Del Rosal, L. Maron, J. Trébosc, G.Tricot, R. M. Gauvin, L. Delevoye, Heteronuclear NMR spectroscopy

as a surface-selective technique: A unique look at the hydroxyl groups of γ-alumina, *Chem. Eur. J.* 20, 4038–4046, 2014.

2.35 D. Lee, N. T. Duong, O. Lafon, G. De Paëpe, Primostrato solid-state NMR enhanced by dynamic nuclear polarisation: Pentacoordinated Al^{3+} ions are only located at the surface of hydrated γ-alumina, *J. Phys. Chem. C* 118, 25065–25076, 2014.

2.36 L. Oliviero, A. Vimont, J.-C. Lavalley, F. Romero Sarria, M. Gaillard, F. Maugé, 2,6-Dimethylpyridine as a probe of the strength of Brønsted acid sites: Study on zeolites. Application to alumina, *Phys. Chem. Chem. Phys.* 7, 1861–1869, 2005.

2.37 T. Onfroy, G. Clet, M. Houalla, Quantitative IR characterisation of the acidity of various oxide catalysts, *Mesop. Micropor. Mater.* 82, 99–104, 2005.

2.38 M. Digne, P. Raybaud, P. Sautet, D. Guillaume, H. Toulhoat, Atomic scale insights on chlorinated γ-alumina surfaces, *J. Am. Chem. Soc.* 130, 11030–11039, 2008.

2.39 P. W. Tasker, The stability of ionic crystal surfaces, *J. Phys. C Solid State Phys.* 12, 4977–4984, 1979.

2.40 H. P. Pinto, R. M. Nieminen, S. D. Elliott, *Ab initio* study of γ-Al_2O_3 surfaces, *Phys. Rev. B* 70, 125402, 2004.

2.41 A. Vijay, G. Mills, H. Metiu, Structure of the (001) surface of γ alumina, *J. Chem. Phys.* 117, 4509–4516, 2002.

2.42 L. Kovarik, A. Genc, C. Wang, A. Qiu, C. H. F. Peden, J. Szanyi, J. H. Kwak, Tomography and high-resolution electron microscopy study of surfaces and porosity in a plate-like γ-Al_2O_3, *J. Phys. Chem. C* 117, 179–186, 2013.

2.43 F. Hoffmann, M. Cornelius, J. Morell, M. Fröba, Silica-based mesoporous organic-inorganic hybrid materials, *Angew. Chem. Int. Ed.* 45, 3216–3251, 2006.

2.44 Y. Wang, N. Lang, A. Tuel, Nature and acidity of aluminum species in AlMCM-41 with a high aluminum content (Si/Al = 1.25), *Micropor. Mesopor. Mater.* 93, 46–54, 2006.

2.45 M. Behrens, Coprecipitation: An excellent tool for the synthesis of supported metal catalysts — From the understanding of the well known recipes to new materials, *Catal. Today* 246, 46–54, 2015.

2.46 Z. Peng, F. Somodi, S. Helveg, C. Kisielowski, P. Specht, A. T. Bell, High resolution *in situ* and *ex situ* TEM studies on graphene formation and growth of Pt nanoparticles, *J. Catal.* 286, 22–29, 2012.

2.47 Mutus Laboratory, University of Windsor, Ontario, Canada.

2.48 J. H. Kwak, J. Z. Hu, D. H. Kim, J. Szanyi, C. H. F. Peden, Pentacoordinated Al^{3+} ions as preferential nucleation sites for BaO on γ-Al_2O_3: An ultra-high-magnetic field ^{27}Al MAS NMR study, *J. Catal.* 251, 189–194, 2007.

2.49 Y. Yu, B. Fonfé, A. Jentys, G. L. Haller, J. A. R. van Veen, O. Y. Gutiérrez, J. A. Lercher, Bimetallic Pt–Pd/silica–alumina hydrotreating catalysts — Part I: Physicochemical characterisation, *J. Catal.* 292, 1–12, 2012.

Questions

2.1 Describe the four steps needed to produce γ-Al_2O_3 from bauxite.

2.2 Bayerite and gibbsite are the most stable forms of aluminium (oxy)hydroxides at low temperature, but why are boehmite and diaspore more stable at higher temperature?

2.3 Why is boehmite and not diaspore formed upon heating of bayerite, even though diaspore is slightly more stable than boehmite (Fig. 2.15)?

2.4 The hydroxides, oxides and oxyhydroxides of iron and aluminium often have the same structures. For instance, goethite (α-FeO(OH)), lepidocrocite (γ-FeO(OH)), hematite (α-Fe_2O_3), maghemite (γ-Fe_2O_3, present in tropical soils and on the surface of Mars) and ferrihydrite (Fe_5O_7(OH)) have the same structures as diaspore (α-AlO(OH)), boehmite (γ-AlO(OH)), α-Al_2O_3, γ-Al_2O_3 and tohdite (Al_5O_7(OH)), respectively. Why are the structures of iron and aluminium hydroxides, oxides and oxyhydroxides the same?

2.5 What is the structure of magnetite (Fe_3O_4) and why does an equivalent Al compound not exist?

2.6 One wants to prepare 0.5 g of a 1 wt.% Pt on Al_2O_3 catalyst. How many ml of an aqueous solution of H_2PtCl_6 should one then add to 0.5 g Al_2O_3 and what should the Pt concentration of this solution be? The pore volume of Al_2O_3 is $0.5\,mL{\cdot}g^{-1}$ and the atomic weight of Pt is 195.

2.7 Suppose the catalytic muffler under a car contains 1 kg of a 1 wt.% Pt/Al_2O_3 catalyst, with 100% Pt dispersion. How many accessible Pt atoms are present in this muffler and how large is the Pt surface area, assuming that one Pt atom has a surface area of 6 $Å^2$?

2.8 When the IR spectra of zeolites are measured in the presence of D_2 gas the OH vibrations at about $3{,}750\,cm^{-1}$ disappear and new bands appear at about $2{,}700$–$2{,}600\,cm^{-1}$. Why?

2.9 The intensity of NMR signals is weaker than the intensity of IR signals. How much smaller is the energy difference measured on a 400 MHz NMR spectrometer than the energy difference measured by IR at $3{,}000\,cm^{-1}$?

2.10 Calculate the diameter of the pseudoboehmite particles from its XRD pattern (Fig. 2.8).

Chapter 3

Adsorption

3.1 Physisorption

In order to react, molecules must first come into contact with each other. In heterogeneous catalysis the molecules must first adsorb on the catalyst surface. Molecules will adsorb only when they are attracted to the surface, i.e. when the enthalpy of adsorption is negative; otherwise the loss of entropy, when molecules adsorb and lose mobility, will prevent adsorption. Adsorption is an essential step in a catalysed reaction because the adsorbed molecules form intermediate surface complexes that are more likely to react. In adsorbed molecules, the bonds are weaker and, thus, reactions can occur more easily. Two interactions determine adsorption: physisorption and chemisorption. Even though the interaction energy is weak ($< 40\,\mathrm{kJ{\cdot}mol^{-1}}$), physisorption plays an important role in nature. For instance, the van der Waals attraction between surfaces and the hairs on the feet of geckos enables them to climb vertical surfaces.

When molecules approach a surface, they already undergo physisorption at long distance. At short distance they interact much more strongly by chemical forces (chemisorption). However, the molecules must lose energy in order to be trapped at the surface in the physisorption or in the chemisorption well (Fig. 3.1). If the molecules do not lose energy during collision with the surface they cannot be adsorbed and will reflect.

Physisorption is caused by interactions among permanent, induced and transient electric dipoles. The energy of most of these interactions can be calculated according to Coulomb's law. For instance, q_1q_2/r is the interaction between two molecules with charges q_1 and q_2 and $-\mu_1\mu_2(1{-}3\cos^2\theta)/r^3$ is the interaction between two molecules with dipole

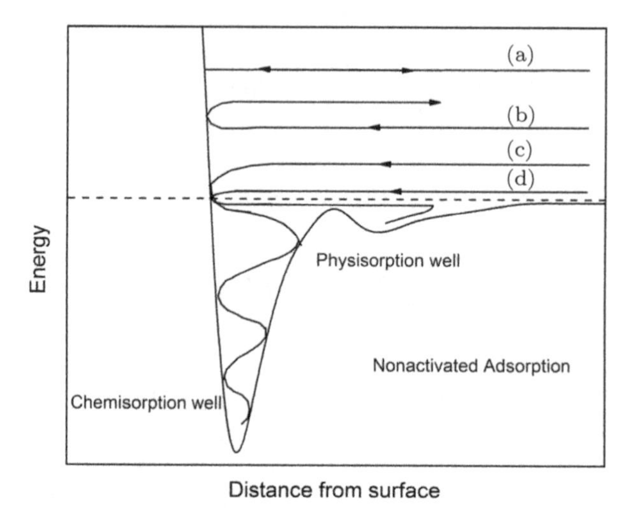

Fig. 3.1 One-dimensional representation of non-activated adsorption. (a) elastic scattering, (b) inelastic scattering, (c) chemisorption, (d) physisorption.

moments μ_1 and μ_2, in both cases at distance r between the molecules. The latter interaction energy depends on the orientation of the dipoles. Assuming that molecules rotate fast, that the statistical chance of an orientation depends on the Boltzmann factor $\exp(-E/kT)$ and that the energy is small compared to kT (so that $\exp(-E/kT) \approx 1 - E/kT$), the average dipole–dipole interaction energy is $-2/3 \cdot \mu_1^2\mu_2^2\, kT/r^6$. Many organic molecules have no dipole moment and are not ions and for these molecules the interaction between their induced dipoles must be calculated with the aid of quantum mechanics. The corresponding interaction energy $-3/2[I_1I_2/(I_1 + I_2)]\alpha_1\alpha_2/r^6$ between two molecules (in which I_i is the ionisation potential of molecule i and α_i the polarisability of molecule i) is referred to as London or dispersion energy.

All above interactions are referred to as **van der Waals interactions** (J. D. van der Waals, Nobel Prize in Physics, 1910). Van der Waals interactions are always attractive, with a $-1/r^6$ dependence on distance: molecules move as close as possible to each other. At short distance, however, repulsive (quantum mechanical) forces play a role and depend strongly ($\propto 1/r^{12}$) on distance. Together with the van der Waals interaction, this gives the Lennard–Jones energy

$$E = \frac{A}{r^{12}} - \frac{B}{r^6}. \tag{3.1}$$

The Lennard–Jones energy has a minimum (Fig. 3.1), called the physisorption well.

Van der Waals interactions determine the boiling points of molecules. Noble gas atoms and small symmetric molecules, which have no charge or dipole moment, interact weakly: He 0.2 kJ·mol^{-1}, H_2 1.5 kJ·mol^{-1}, N_2 7.4 kJ·mol^{-1}, CH_4 14.6 kJ·mol^{-1} (the larger the molecule, the larger the polarisability and, thus, the larger the London energy is). As a consequence, these molecules have low boiling points. Liquid He was produced for the first time in 1908. Amongst others, it is used in superconducting magnets, a component of modern nuclear magnetic resonance (NMR) spectrometers (cf. [2.16]). Van der Waals forces are more or less additive, meaning that contributions by the atoms in molecules can be added together. As a consequence, larger molecules interact more strongly and have higher boiling points. Thus, in a series of hydrocarbons C_nH_{2n+2}, the boiling point increases almost linearly with carbon number and branched hydrocarbons have lower boiling points than linear hydrocarbons (pentane 36.1 °C, methylbutane 27.8 °C and dimethylpropane 9.5 °C). Branched hydrocarbons are smaller and their atoms cannot get as close together as in unbranched hydrocarbons and the interactions are therefore weaker. Molecules with a dipole moment, such as NH_3 and H_2O, have larger interaction energies (29.6 and 47.3 kJ·mol^{-1}, respectively).

Van der Waals interactions between molecules are weak compared to the binding energy between atoms in molecules. For instance, the binding energy of the H atoms in H_2 is 435 kJ·mol^{-1} and that of the C–H bond in CH_4 is 418 kJ·mol^{-1}. Physisorption of molecules on surfaces is, thus, also weak. Physisorption is not specific, it occurs between all adsorbents and adsorbates. Hence, although physisorption is the reason why molecules approach a surface, it does not explain catalysis, the breaking and building of specific chemical bonds.

3.1.1 *Adsorption on Surfaces*

A molecule can hit a surface elastically (without exchange of energy) or inelastically (Fig. 3.1). In an inelastic collision, the molecule exchanges energy with the surface, but it can leave the surface again if it still has sufficient kinetic energy perpendicular to the surface. If the adsorption process is dynamic, we can use statistics to define the average **sticking time** τ of a molecule at the surface and the average **collision number** n of molecules that hit the surface per second and per surface area. The **surface**

coverage σ, the average number of molecules adsorbed per unit surface area then is:

$$\sigma = n\tau s, \tag{3.2}$$

where s is the **sticking probability**, the probability that a molecule that hits the surface will adhere to the surface. For an ideal gas with density D and molar mass M one can derive [3.1]

$$n = D\sqrt{\frac{kT}{2\pi m}} = \frac{N_A P}{\sqrt{2\pi \mathrm{MRT}}}. \tag{3.3}$$

The **Hertz–Knudsen equation** (3.3) shows that the number of collisions per unit surface area and per unit time is proportional to the pressure and inversely proportional to the square root of the temperature. From the Hertz–Knudsen equation, it can be calculated that, at 20 °C and 1 atm (101.3 kPa), H_2 molecules collide 11×10^{23} times per second with a surface of 1 cm^2 and N_2 molecules 3×10^{23} times. This demonstrates that the number of gas molecules colliding with a surface is enormous and that the frequency, with which molecules collide with a surface, is never a limiting factor for the coverage of a surface with molecules.

The average residence time τ of molecules at a surface depends on the probability that the molecule desorbs from the surface. In turn, this probability depends on the heat of adsorption Q, the interaction between the molecule and the surface, which must be provided so that the molecule can leave the surface. The desorption frequency ν is described by an Arrhenius-type equation, because the probability that a molecule collects energy Q is proportional to $\exp(-Q/RT)$. This gives $\nu = \nu_0 \exp(-Q/RT)$, where ν_0 is a frequency factor that is roughly equal to $\nu_0 = kT/h = 0.6 \times 10^{13}\,\mathrm{s}^{-1}$ at room temperature. With $\tau = 1/\nu$, the residence time is:

$$\tau = \tau_0 \exp \frac{Q}{RT}. \tag{3.4}$$

Note the positive sign of Q in equation (3.4) ($Q = -\Delta H_{\mathrm{ads}}$). Because of the exponential dependence, the residence time increases strongly with the heat of adsorption and decreases strongly with temperature. For instance, assuming that $\tau_0 = 10^{-13}\,\mathrm{s}$, τ is of the same order of magnitude as τ_0 ($10^{-13}\,\mathrm{s}$) at room temperature for $Q < 4\,\mathrm{kJ \cdot mol^{-1}}$. For $Q = 6\,\mathrm{kJ \cdot mol^{-1}}$, which is about the heat of physisorption of H_2 on a surface, τ is about $10\tau_0$ ($10^{-12}\,\mathrm{s}$) and weak adsorption occurs. For a heat of adsorption of about $16\,\mathrm{kJ \cdot mol^{-1}}$, as for Ar, O_2, N_2 and CO on several surfaces, $\tau \approx 10^{-10}\,\mathrm{s}$

and a small portion of the surface will be covered at room temperature and 1 atm. At higher heat of adsorption, τ increases strongly, as does σ.

The **degree of coverage** of a surface is defined as $\theta = \sigma/\sigma_0$, where σ is the number of molecules adsorbed on unit surface area and σ_0 is the number of molecules that can cover a unit of surface area in a monolayer. The degree of coverage is the number of monolayers that is adsorbed on a surface. For N_2 molecules, σ_0 is about 10^{15} molecules·cm^{-2}. At room temperature and 1 atm, $\theta = 300$ for $Q = 40\,\text{kJ}\cdot\text{mol}^{-1}$ (as for hexane on Pt) and $\theta = 10^9$ for $Q = 80\,\text{kJ}\cdot\text{mol}^{-1}$ (as for dissociative chemisorption of H_2 on Pt). Values above 1 are unrealistic, because they would require more than monolayer coverage and the heat of adsorption of hexane on hexane, and of H_2 on H, is much lower than of hexane on Pt and H on Pt, respectively. However, these calculated high degrees of coverage demonstrate the clear tendency to cover the surface. At the higher temperature of 600 K and a pressure of 1.3×10^{-4} Pa (10^{-6} Torr), the situation is different. For $Q = 40\,\text{kJ}\cdot\text{mol}^{-1}$, $\theta = 8 \times 10^{-11}$ and for $Q = 80\,\text{kJ}\cdot\text{mol}^{-1}$ $\theta = 10^{-7}$, indicating that under these conditions molecules can be removed from a surface.

3.1.2 *Langmuir Adsorption Isotherm*

Substituting Eqs. (3.3) and (3.4) into Eq. (3.2) gives

$$\sigma = n\tau s = \frac{N_A P}{\sqrt{2\pi \text{MRT}}} \tau_0 e^{\frac{Q}{RT}}\, s. \tag{3.5}$$

Because M, τ_0 and s are constants for the adsorption of a certain molecule on a certain surface, σ is a function of p and T: $\sigma = k_0 \frac{P}{\sqrt{T}} \exp \frac{Q}{RT}$.

At constant temperature this gives $\sigma = k_1 P$. This most simple form of adsorption isotherm (constant T) predicts that surface coverage is proportional to pressure when the molecules behave ideally in the gas phase as well as on the surface and when Q is constant and does not depend on coverage. This might only be true at low coverage, when the surface is homogeneous and adsorbed molecules do not hinder each other. Another problem is that, after initial coverage of the surface, gas molecules may collide with adsorbed molecules. If the interaction energy Q of a molecule with an adsorbed molecule is low, then collision will not lead to adsorption. Langmuir therefore corrected the isotherm $\sigma = k_1 P$ by assuming that only the collision of molecules with the bare surface, but not with already adsorbed molecules, leads to adsorption [3.2]. If that is true, the sticking probability must be corrected for the degree of coverage that has already taken place; thus $s = 1 - \sigma/\sigma_0$, $\sigma = n\tau(1 - \sigma/\sigma_0)$ and $\sigma = \sigma_0 n\tau/(\sigma_0 + n\tau)$.

This gives:

$$\theta = \frac{\sigma}{\sigma_0} = \frac{\frac{n\tau}{\sigma_0}}{1 + \frac{n\tau}{\sigma_0}} \quad \text{or} \quad \theta = \frac{k_L P}{1 + k_L P}, \tag{3.6}$$

with $k_L = \frac{n\tau}{\sigma_0 P} = \frac{N_A}{\sqrt{2\pi \mathrm{MRT}}} \frac{\tau}{\sigma_0}$.

The coverage first increases with pressure but is saturated at high pressure, because Langmuir assumed that the maximum coverage of the surface is a monolayer (Fig. 3.2). Langmuir derived the **Langmuir equation** (3.6) in 1916 while working in the laboratory of General Electric, USA, studying gas adsorption on tungsten wires in order to produce better light bulbs. In 1932, he received the Nobel Prize in Chemistry for his work on the adsorption of molecules on surfaces.

At low pressure, the Langmuir equation is identical to the simple adsorption isotherm equation $\theta = k_1 P$. At high pressure the Langmuir equation leads to the limiting coverage of $\theta = 1$, due to the assumption that only collisions with the bare surface can lead to adsorption. The International Union of Pure and Applied Chemistry (IUPAC) refers to this adsorption isotherm as a class I isotherm. Other classes will be discussed in Section 3.1.3.

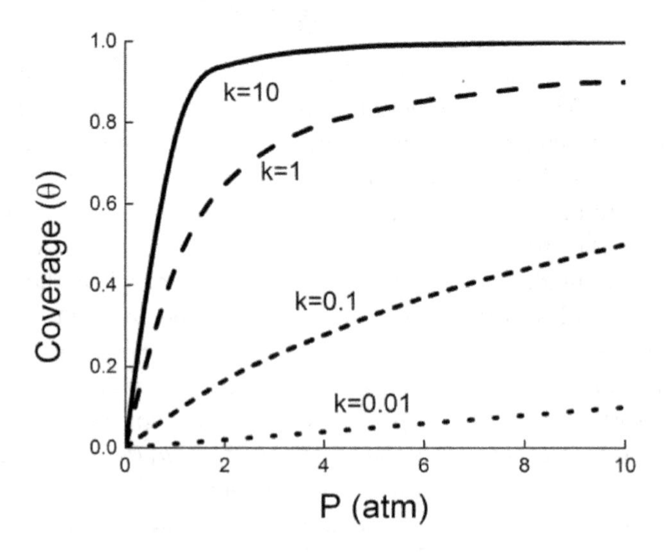

Fig. 3.2 Surface coverage as a function of pressure in a Langmuir isotherm.

I. Langmuir

Measurement of gas adsorption on a surface is based on Boyle's law $N = PV/RT$. By measuring the pressure P_1 of a gas in a volume V_1 at temperature T, the number N of gas molecules is known. When one opens the valve to an evacuated volume V_2 that contains the material, whose surface area one wants to determine, the pressure in the total volume $V_1 + V_2$ will be $P = P_1V_1/(V_1 + V_2)$ if no adsorption takes place but will be lower when adsorption takes place. From this pressure decrease one can calculate the number of gas molecules that was adsorbed on the surface of the material. By dividing this number by the number of molecules that can cover the surface in a monolayer one obtains the surface coverage θ. The addition of varying amounts of gas to the adsorbate, followed by measurement of the equilibrium pressures, gives a series of V, P values. From these values the Langmuir constant k_L is calculated.

When two molecules A and B adsorb on a surface, the changes in the coverages of these molecules at the surface are equal to

$$\frac{d\theta_A}{\mathrm{dt}} = k_1 P_A \theta_v - k_{-1}\theta_A$$

$$\frac{d\theta_B}{\mathrm{dt}} = k_2 P_B \theta_v - k_{-2}\theta_B$$

A B

↑↓ ↓↑

___A∗ _____ B∗___

In these equations, θ_A and θ_B are the coverages of the surface by A and B, respectively, θ_v is the fraction of vacant sites (the fraction of the surface that is not covered by A and B), and k_i and k_{-i} ($i = 1$ or 2) are the rate constants for adsorption and desorption of molecule i (A or B), respectively.

In equilibrium, $\frac{d\theta_A}{dt} = \frac{d\theta_B}{dt} = 0$ and

$$\frac{\theta_A}{1-\theta_A-\theta_B} = \frac{k_1 P_A}{k_{-1}} = K_A P_A \qquad \theta_A = K_A P_A (1-\theta_A-\theta_B),$$

$$\frac{\theta_B}{1-\theta_A-\theta_B} = \frac{k_2 P_B}{k_{-2}} = K_B P_B \qquad \theta_B = K_B P_B (1-\theta_A-\theta_B).$$

$K_A = k_1/k_{-1}$ and $K_B = k_2/k_{-2}$ are the equilibrium constants for the adsorption of A and B at the catalyst surface, respectively.

As a result, the Langmuir coverages θ_A of A and θ_B of B at the surface are

$$\theta_A = \frac{K_A P_A}{1 + K_A P_A + K_B P_B} \quad \text{and} \quad \theta_B = \frac{K_B P_B}{1 + K_A P_A + K_B P_B}. \tag{3.7}$$

When many molecules i adsorb at the surface, the Langmuir coverage of molecule A is

$$\theta_A = \frac{K_A P_A}{1 + \Sigma K_i P_i}. \tag{3.8}$$

When $P_A = P_B$ in the gas phase, the ratio of the surface coverages of A and B is equal to

$$\frac{\theta_A}{\theta_B} = \frac{K_A}{K_B} = \frac{k_1/k_{-1}}{k_2/k_{-2}} = \frac{K_1^0}{K_2^0} \exp\frac{Q_A - Q_B}{RT} \approx \exp\frac{\Delta Q}{RT}. \tag{3.9}$$

In Eq. (3.9), $k_i/k_{-i} = K_i = K_i^0 \exp(Q_i/RT)$, where the heat of adsorption Q_i of molecule i is equal to the difference in the activation energy of adsorption and desorption (Fig. 3.3).

The heat of adsorption is defined as $Q = -\Delta H_{\text{ads}}$; the enthalpy of adsorption is negative but the heat of adsorption is always positive. The pre-exponential factors K_i^0 are often of a similar magnitude and, in a semi-quantitative calculation of θ_A/θ_B, K_1^0/K_2^0 can be regarded as being equal to 1 in Eq. (3.9). Because of the exponential dependence, even a small difference in the heat of adsorption leads to a large difference in coverage. For example, at $P_A = P_B$ and $\Delta Q = 40\,\text{kJ·mol}^{-1}$, $\theta_A/\theta_B = 1.4 \times 10^7$ at 20 °C and even at 400 °C the ratio is high ($\theta_A/\theta_B = 1.3 \times 10^3$). This demonstrates that a surface is covered preferentially by the molecule with

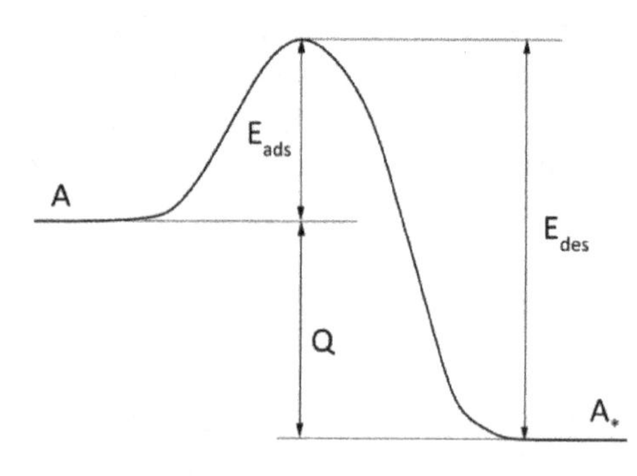

Fig. 3.3 Energy diagram of adsorption of molecule A at a catalytic site *.

the highest heat of adsorption. Furthermore, surfaces are easily poisoned, because, for $\theta_A/\theta_B = 1$ and $\Delta Q = 40\,\text{kJ}\cdot\text{mol}^{-1}$, the ratio $P_A/P_B = K_B/K_A = 0.7\times10^{-7}$ at 20 °C. Thus, even a few ppm of poison are enough to poison the surface when its adsorption is stronger than that of the reactants!

3.1.3 *Multilayer Adsorption, BET*

The Langmuir equation is widely applied to chemisorption when adhesion (binding of molecules to a surface) is much stronger than cohesion (binding between molecules). It also finds application to the adsorption of molecules from solution by solid surfaces (treatment of wastewater). However, when adhesion and cohesion do not differ significantly, the Langmuir equation is of limited value because then the adsorption of molecules on top of already adsorbed molecules must be taken into account.

Photo of S. Brunauer copyright of the American Chemical Society, Jan. 1, 1987. Photo of P. Emmett from the Paul Emmett Papers, 1918–2001, OSU Libraries Special Collections & Archives Research Centre.

Brunauer, Emmett and Teller therefore extended the Langmuir model from monolayer to multilayer adsorption [3.3]. Figure 3.4 (left) indicates how molecules are adsorbed at different heights. A model that can describe multi adsorption mathematically is shown in Fig. 3.4, right.

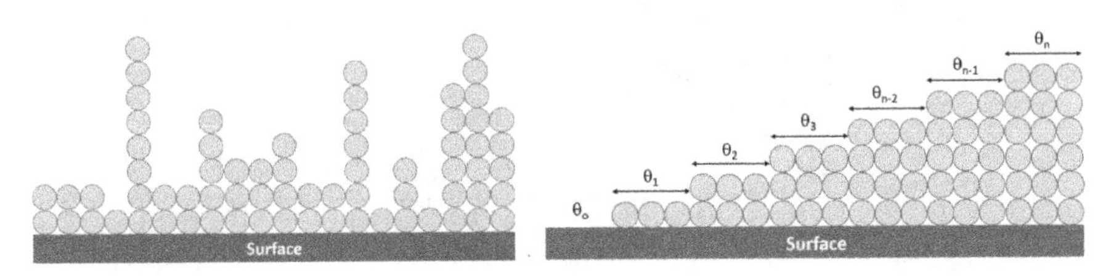

Fig. 3.4 BET model of multilayer coverage of a surface.

Assuming that there are no lateral interactions between adsorbed molecules, Brunauer, Emmett and Teller derived the following BET equation (3.10) (see [3.4] for derivation):

$$\theta = \frac{cP'}{(1-P')(1+(c-1)P')} \quad \text{with} \quad P' = P/P_0, \tag{3.10}$$

where P_0 is the saturation pressure of the adsorbate and $c = \tau_0/\tau_1$ the ratio of the residence times of a molecule adsorbed on the bare surface and a molecule adsorbed on already adsorbed molecules. Figure 3.5 gives the BET adsorption isotherms for three values of c.

The BET equation is very important to determine the surface area of powders, especially of catalyst supports. With automated equipment it is possible to measure adsorption isotherms and determine the constant c. If the heat Q_i of adsorption of a molecule on already adsorbed molecules is equal to the heat of evaporation Q_L of liquid molecules and if the frequency factors of both types of adsorbates are equal, then

$$c = \tau_0/\tau_1 = \left(\tau_0^0 \cdot \exp\frac{Q_A}{RT}\right) \Big/ \left(\tau_1^0 \cdot \exp\frac{Q_L}{RT}\right) \sim \exp\frac{Q_A - Q_L}{RT}. \tag{3.11}$$

When adsorption on the bare surface is stronger than on already adsorbed molecules ($Q_A > Q_L$), then $c > 1$, giving type II isotherms (Fig. 3.6). Type II isotherms first increase strongly with pressure, but the increase slows down at increasing pressure, because multi adsorption is not favoured to the same extent as mono adsorption. Type II isotherms are, thus, convex

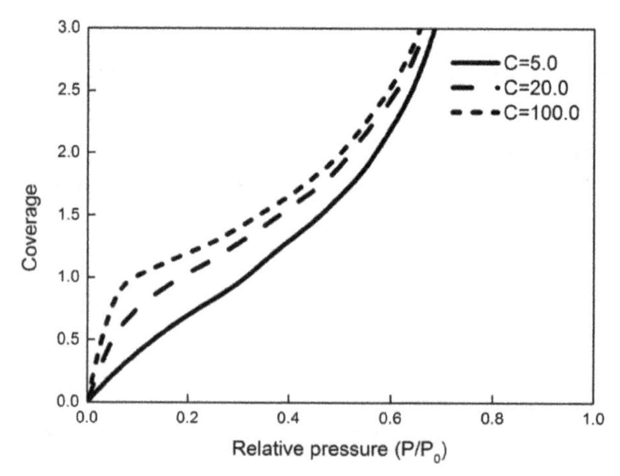

Fig. 3.5 BET adsorption isotherms for $c = 5$, 20 and 100.

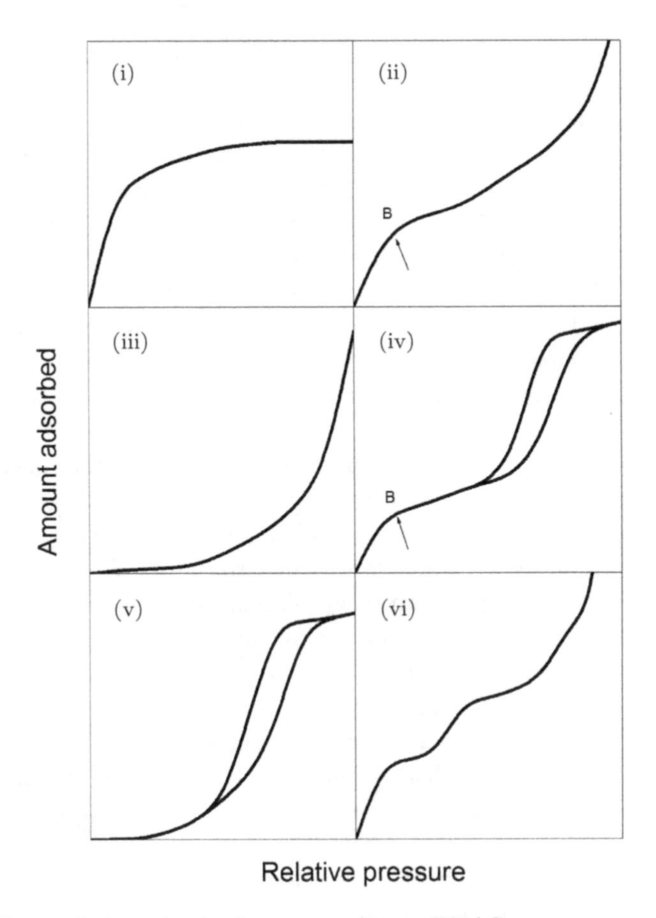

Fig. 3.6 Types of adsorption isotherms according to IUPAC.

at lower pressure. Multi adsorption becomes important above the inflection point B (Fig. 3.6), when condensation begins and the coverage increases sharply with pressure, leading to a concave isotherm. When adhesion is weaker than cohesion, then $Q_A < Q_L$ and $c < 1$. This occurs when hydrophobic molecules (e.g. benzene) adsorb on glass, leading to type III concave adsorption isotherms over the whole pressure range, because adsorption is initially weak but increases with pressure (Fig. 3.6).

As expected, the Langmuir equation is included in the BET equation, because for $P' \ll 1$ and $c \gg 1$ one obtains $cP' = cP/P_0 = (\tau_0/\tau_1)(\beta\tau_1/\sigma_0)P = (\tau_0\beta/\sigma_0)P = k_L P$ and $\theta = cP'/(1 + cP') = k_L P/(1 + k_L P)$.

Like the Langmuir model, the BET model is based on the assumption that the surface is homogeneous and that no lateral interactions occur between adsorbed molecules. This leads to the physically absurd situation that molecules stack up in columns and do not form islands of molecules [3.5], as shown in Fig. 3.4 (left side). Furthermore, the heat of adsorption is assumed to be constant from the second adsorption layer onwards. For these reasons, the BET equation should not be applied to the region $P/P_0 > 0.3$. The region $P/P_0 < 0.05$ should not be considered either, because the surface may contain defects that have a high heat of adsorption and strongly adsorb the first molecules admitted to the surface. At intermediate pressure ($0.05 < P/P_0 < 0.3$) the heat of adsorption may vary only slightly and, thus, the BET isotherm gives acceptable results.

At higher relative pressure, deviations between the experiment and the BET equation occur because, in the derivation of the BET equation, it is assumed that the adsorption becomes infinite when P reaches the saturation pressure of the adsorbate. However, when the heat of adsorption of the second and higher adsorption layers is different from the heat of evaporation Q_L of the liquid molecules, then the residence time τ_1 will differ from the expected value and adsorption will increase more sharply or less sharply. Furthermore, small pores can contain only a limited number of layers, whereas an infinite number of layers is assumed in the derivation of the BET equation. Despite its shortcomings, the BET method still enjoys great popularity, mainly because in the region $0.05 < P/P_0 < 0.3$ the advantages outweigh the disadvantages.

Any gas that can be liquefied can be used for BET measurements; in practice, however, N_2 is used in most cases. A standard measurement of the surface area of a porous material consists of the automated determination of the adsorption isotherm of N_2 at 77 K (−196 °C) between 0 and 1 atm and the calculation of V_m, the gas volume that represents monolayer coverage. The surface area SA is calculated from V_m: SA ($m^2 \cdot g^{-1}$) = 4.36 V_m (cm^3 gas) for N_2. The BET method enables the calculation of surface areas up to 1,000 $m^2 \cdot g^{-1}$. For small surface areas a gas with a pressure lower than N_2 at −196 °C (Ar, Kr or CH_4) should be selected.

After measuring a BET isotherm by increasing the pressure of N_2, it is also possible to measure the desorption isotherm by decreasing the pressure from high to low. **Hysteresis** of adsorption is often observed, with the desorption isotherm above the adsorption isotherm (Fig. 3.7). Hysteresis occurs when filling of the pores does not occur in the same way as emptying of pores. Filling occurs when an increasing number of layers

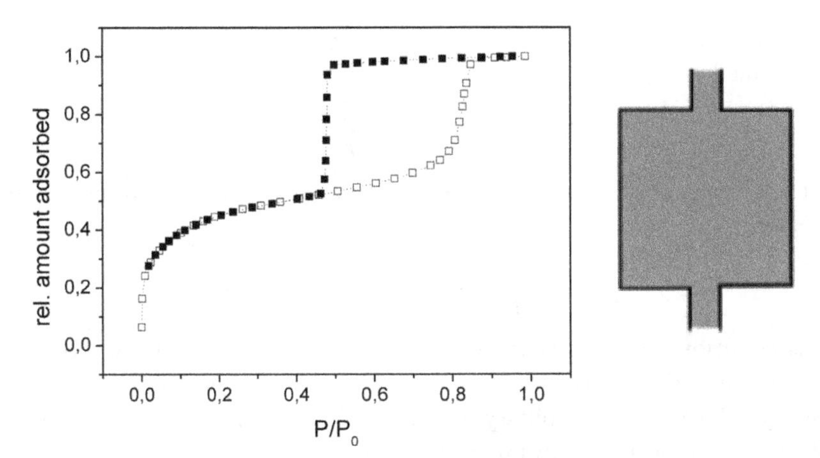

Fig. 3.7 Adsorption (open squares) and desorption (filled squares) isotherms of N_2 at $-196\,°C$ in porous silica glass, consisting of spherical cavities with a diameter of 20 nm connected by channels, 3 nm in diameter.

is deposited on all the surfaces (walls of the pores). Emptying, however, does not occur by removing one layer after another from all the surfaces simultaneously but starts at the widest pores (with the lowest curvature) and ends with removal of the molecules from the narrow pores. Hysteresis can occur when adsorption on the bare surface is stronger than on already adsorbed molecules (changing type II behaviour to type IV, cf. Fig. 3.6) as well as when adsorption on the bare surface is weaker than on already adsorbed molecules (changing type III behaviour to type V). A type VI isotherm (Fig. 3.6) is characterised by steps that may reflect adsorption in more than one layer. It may occur when the first-layer has a specific registry with the surface or has an interaction that gives rise to a second layer.

Hysteresis is explained quantitatively by the Kelvin equation. When molecules adsorb on a curved surface with radii r_1 and r_2, the gaseous adsorbate is in equilibrium with the layer adsorbed on the curved surface when the free enthalpy is constant or when dG is zero:

$$dG = \mu_{\text{ad}}\, dN - \mu_{\text{G}}\, dN + \gamma\, dS = 0, \quad \text{or}$$

$$dS/dN = (\mu_{\text{ad}} - \mu_{\text{G}})/\gamma.$$

$dS/VdN = \frac{1}{r_1} + \frac{1}{r_2}$, with μ the chemical potential and V the molar volume.

With $\mu_G = \mu_G^0 + RT \ln \frac{P}{P_0}$ and $\mu_{\text{ad}} = \mu_L = \mu_G^0$, the **Kelvin equation** is obtained:

$$RT \ln \frac{P}{P_0} = -\gamma V \left(\frac{1}{r_1} + \frac{1}{r_2} \right). \tag{3.12}$$

For a cylinder (capillary) with radius r, the radii r_1 and r_2 are $r_1 = r$ and $r_2 = \infty$, thus

$$\ln \frac{P}{P_0} = -\frac{\gamma V}{RTr} \quad \text{or} \quad P = P_0 \exp\left(-\frac{\gamma V}{RTr} \right). \tag{3.13}$$

This equation indicates that molecules with pressure P_i in the gas phase are in equilibrium with molecules on the surface when the radius of the adsorbed layer on the capillary wall is r_i. When more and more molecules adsorb on the wall, the radius r decreases. However, P decreases when r decreases and, thus, the pressure required to maintain equilibrium between the gas and the adsorbed phase decreases. As a consequence, when the pressure of the gas is sufficient to start adsorption on the wall of the cylindrical pore with radius r, the adsorption will continue until the whole pore is filled. Concave surfaces will be filled one by one. First the surfaces with the largest curvature (smallest r) will become full at a certain pressure. Surfaces with a somewhat larger r will fill at a slightly higher pressure. This is what happens during an increase in gas adsorption above about $P/P_0 = 0.3$.

When all the pores are full, desorption starts at the end of the cylindrical capillary (pore mouth) and the liquid surface takes the shape of a half sphere with $r_1 = r_2 = r$. For desorption, we thus get $\ln(P_{\text{des}}/P_0) = -2\gamma V/RTr$. Thus, $P_{\text{des}}/P_0 = (P_{\text{ads}}/P_0)^2$ and $P_{\text{des}} < P_{\text{ads}}$ (because $P/P_0 < 1$). Desorption therefore starts at lower pressure than adsorption and the desorption curve is above the adsorption curve. When measuring the desorption isotherm of porous systems with many flat surfaces and a few strongly curved surfaces (as in so-called ink bottle pores, (see Fig. 3.7, right)), lowering the pressure does not lead initially to strong desorption. The liquid always contracts in the smallest of the ink bottle pores and for this radius r desorption does not occur yet, according to the equation $\ln(P_{\text{des}}/P_0) = -2\gamma V/RTr$. Finally, when the pressure is so low that desorption of the narrow pores should occur, according to the desorption equation, almost total desorption occurs and the desorption isotherm drops sharply. Warning! Although such a sharp drop is often observed for N_2 adsorption at $P/P_0 = 0.42$, this is not due to ink bottle pores, but is an artifact of N_2.

3.2 Surface Diffusion

Until now we assumed that a molecule can adsorb (chemisorb) only at the position (site) where it collides with the surface, but in reality the situation is more complex. It was known for a long time that the probability for desorption is not proportional to the surface coverage but increases much more slowly. As an explanation, Kisliuk proposed that diffusion of molecules over the surface increases the sticking coefficient [3.6]. A molecule can hit the surface and be physisorbed at the site of impact. To become physisorbed, the molecule does not have to transfer all the translational energy that it had in the gas phase to the surface. The molecule only needs to transfer its translational energy perpendicular to the surface and can maintain the translational energy parallel to the surface and, thus, can diffuse over the surface. After several diffusion steps it may lose more energy to the surface and become locally physisorbed (Fig. 3.8). After more loss of energy, it becomes chemisorbed but is still in an excited state. After losing more energy it finally reaches the bottom of the chemisorption well (see Fig. 3.1).

Because physisorption energy is low, the activation energy for diffusion to a neighbouring position will also be low. A physisorbed molecule has, therefore, three possibilities. It can desorb or adsorb, and it can migrate (diffuse) to another position. The sum of the probabilities for adsorption p_a, desorption p_d and migration p_m at site 1 are $p_{a1} + p_{d1} + p_{m1} = 1$.

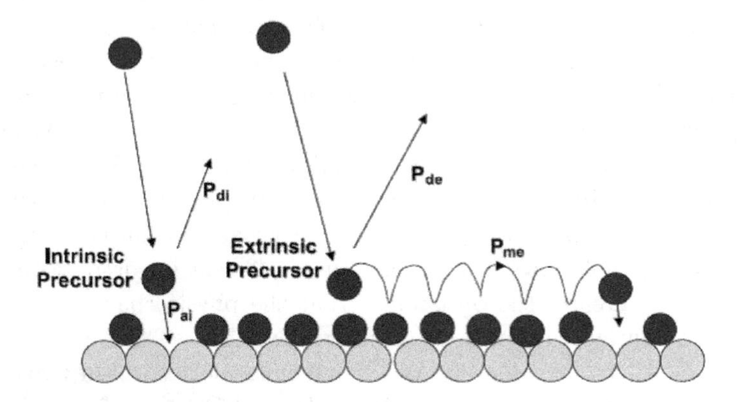

Fig. 3.8 Adsorption to a precursor state over empty (intrinsic) or filled (extrinsic) sites. The intrinsic state can adsorb ($p_{\rm ai}$), desorb ($p_{\rm di}$) or diffuse. The extrinsic state can desorb ($p_{\rm de}$) or diffuse ($p_{\rm me}$) to another site before adsorbing at an empty site [3.7]. Copyright (2010) IOP Publishing.

The sticking probability s is the sum of all probabilities that the molecule adsorbs on sites $1, 2, 3, \ldots$, multiplied by the migration probabilities:

$$s = p_{a1} + p_{m1}p_{a2} + p_{m1}p_{m2}p_{a3} + \cdots.$$

If p_{ai} and p_{mi} are independent of i, then $s = p_a(1 + p_m + p_m^2 + \cdots) = p_a \frac{1}{1-p_m} = \frac{p_a}{p_a+p_{d1}}$.

The probability of desorption depends on the desorption of molecules from the bare surface and of molecules from adsorbed molecules. $p_{d1} = p_d(1-\theta) + p'_d\theta$, thus

$$s = \frac{p_a f(\theta)}{p_a f(\theta) + p_d(1-\theta) + p'_d\theta}.$$

For a clean surface, $f(\theta) = 1$ and $s_0 = p_a/(p_a + p_d)$. If each surface position can accommodate one molecule, then $f(\theta) = 1 - \theta$, as in the derivation of the Langmuir equation, and

$$\begin{aligned} s &= \frac{p_a(1-\theta)}{p_a(1-\theta) + p_d(1-\theta) + p'_d\theta} = \frac{p_a(1-\theta)}{(p_a + p_d)(1-\theta) + p'_d\theta} \\ &= \frac{s_0(1-\theta)}{(1-\theta) + \frac{p'_d\theta}{p_a+p_d}} \\ \frac{s}{s_0} &= \frac{(1-\theta)}{(1-\theta) + K\theta}, \end{aligned} \tag{3.14}$$

where $K = p'_d/(p_a + p_d)$ is the precursor desorption coefficient, the ratio of the number of molecules desorbing from adsorbed molecules to the number of molecules adsorbing at and desorbing from the bare surface.

Figure 3.9 shows that the sticking probability decreases linearly with coverage ($s/s_0 = 1 - \theta$) for pure Langmuir adsorption without precursor adsorption ($K = 1$, high coefficient for precursor desorption). When the probability of desorption from the second layer is small (K is small), the probability of diffusion becomes high. In that case, the precursor state keeps the sticking probability close to 1, even at substantial coverage of the surface by molecules. The smaller the coefficient for desorption of the precursor, the greater the probability that the physisorbed molecule will stick to the surface. Migration (diffusion) over the surface increases the sticking coefficient considerably. When one molecule adsorbs at two surface positions, as in dissociative adsorption of H_2 and O_2, then $f(\theta) = (1-\theta)^2$ and $s/s_0 = (1-\theta)/[(1 - s_0\theta) + K\theta/(1-\theta)]$. In that case too, the sticking coefficient decreases only slowly with coverage, similar to the sticking coefficient in Fig. 3.9.

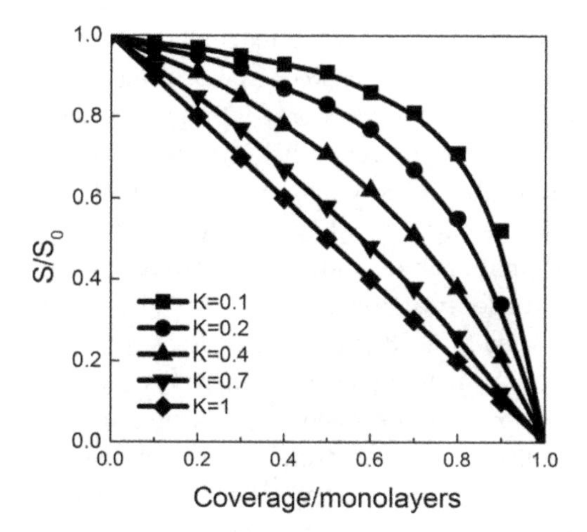

Fig. 3.9 The effect of the precursor parameter on the adsorption curve; $K = 1$ corresponds to the absence of a precursor effect and Langmuirian kinetics [3.7]. Copyright (2010) IOP Publishing.

3.3 Chemisorption

3.3.1 *Chemical Bonding*

Chemisorption is the strong binding of molecules and atoms with surfaces. Metallic bonds, as in pure metals, are responsible for the chemisorption of alkali metals on metals (Cs on W) (see Section 5.4 for metal bonding). Ionic bonds are responsible for the chemisorption of oxygen and halogens at metal surfaces, when surface metal atoms are oxidised to metal cations and the oxygen and halogen atoms become anions (O_2 on Ni). Covalent bonds describe the bonding of hydrogen on metals (H_2 on Pt), CO on metals (CO on Ni) and alkenes on metal oxides (propene on bismuth molybdate). The strongest contributions to covalent bonding between two molecules are made by the frontier orbitals, the highest occupied molecular orbitals (**HOMO**) of one molecule and the lowest unoccupied molecular orbitals (**LUMO**) of the other molecule. This interaction can be seen as an acid-base interaction between an orbital with fewer electrons (LUMO, the acceptor orbital) and an orbital with more electrons (HOMO, the donor orbital).

The bonding between atoms in a molecule as well as between molecules is strongest when the interaction between orbital A of one atom and of orbital B of the other atom is strongest. The interaction is proportional to $(H_{AB})^2/\Delta E_{AB}$, where H_{AB} is the exchange interaction $\int A\frac{e}{r}B\,dx$ and ΔE_{AB} is the energy difference between the orbitals. Because the exchange interaction $\int A\frac{e}{r}B\,dx$ is roughly proportional to the overlap $\int AB\,dx$ between both orbitals, the interactions between orbitals are strongest when the orbitals overlap substantially and ΔE_{AB} is small.

When two orbitals interact, they form two new orbitals, which are a mixture of the original orbitals and, thus, surround both atoms [3.8]. They are referred to as **molecular orbitals** (MOs). The energy $E^- = E_c^0 - \Delta E$ of the new lower level (Fig. 3.10, right) decreases by ΔE and the energy $E^+ = E_b^0 + \Delta E'$ of the new higher level increases by $\Delta E'$.

If before the interaction between the orbitals on atom A and on atom B the level E_c^0 contains two electrons while the level E_b^0 is empty, then following the interaction the new lower level with energy E^- contains two electrons and the new higher level with energy E^+ is empty. In that case, the interaction leads to an energy gain $2\Delta E = 2(E_c^0 - E^-)$ and, thus, to bonding of the atoms (or molecule fragments) to which the orbitals A and B belong. This MO is referred to as a **bonding orbital**.

If, on the other hand, each of the original levels E_c^0 and E_b^0 is filled with two electrons, then each of the two new levels will also contain two electrons. This would lead to an energy gain of $2\Delta E$ and an energy loss of $2\Delta E'$ (Fig. 3.10, right). Since quantum mechanics teaches us that the antibonding MO increases more in energy than the bonding MO decreases ($\Delta E' > \Delta E$), the net result is energy loss. Thus, the MO that has gone up in energy relative to the energy E_b^0 leads to a loss in energy when it is filled with electrons and to a repulsive interaction between the atoms (or molecule fragments). This MO is referred to as an **antibonding orbital**.

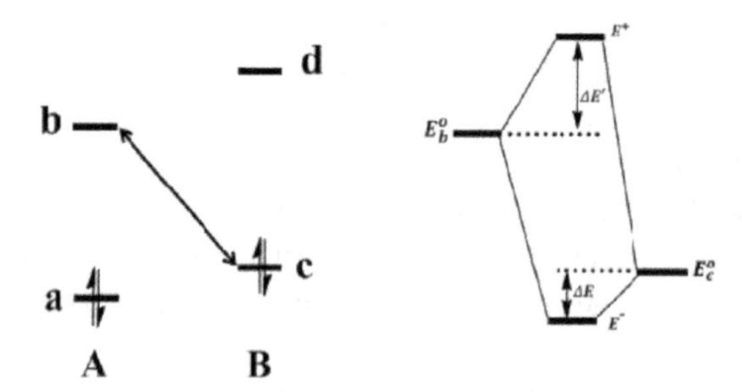

Fig. 3.10 Interaction between orbitals b and c leads to two new MOs, a decrease in energy of the lower orbital by ΔE and an increase in the energy of the higher orbital by $\Delta E'$.

The interactions between orbitals can be summarised as follows. Interactions of filled and empty orbitals (two-electron interactions) are attractive, while interactions of filled orbitals (four-electron interactions) are repulsive. Interactions of empty orbitals have no effect; as long as there are no electrons, there is no change in energy.

When several filled and empty orbitals (levels) exist on both interacting atoms A and B (Fig. 3.10, left), all three types of interactions (attractive, repulsive and no effect) will occur. Interactions between the filled orbital a from A and the empty orbital d from B and between the filled orbital c from B and the empty orbital b from A are attractive. The lower MO that is formed by the interaction is filled with two electrons and gains energy, while the upper MO is empty and, thus, cannot lose energy. Therefore, a–d and b–c two-electron interactions lead to bonding. An interaction between orbitals b and d has no energetic consequences, since the orbitals are empty. The four-electron interaction between orbitals a and c is repulsive, because the total energy is greater than that of the separate isolated levels.

In an atom or molecule there can be many more filled and empty orbitals than indicated in Fig. 3.10 and, as a consequence, there are many attractive and repulsive interactions. Fortunately, the interactions between orbitals that overlap substantially and which are similar in energy dominate because the interaction between orbitals is proportional to $(H_{AB})^2/\Delta E_{AB}$. Orbitals overlap substantially when they extend in space. This is the case for orbitals of the least bound electrons (the upper levels) of an atom or molecule, the valence orbitals. Orbitals of strongly bonded electrons, with low energy, are drawn close to the nucleus and do not extend into space. The overlap of such contracted orbitals is small. As a consequence, orbitals in the valence region determine chemical reactivity: the highest filled and lowest empty orbitals.

The empty (acceptor) level b of fragment A is closer in energy to the filled (donor) level c of fragment B than to levels below level c. If the overlap $\int bh\,dr$ of the orbitals h belonging to such levels is smaller than $\int bc\,dr$, then the numerator and denominator of $(H_{AB})^2/\Delta E_{AB}$ set the b(A)–c(B) interaction apart as an important, perhaps the most important interaction. The HOMO, or a small subset of higher-lying occupied MOs, and the LUMO, or a subset of unoccupied MOs, often dominate the interaction between two molecules. These are referred to as the frontier orbitals, the valence orbitals of the molecule and the orbitals most easily perturbed in any molecular interaction. They control the chemistry of the molecules [3.8].

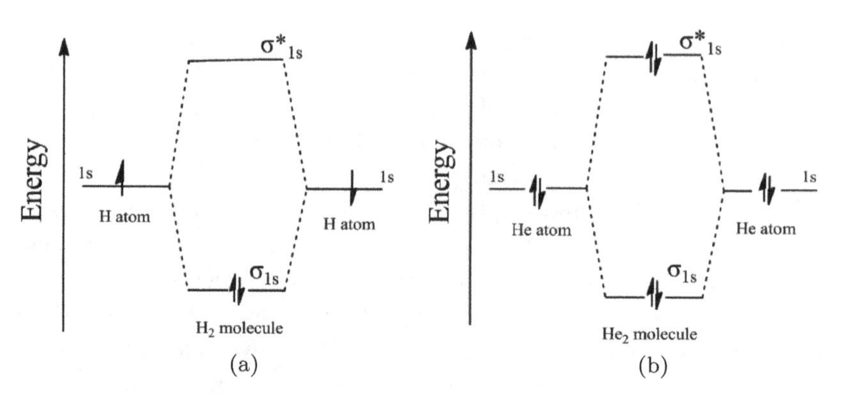

Fig. 3.11 Molecular orbitals and MO energies of (a) H_2 and (b) He_2.

This qualitative knowledge can explain the stability of molecules. Why does H_2 exist and why does He_2 not exist? The energy difference ΔE_{AB} between the orbitals of the fragments A and B is smallest when A and B are the same atoms or molecules, then $\Delta E_{AB} = 0$. This explains why the bonding between two H atoms is strong, as seen in Fig. 3.11, which indicates what happens to the energy when a hydrogen atom A approaches a hydrogen atom B. The 1s atomic orbital (AO) on atom A and the 1s AO on atom B form two MOs. One MO is lower in energy than the original AOs and is referred to as the bonding orbital, because it leads to a gain in energy when it is filled with electrons. The second MO is higher in energy than the energy of the original 1s orbitals. This MO is the antibonding orbital, leading to a loss of energy when filled with electrons.

When two H atoms (each having one electron) are combined to a H_2 molecule, the two electrons can be put in the lowest MO, because the **Pauli exclusion principle** allows putting two electrons in one orbital when they have a different spin α and β. Pauli discovered the exclusion principle in 1925, when he was 25 years old, and received the Nobel Prize in Physics in 1945. The lowest MO of H_2 is the bonding MO and, thus, the combination of two H atoms leads to a gain in energy (Fig. 3.11, left). Two H atoms form a stable bond in the H_2 molecule. If an attempt is made to combine two He atoms to a He_2 molecule, then four electrons must be transferred to the MOs of the He_2 molecule, because each He atom has two electrons. As Fig. 3.11 (right) shows, two electrons must go into the bonding MO and two into the antibonding MO. Because the antibonding MO increases more in energy than the bonding MO decreases, the energy increases when the two He atoms are combined to a molecule, meaning that two He atoms repel each other.

In the same way one can construct the MOs and the energy levels of other molecules, for instance of carbon monoxide. The CO molecule is important in catalysis, not only in catalytic reactions (such as methanation and Fischer–Tropsch reactions, cf. Section 6.4.1), but also as a probe molecule to characterise catalyst surfaces. The bonding of the C and O atoms in CO and the bonding of CO to a metal atom or cation are, therefore, important issues. Figure 3.12 indicates how the 1s, 2s and 2p AOs of the C and O atoms combine to MOs. The 1s AOs are not important in the chemical binding of the C and O atoms, because they are small and do not overlap. Thus, the interaction energy is very low. The valence-shell carbon and oxygen 2s and 2p AOs combine to MOs of two types: σ orbitals, which are symmetric around the CO axis, and π orbitals, which have a plane through the CO axis as plane of antisymmetry (positive part of the orbital on one side and negative part of the orbital on the other side of the plane). As a consequence, the electron density of the π orbitals is zero in the CO axis, while the σ orbitals have a maximum electron density in the CO axis. With six electrons from the C atom ($1s^2$, $2s^2$, $2p^2$) and eight from the O atom ($1s^2$, $2s^2$, $2p^4$), the 1σ, 2σ, 3σ, 4σ, 1π and 5σ MO orbitals of CO will be filled (Fig. 3.12). The bonding contributions of the 1σ, 2σ and 3σ orbitals are small, but the contributions of the 4σ, 1π and 5σ orbitals are

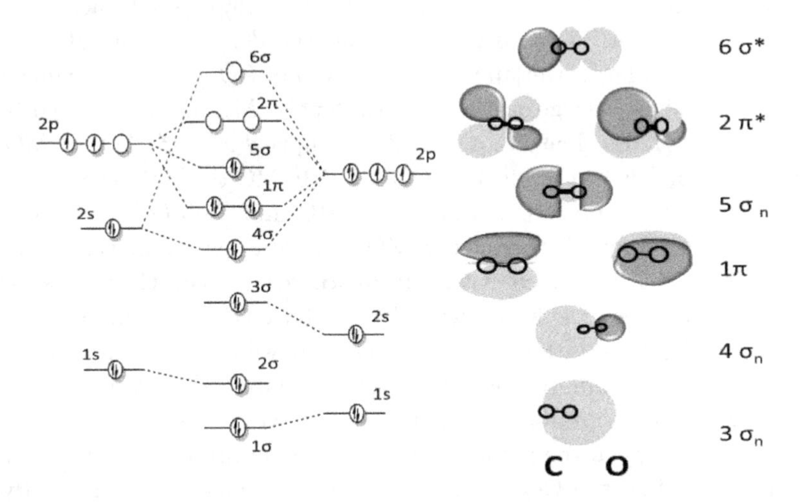

Fig. 3.12 Molecular orbitals and molecular orbital energies of CO.

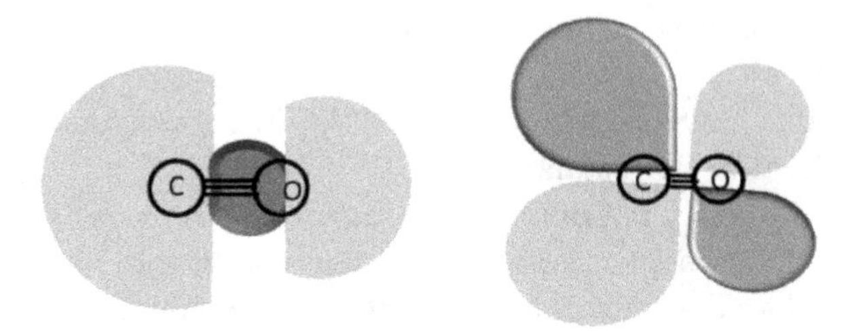

Fig. 3.13 5σ bonding HOMO (left) and 2π* antibonding LUMO (right) of CO.

strong and bonding. Therefore, bonding in the CO molecule is strong. The HOMO of the CO molecule is the 5σ MO, which is mainly a combination of the C 2p$_z$ and O 2p$_z$ orbitals (z axis is along the CO axis). Because the carbon levels are slightly higher than the oxygen levels, the 5σ MO has a larger contribution from the C atom than from the O atom. This is indicated by a larger cloud in Fig. 3.13. The LUMO is the 2π* MO, which actually consists of two MOs with the same energy (the label π indicates twofold degeneracy and the label * an antibonding orbital). One MO is a combination of the C 2p$_x$ + O 2p$_x$ AOs and the other is a combination of the C 2p$_y$ + O 2p$_y$ AOs (the x and y axes are perpendicular to the CO axis). Both orbitals have exactly the same shape, but are rotated 90° relative to each other. The 2π* LUMO has a larger contribution from the C atom than from the O atom (Fig. 3.13).

The HOMO and LUMO of CO have the ideal shape to interact strongly with a metal atom. Figure 3.14 shows that the $d(x^2 - z^2)$ AO of a metal atom can have a large overlap with the 5σ MO of CO (Fig. 3.13), while the $d(xz)$ AO can have a large overlap with the 2$\pi^*(x)$ MO and the $d(yz)$ orbital with the 2$\pi^*(y)$ MO. The interaction of the empty $d(x^2 - z^2)$ AO (LUMO on the metal atom) with the filled 5σ CO MO (HOMO on CO) leads to a new M–C–O orbital, a mixture of the 5σ CO MO and the $d(x^2 - z^2)$ orbital. When two electrons fill this new σ MO, electron density moves from the CO to the metal atom: **electron donation or σ donation**. The other interaction, between the filled metal $d(xz)$ orbital (HOMO) and the empty CO 2$\pi^*(x)$ MO (LUMO), leads to an electron flow in the opposite direction, from the metal atom to the CO ligand: **π back donation or back bonding**. Both interactions, σ donation and π back donation, contribute to the bonding between the metal atom and the C atom of the CO molecule. At the same time the CO bond is weakened, because electron density is taken from the 5σ CO MO, which is bonding in CO, and electrons are

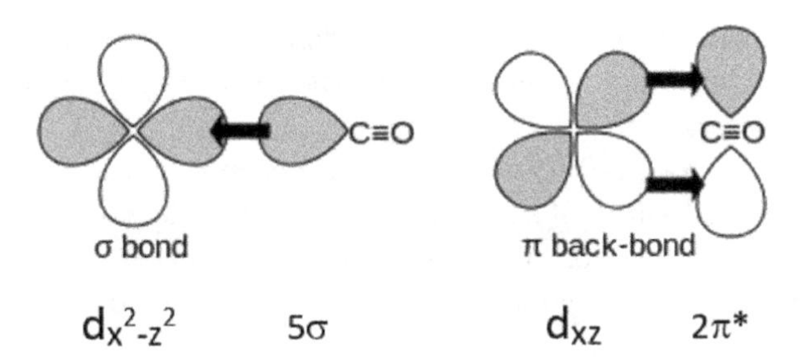

Fig. 3.14 Interactions between the metal $d(x^2 - z^2)$ AO and the CO 5σ MO and between the metal $d(xz)$ AO and the CO $2\pi^*$ MO.

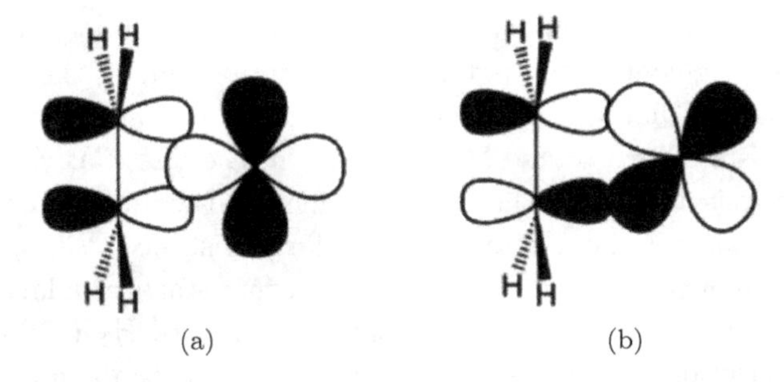

Fig. 3.15 Dewar–Chatt–Duncanson model of the interaction of a metal atom with ethene. (a) σ-donation via the filled ethene π MO and (b) π-back donation via the empty π^* MO of ethene.

donated to the CO $2\pi^*(x)$ MO, which is antibonding in CO. Less bonding and more antibonding leads to a weaker CO bond, which manifests itself in a lowered frequency of the CO stretching IR vibration. At the same time new M–C IR vibrations occur at low frequency. The above description of the bonding of CO to a metal is referred to as the **Blyholder model** [3.9].

Similar σ donation and π back donation occur in the interactions between the d orbitals of a transition-metal atom and the π electrons of the double bond of an alkene molecule (Fig. 3.15). The two 2p AOs on the two carbon atoms form a bonding π MO and an antibonding π^* MO. The bonding alkene π MO interacts with a metal d orbital that is symmetric relative to the M-alkene axis (Fig. 3.15a), while the antibonding π^* MO interacts with a metal d orbital that is antisymmetric relative to the M-alkene axis (Fig. 3.15b). The former interaction leads to σ donation and the latter to π back donation. Both contribute to bonding of the alkene to the metal atom and, at the same time, to weakening of the double bond in the alkene molecule. This model of σ donation and π back donation in metal-alkene complexes is referred to as the **Dewar–Chatt–Duncanson model.**

The formation of new bonds and the weakening of old bonds due to shifts in electron density form the basis of catalytic reactions. The new bonds allow chemisorption of molecules on metal atoms at metal surfaces or on metal ions in metal complexes and the weaker bonds in the molecules prepare the molecules for reaction by lowering the activation energies.

3.3.2 *Dissociative Chemisorption*

Chemisorption can be associative when the molecule adsorbs and stays intact or dissociative when the molecule adsorbs and dissociates in atoms or molecule fragments. The potential energy diagram of the dissociative chemisorption of diatomic molecules such as H_2, N_2 and O_2 on a metal surface consists of two parts. The first step in the dissociative adsorption of H_2 is physisorption of the H_2 molecule on the metal surface. This leads to a weak minimum at a relatively large distance (Fig. 3.16a). The chemisorption of two H atoms at the metal surface starts at a large distance at high energy because of the dissociation energy of H_2 (436 kJ·mol^{-1}). The chemisorption curve has a minimum at a shorter distance than the physisorption curve. The depth of the minimum below the zero level

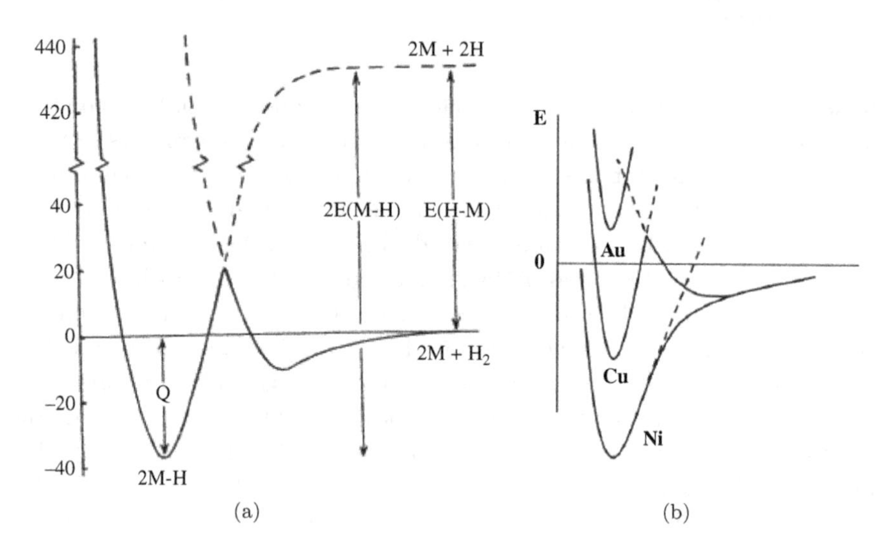

Fig. 3.16 (a) Potential energy (in kJ·mol^{-1}) of dissociative adsorption of H_2 on Cu, and (b) potential energy (in eV) along the minimum-energy reaction path for H_2 dissociation on the (111) surfaces of Ni, Cu and Au.

(H_2 molecule far away from the surface) is the heat of adsorption Q of H_2 ($Q = -\Delta H$ is positive for exothermic adsorption). Q depends on the M–H interaction and is about 120 kJ·mol^{-1} for Ni, 90 kJ·mol^{-1} for Pt, 30 kJ·mol^{-1} for Cu and −50 kJ·mol^{-1} for Au [3.10] as in Fig. 3.16b. When the minimum of the chemisorption curve is deep, the crossing point between the physisorption and chemisorption curves is below zero energy. As a result, the dissociation of H_2 on Pt occurs without an activation barrier. For Cu (Fig. 3.16a), the crossing point is at $E > 0$ and H_2 adsorption on Cu is an activated process. When $Q < 0$ (endothermic adsorption, the chemisorption minimum is above zero energy), the adsorption of H_2 is both activated and endothermic, as is the case for Au. Adsorption of H_2 leads to a very low coverage of Au.

Chemisorption of hydrogen is used to measure the metal surface area of metal on support catalysts. Most metals have a H–M binding energy that is larger than half the dissociation energy of H_2, so that H_2 adsorbs dissociatively in two H atoms (Fig. 3.16). Monolayer (Langmuir) adsorption of H atoms occurs on the metal surface, not only because H atoms have a much stronger interaction with the metal surface than with other H atoms (no multilayer formation), but also because the van der Waals interaction of H_2 molecules or H atoms with the support, on which the metal particles are located, is weak. Therefore, by measuring the amount of H_2 that a metal-on-support catalyst adsorbs, it is possible to calculate the H/M ratio, the number of adsorbed H atoms relative to the total number of metal atoms in the sample. When each H atom binds to one surface metal atom, H/M is equal to M_s/M, the ratio of the number of metal atoms on the surface and the total number of metal atoms. This M_s/M ratio is called the **dispersion** of the metal.

Not only dissociative chemisorption of H_2, but also of O_2 and N_2 is very important in catalysis. Chemisorption of O_2 is the basis of corrosion of metals and of catalytic oxidation reactions of hydrocarbons to oxygen-containing molecules and of harmful CO and hydrocarbons in exhaust gas to CO_2 (Chapter 9). Dissociation of N_2 is the basis of the synthesis of ammonia (Section 6.5), which finds its main application in the production of fertiliser required to feed the world population. The bond energy of N_2 (945 kJ·mol^{-1}) is much higher than that of H_2 (436 kJ·mol^{-1}) and of O_2 (497 kJ·mol^{-1}). The adsorption and dissociation of N_2, with physisorption of a molecular precursor and chemisorption of adsorbed atoms, occurs similar to H_2 (Fig. 3.16). Because of the high bond energy of N_2, the minimum of the chemisorption curve for Fe and Ru (the best ammonia synthesis catalysts) is not far below zero energy and the crossing point of the physisorption and chemisorption curves of N_2 for these metals is located well above zero energy. Thus, N_2 dissociation has a high activation energy.

References

3.1 The probability that an ideal gas molecule has a velocity v_x is $p(v_x) = \frac{1}{\sqrt{2\pi\alpha^2}} \exp\frac{-v_x^2}{2\alpha^2}$, where $\alpha^2 = \int_{-\infty}^{\infty} v_x^2 p(v_x) dv_x$ is the average of v_x^2, because $\int_{-\infty}^{\infty} x^2 \exp(-x^2)\, \mathrm{d}x = 0.5\sqrt{\pi}$.

For a gas of density D, the number of molecules with velocity v_x in the x direction perpendicular to a surface S that hit the surface S per second, is equal to $DSv_x p(v_x) = DSv_x \frac{1}{\sqrt{2\pi\alpha^2}} \exp\frac{-v_x^2}{2\alpha^2}$.

The total number of molecules hitting the surface with any velocity per second is, thus, $n = \int_0^{\infty} F\, dv_x$, with $F = DSv_x \frac{1}{\sqrt{2\pi\alpha^2}} \exp\frac{-v_x^2}{2\alpha^2 v}$.

Because $E_{\mathrm{kin}} = 0.5\, mv_x^2 = 0.5\, \mathrm{kT}$, $\alpha^2 = v_x^2 = \mathrm{kT/m}$ and $D = N/V = N_A P/RT$ (N_A Avogadro constant), we get $n = D\sqrt{\frac{kT}{2\pi m}} = \frac{N_A P}{\sqrt{2\pi \mathrm{MRT}}}$.

3.2 I. Langmuir, The constitution and fundamental properties of solids and liquids. Part I. Solids, *J. Am. Chem. Soc.* 38, 2221–2295, 1916.

3.3 S. Brunauer, P. Emmett, E. Teller, Adsorption of gases in multimolecular layers, *J. Am. Chem. Soc.* 60, 309–319, 1938.

3.4 In thermal equilibrium, the residence time τ_i of a molecule at layer i is equal to the inverse of the rate of desorption ν_{i-1} from layer $i-1 : \tau_i = 1/\nu_{i-1}$. With $s_i = \theta_{i-1}$ this gives $n\theta_{i-1} = \nu_{i-1}\sigma_0\theta_i$ or $\theta_i = (n\tau_i/\sigma_0)\theta_{i-1}$. Thus, $\theta_i = (n\tau_i/\sigma_0)(n\tau_{i-1}/\sigma_0)(n\tau_{i-2}/\sigma_0)\ldots(n\tau_1/\sigma_0)(n\tau_0/\sigma_0)\theta_0$.

If the residence time of a molecule in all layers apart from layer 0 is equal, $\tau_i = \tau_1$ for $i \geq 1$, then $\theta_i = (n\tau_1/\sigma_0)^{i-1}(n\tau_0/\sigma_0)\theta_0 = (n\tau_1/\sigma_0)^i(\tau_0/\tau_1)\theta_0$. The total coverage θ of the surface thus is $\theta = \sum_1^{\infty} i\theta_i = (\tau_0/\tau_1)\theta_0 \sum_1^{\infty} i(n\tau_1/\sigma_0)^i = c\theta_0 \sum_1^{\infty} ix^i$ and $\theta_0 = 1 - \sum_1^{\infty} \theta i = 1 - c\theta_0 \sum_1^{\infty} x^i$ and thus $\theta_0 = 1/(1 + c\sum_1^{\infty} x^i)$,

$$\theta = c\sum_1^{\infty} ix^i \Big/ \left(1 + c\sum_1^{\infty} x^i\right) = cx/(1-x)(1-x+cx),$$

because $\sum_1^{\infty} x^i = x/(1-x)$ and $\sum_1^{\infty} ix^i = x/(1-x)^2$.

In these formulae $x = n\tau_1/\sigma_0$ and $c = \tau_0/\tau_1$, $n = N_A P/\sqrt{(2\pi \mathrm{MRT})} = \beta P$ with $\beta = N_A/\sqrt{(2\pi \mathrm{MRT})}$

$$x = n\tau_1/\sigma_0 = \beta P\tau_1/\sigma_0 = P/q \quad \text{with } q = \sigma_0/\beta\tau_1,$$

$$\theta = V/V_m = cx/(1-x)(1-x+cx) = cP/[(q-P)(1+(c-1)P/q].$$

This equation indicates that V/V_m should become infinite when P approaches q. This happens when condensation of the molecules at the surface occurs. The physical meaning of q thus is $q = P_0$, the saturation pressure of the adsorbate. Therefore

$$\theta = cP'/(1 - P')(1 + (c - 1)P') \quad \text{with } P' = P/P_0.$$

3.5 S. Brunauer, About some critics of the BET theory, *Langmuir* 3, 3–4, 1987.
3.6 P. Kisliuk, The sticking probabilities of gases chemisorbed on the surfaces of solids, *J. Phys. Chem. Solids* 3, 95–101, 1957.
3.7 M. Bowker, The role of precursor states in adsorption, surface reactions and catalysis, *J. Phys. Condens. Matter* 22, 263002, 2010.
3.8 R. Hoffmann, A chemical and theoretical way to look at bonding on surfaces, *Rev. Mod. Phys.* 60, 601–628, 1988.
3.9 G. Blyholder, Molecular orbital view of chemisorbed carbon monoxide, *J. Phys. Chem.* 68, 2772–2777, 1964.
3.10 B. Hammer, J. K. Nørskov, Why gold is the noblest of all the metals, *Nature* 376, 238–240, 1995.

Questions

3.1 Why are the boiling points of branched alkanes lower than the boiling points of linear alkanes?

3.2 Why is cleaning of a surface carried out at elevated temperature and low pressure?

3.3 In Section 3.3.1 we learned that two-electron interactions are attractive, while four-electron interactions are repulsive. What about three-electron interactions? What is the lightest three-electron two-atom molecule?

3.4 Adsorption of H_2 leads to a very small coverage of Au. Despite this, it is still possible to deposit H atoms on the surface of Au. How?

3.5 The bond energy of N_2, O_2 and H_2 are 945, 497 and 436 $kJ{\cdot}mol^{-1}$, respectively. Explain these differences in bond energy.

3.6 The σ donation and the π back donation between an alkene and a metal atom lead to the bonding of the alkene to the metal atom and to the weakening of the double bond in the alkene molecule. How can you prove the existence of the new bond and the weakening of the C=C bond?

3.7 One g Pt/Al_2O_3 catalyst with 0.3 wt% Pt loading adsorbs 6 $\mu mol\,H_2$. Calculate the Pt dispersion.

Chapter 4

Kinetics

4.1 Langmuir–Hinshelwood Model

The rate of a reaction in the homogeneous gas or liquid phase is proportional to the product of the reactant concentrations c_i: $r = k\Pi c_i^{ni}$, where k is the rate constant and n_i the exponent of the concentration c_i. Catalytic reactions do not take place in the gas or liquid phase; in heterogeneous catalysis they take place at the surface of a catalyst and in homogeneous catalysis on the metal atom of an organometallic complex. The concentrations of the reactants at the surface or on a metal atom differ from those in the gas and liquid phases. Because it is very difficult to measure the concentrations of the reactants at the surface or on a metal atom, their values are often derived from the concentrations in the gas or liquid phase by means of equilibrium isotherms. The Langmuir isotherm (Section 3.1.2) is often chosen to deduce the **Langmuir–Hinshelwood** kinetic rate equation.

4.1.1 *Monomolecular Reaction*

In a surface-catalysed reaction of A to B, first A adsorbs at the surface and B desorbs after reaction.

$$\mathrm{A} + * \underset{k_1}{\overset{k_{-1}}{\leftrightarrows}} \mathrm{A}* \underset{k_2}{\overset{k_{-2}}{\leftrightarrows}} \mathrm{B}* \underset{k_3}{\overset{k_{-3}}{\leftrightarrows}} \mathrm{B} + *$$

$$\begin{array}{ccc} \mathrm{A} & & \mathrm{B} \\ \uparrow\downarrow & & \downarrow\uparrow \\ \_\_\ \mathrm{A}*\ \_\_ & \leftrightarrows & \_\_\ \mathrm{B}*\ \_\_ \end{array}$$

with the following rate equations:

$$\frac{d\theta_v}{dt} = -k_1 P_A \theta_v + k_{-1}\theta_A + k_3\theta_B - k_{-3}P_B\theta_v = -r_1 + r_3,$$

$$\frac{d\theta_A}{dt} = k_1 P_A \theta_v - k_{-1}\theta_A - k_2\theta_A + k_{-2}\theta_B = r_1 - r_2,$$

$$\frac{d\theta_B}{dt} = k_2\theta_A - k_{-2}\theta_B + k_3\theta_B - k_{-3}P_B\theta_v = r_2 - r_3.$$

In these equations the rate constants k_i can be written as $k_i = k_{io} \exp - E_i/RT$, where E_i is the activation energy for step i. The rate constants of the forward and backward reactions are connected by Q_i, the heat of adsorption: $k_i/k_{-I} = K_1^0 \exp Q_i/RT$ (Section 3.1.2). In **microkinetics**, such a set of rate equations is solved by computation. One determines the boundary conditions for the system (partial pressures, surface concentrations and temperature), one determines the rate constants k_{io} and activation energies A_i of all kinetic steps from transition state theory or other theories and one solves the set of differential equations. In this particular case of a monomolecular reaction from A to B the initial conditions could be $P_A = P_{Ao}$ and $P_B = 0$, $\theta_A = \theta_B = 0$ and T constant. In simple microkinetic models, as in this monomolecular reaction of A to B on one catalytic site, the surface is represented by a single type of adsorption site and each adsorbate can only occupy one site. Solving the set of differential equations gives the surface coverages as a function of time and the rates of the individual elementary reaction steps. Microkinetic modelling has gained great popularity because it allows to calculate surface coverages, apparent activation energies and reaction orders under many conditions.

Microkinetics has the advantage that it solves the rate equations in an exact manner but the disadvantage that it needs substantial computational effort and does not give insight. When certain conditions are met, however, it is not necessary to turn to a computer to solve the rate equations. For instance, because the number of catalytically active sites is often much smaller than the number of molecules of the reactant and product, the change in the concentrations of the reaction intermediates A$*$ and B$*$ is small relative to the concentrations of the reactants and products. That means that we can use the steady-state approximation $d\theta_v/dt = d\theta_A/dt = d\theta_B/dt = 0$ for the major part of the reaction; only initially and at the end the steady-state approximation does not hold. If one of the three steps shown above for the monomolecular reaction is much slower than the other steps, it determines the rate. In our case, three **rate-determining steps** are possible:

1. The surface reaction is rate-determining.
2. Adsorption of A is rate-determining.
3. Desorption of B is rate-determining.

4.1.1.1 *Surface reaction is rate-determining*

When the surface reaction is rate-determining, the forward and backward rates of reactions 1 (adsorption equilibrium of A) and 3 (desorption equilibrium of B) are much larger than the corresponding rates of reaction 2 (Fig. 4.1). This means that adsorption and desorption are fast relative to the surface reactions and that they are almost at equilibrium, because the difference in the forward and backward rates is small relative to the rates (Fig. 4.1). Therefore, the equilibrium Langmuir equation can be used to calculate the coverage of the surface with A and B.

The net reaction rate is $r = r_2$ (forward) $- r_2$ (backward) $= k_2\theta_A - k_{-2}\theta_B$.

With Eq. (3.7) for θ_A and θ_B this gives

$$r = \frac{k_2K_AP_A - k_{-2}K_BP_B}{1 + K_AP_A + K_BP_B} = \frac{k_2K_A(P_A - K_{\text{eq}}^{-1}P_B)}{1 + K_AP_A + K_BP_B}, \tag{4.1}$$

where $K_{\text{eq}} = k_2K_A/k_{-2}K_B$ is the equilibrium constant of the overall reaction A $\leftrightarrows$ B. At low conversion, $K_{\text{eq}}^{-1}P_B \ll P_A$ and Eq. (4.2) simplifies to

$$r = \frac{k_2K_AP_A}{1 + K_AP_A + K_BP_B}. \tag{4.2}$$

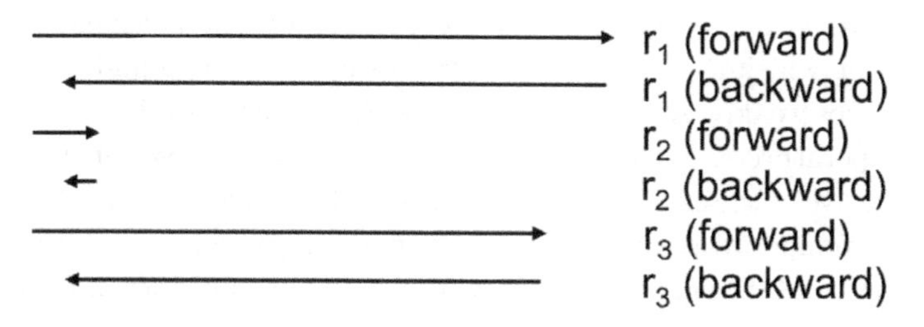

Fig. 4.1 Magnitude of the rates of the elementary steps in the monomolecular reaction $A + * \leftrightarrows A* \leftrightarrows B* \leftrightarrows B + *$. The lengths of the arrows of the forward and backward rates represent their magnitudes.

There are three sub cases:

1a. A and B adsorb weakly.
1b. A adsorbs strongly and B weakly.
1c. A adsorbs weakly and B strongly.

(1a) When both **A and B adsorb weakly** on the catalyst surface, $K_A P_A \ll 1$ and $K_B P_B \ll 1$ and $r = N k_2 K_A P_A$. The rate constant k_2 is equal to $k_2 = k_0 \exp(-E_0/RT)$ (Arrhenius equation), where E_0 is the activation energy of the reaction at the catalyst surface. With $K_A = K_A^o \exp(Q_A/RT)$ this gives

$$r = k_0 K_A^o P_A \cdot \exp[-(E_0 - Q_A)/RT]. \tag{4.3}$$

The reaction rate is proportional to P_A; i.e. **first order** in A. Because $E_{\text{app}} = \frac{d \ln r}{d \frac{-1}{RT}} = E_0 - Q_A$, the apparent activation energy is

$$E_{\text{app}} = E_0 - Q_A. \tag{4.4}$$

(1b) When the adsorption of A is strong and the adsorption of B is weak (**reactant inhibition**), then $K_A P_A \gg 1$, $K_B P_B \ll 1$ and

$$r = k_2 = k_0 \exp(-E_0/RT). \tag{4.5}$$

The reaction does not depend on the concentration of A in the gas phase (**zero order** in A), and E_{app} is equal to the activation energy of the surface reaction

$$E_{\text{app}} = E_0. \tag{4.6}$$

Cases 1a and 1b are extremes of the same monomolecular reaction, during which the surface reaction is rate-determining and the product adsorbs weakly. At high temperature, K_A is low (case 1a), while at low temperature K_A is high (case 1b). The apparent activation energy (the negative of the slope of the curve in Fig. 4.2) increases from $E_0 - Q_A$ at high temperature to E_0 at low temperature. Similarly, case 1a changes to 1b (and the activation energy from $E_0 - Q_A$ to E_0) when P_A changes from low to high pressure, as we will see in the hydrogenation of ethene (Section 6.2).

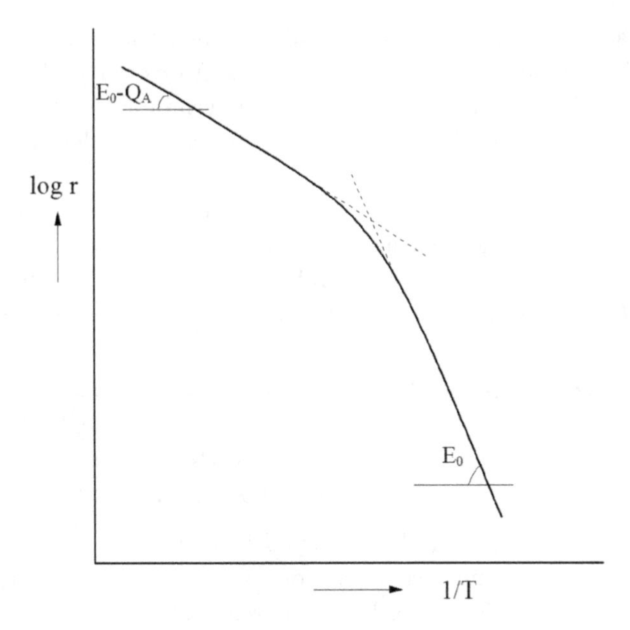

Fig. 4.2 Log r as a function of the reciprocal temperature.

(1c) When the adsorption of the product is strong (**product inhibition**), $K_AP_A \ll 1$ and $K_BP_B \gg 1$ and

$$r = k_2\frac{K_AP_A}{K_BP_B} = k_0 \cdot \frac{K_A}{K_B} \cdot \exp\left[\frac{E_0 - Q_A + Q_B}{RT}\right] \cdot \frac{P_A}{P_B}. \tag{4.7}$$

The reaction is first order in A and minus first order in B. The apparent activation energy is

$$E_{\text{app}} = E_0 - Q_A + Q_B \approx E_0 + Q_B. \tag{4.8}$$

4.1.1.2 *Adsorption of the reactant or product is rate-determining*

When the adsorption of the reactant is rate-determining, all the other reaction steps are fast relative to the rate of adsorption; A∗ reacts fast to B∗ and B desorbs fast from the catalyst surface. As a consequence, the coverage of the surface is very small and $\theta_v \approx 1$. Thus,

$$r = r_1 = k_1P_A\theta_v = k_1P_A. \tag{4.9}$$

The reaction is first order in A and the apparent activation energy is equal to the activation energy of adsorption. An example of a reaction in which adsorption of the reactant is rate-determining is the synthesis of ammonia (Section 6.5). The rate of NH_3 production is determined by the dissociative adsorption of N_2.

When product desorption is slow, all the reactions preceding the desorption are fast and in equilibrium and the surface is covered by A and B. Because of the fast pre-equilibria, $r_1(\text{forward}) - r_1(\text{backward}) = k_1 P_A \theta_v - k_{-1}\theta_A = 0$ and $r_2(\text{forward}) - r_2(\text{backward}) = k_2\theta_A - k_{-2}\theta_B = 0$. Thus,

$$\theta_A = K_A P_A \theta_v, \quad \theta_B = K_2 \theta_A \quad \text{and} \quad \theta_v = \frac{1}{1 + (1 + K_2) K_A P_A},$$

$$r = r_3 = k_3 \theta_B = k_3 K_2 \theta_A = k_3 \frac{K_A K_2 P_A}{1 + (1 + K_2) K_A P_A}. \tag{4.10}$$

The order of the reaction in A is between 1 and 0.

Thus, in the Langmuir–Hinshelwood mechanism for a monomolecular reaction, the reaction is first order in the reactant A unless the adsorption of A is strong, in which case the order decreases to zero. When product inhibition occurs, the order of the reaction is −1 in the product. The total order (sum of all the orders) is always 0 when a reactant or product adsorbs strongly. Low or negative order thus indicates strong adsorption.

4.1.2 *Bimolecular Reaction*

In a monomolecular reaction the reacting molecule must be chemisorbed on the surface if a catalytic reaction is to occur. Catalysed reactions between two molecules can, in principle, occur when only one of the reactants is chemisorbed on the surface — the **Eley–Rideal (ER) mechanism** — or when both reactants are chemisorbed — the **Langmuir–Hinshelwood (LH) mechanism**. In the ER mechanism, one reactant is chemisorbed at the catalyst surface and the other reactant is either in the gas phase and collides with the chemisorbed reactant or is present at the surface in a weak physisorbed state. Although this mechanism is often called ER mechanism, Eley and Rideal actually discussed a different mechanism and the mechanism had already been proposed years before by Langmuir [4.1]. It is usually assumed that catalytic reactions occur by a LH mechanism and not by an ER mechanism, one reason being that, in a LH mechanism, both reaction partners bind to the catalyst surface. The interaction with the surface weakens the bonds between the atoms in the adsorbed molecules, which will facilitate reaction. Only in reactions of a molecule with a radical, as in the hydrogenation of ethene by H atoms or in the reaction of a molecule with NO or O_2 (both are radicals), an ER mechanism might be possible. The other reason for not considering the ER mechanism any further is that the product of the pre-exponential factor and the impingement rate of a gas-surface ER reaction is much lower than the pre-exponential term in the surface LH reaction. Although a LH mechanism is more probable than an ER mechanism, the mechanisms of only a few reactions have been determined experimentally.

Similar equations as for the monomolecular reaction can be derived for a bimolecular reaction of molecules A and B to give molecule C.

$$\mathrm{A} + * \leftrightarrows \mathrm{A}* \quad \mathrm{B} + * \leftrightarrows \mathrm{B}* \quad \text{and} \quad \mathrm{A}* + \mathrm{B}* \leftrightarrows \mathrm{C}* \leftrightarrows \mathrm{C}.$$

If the reaction at the catalyst surface is rate-determining, occurs via a LH mechanism and A, B and C adsorb weakly at the surface ($K_A P_A \ll 1$, $K_B P_B \ll 1$ and $K_C P_C \ll 1$), then

$$r = k\theta_A\theta_B = \frac{k \cdot K_A P_A K_B P_B}{(1 + K_A P_A + K_B P_B + K_C P_C)^2} = k \cdot K_A K_B \cdot P_A P_B. \tag{4.11}$$

The reaction is first order in A and in B and second order in total; the apparent activation energy is

$$E_{\text{app}} = E_0 - Q_A - Q_B. \tag{4.12}$$

If adsorption of A is strong and adsorption of B and C is weak ($K_A P_A \gg 1$, $K_B P_B \ll 1$ and $K_C P_C \ll 1$), then

$$r = k\theta_A\theta_B = k\frac{K_A P_A K_B P_B}{(K_A P_A)^2} = k\frac{K_B P_B}{K_A P_A}, \tag{4.13}$$

and the reaction is first order in B and minus first order in A. The total order is zero.

The above two cases are the extremes of weak and strong adsorption of A. In the intermediate situation, when adsorption of B and C is weak ($K_B P_B \ll 1$ and $K_C P_C \ll 1$):

$$r = k\theta_A\theta_B = k\frac{K_A P_A K_B P_B}{(1 + K_A P_A + K_B P_B + K_C P_C)^2} = k\frac{K_A K_B P_A P_B}{(1 + K_A P_A)^2}. \tag{4.14}$$

At low P_A, there are only few A molecules at the surface and the rate is low but at high P_A pressure A expels B from the surface and the rate is low as well. A high rate requires both A and B to be present at the surface and the rate has a maximum when $d[P_A/(1+K_A P_A)^2]/dt = 0$, when $P_A = K_A^{-1}$.

When the product C adsorbs strongly and the reactants adsorb weakly, then

$$r = k\frac{K_A K_B}{K_P^2}\frac{P_A P_B}{P_C^2}. \tag{4.15}$$

The reaction is first order in A and in B, minus second order in C and the total order is 0. As in the monomolecular reaction, a total order of zero indicates strong adsorption, be it of the reactant or the product.

4.2 Influence of Diffusion

Diffusion brings the reactants to the catalytically actives sites and moves the products away from the sites. The equations derived above are based on the assumption that diffusion is much faster than the chemical reaction and, thus, must not be considered. If this is not the case, then the rate of diffusion influences the overall rate. Consider a cylindrical pore in which a catalytic reaction occurs at the pore wall. When diffusion is fast enough to keep pace with the chemical reaction, the concentrations of the reactants throughout the pore are the same as the concentrations in the gas or liquid phase outside the pore (Fig. 4.3a). Under these conditions, the surface of the catalyst is used efficiently and the total reaction occurs at maximum rate. When the rate of the catalytic reaction at the pore wall increases (e.g. at higher temperature) and the rate of diffusion decreases (longer and narrower pores), the concentrations of the reactants decrease from outside the pore to the centre of the pore (Fig. 4.3b) and the catalyst surface at the pore wall is not fully used. When the reaction rate is very high relative to the rate of diffusion, the concentrations of the reactants in the pore decrease strongly (Fig. 4.3c); only the area right behind the pore mouth is involved effectively in the catalytic reaction.

For a first-order reaction, during which the number of molecules does not change, a rate equation can be derived for a cylindrical pore (with length $2L$ and radius R) perpendicular to the front and back of flat plate-like catalyst particles (Fig. 4.4) by considering the mass balance over a pore

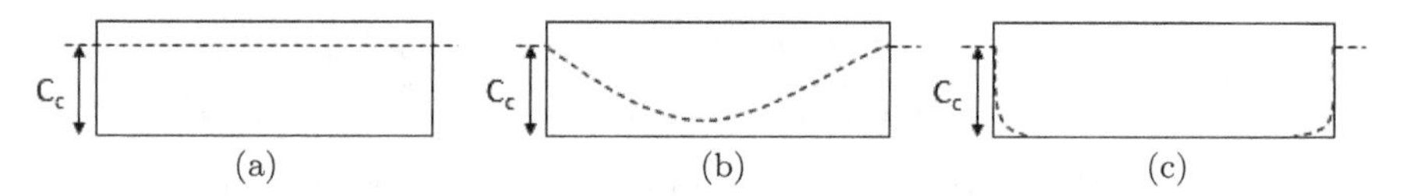

Fig. 4.3 Concentration profile of a reactant in a catalyst pore at (a) $r_{\text{diff}} \gg r_{\text{chem}}$ (low T), (b) at $r_{\text{diff}} \approx r_{\text{chem}}$ (intermediate T), and (c) at $r_{\text{diff}} \ll r_{\text{chem}}$ (high T).

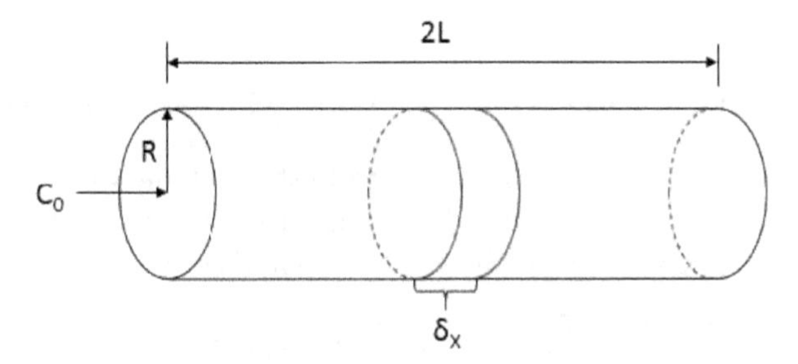

Fig. 4.4 Reaction taking place on the internal surface of a cylindrical pore with length $2L$ and radius R.

element of length δx:

$$\text{input} = \text{output} + \text{chemical reaction or } F_{\text{in}} = F_{\text{out}} + rS, \tag{4.16}$$

where F_{in} is the ingoing flux of matter at position x, F_{out} is the outgoing flux of matter at position $x + \delta x$, rS is the change in the number of molecules during reaction, r is the reaction rate and S is the catalytically active surface area.

Application of Fick's first law at x and $x + dx$ (cf. Fig. 4.4) gives

$$F_{\text{in}} = -DS_0(dc/dx)_x = -\pi R^2 D(dc/dx)_x,$$

and

$$F_{\text{out}} = -DS_0(dc/dx)_{x+dx} = -\pi R^2 D(dc/dx)_{x+dx}.$$

D is the diffusion coefficient and $S_0 = \pi R^2$ is the surface, through which the molecules diffuse. With $rS = k_S cS$ and the catalytically active surface area $S = 2\pi R\delta x$, Eq. (4.16) becomes

$$-\pi R^2 D(dc/dx)_x = -\pi R^2 D(dc/dx)_{x+dx} + k_S c \cdot 2\pi R\delta x \quad \text{or}$$

$$(dc/dx)_{x+dx} - (dc/dx)_x = (2k_S/RD) \cdot c \cdot \delta x.$$

In the limit of $\delta x = 0$

$$\lim \left[\frac{\left(\frac{dc}{dx}\right)_{x+dx} - \left(\frac{dc}{dx}\right)_x}{\delta x} \right]_{dx=0} \equiv (d^2c/dx^2)_x = 2k_S c/RD.$$

If the surface reaction rate constant k_S is replaced by a volume reaction rate constant r_V, then

$$k_V \equiv (S/V)k_S = (2\pi R \cdot 2L)k_S/\pi R^2 \cdot 2L) = 2k_S/R \text{ and } d^2c/dx^2 = k_V c/D.$$

This gives

$$c = A \cdot \exp(mx) + B \cdot \exp(-mx) \quad \text{with} \quad m^2 = k_V/D.$$

The constants A and B are determined from the boundary conditions $c = c_0$ for $x = 0$ and $dc/dx = 0$ for $x = L$, because the pore is open at both ends and is, thus, symmetric for reaction, diffusion and concentrations. This gives

$$c = c_0 \frac{\cosh m(x - L)}{\cosh ml}, \quad \text{where } \cosh \alpha = 0.5(e^{\alpha} + e^{-\alpha}), \tag{4.17}$$

and

$$mL = L(k_V/D)^{0.5} = \phi \text{ (Thiele modulus).} \tag{4.18}$$

The Thiele modulus determines the part of the pore that is involved in the catalytic reaction. The concentration of the reactant inside a catalyst pore decreases fast when the Thiele modulus is large (Fig. 4.5) and the overall reaction rate decreases. The catalytic surface in the pore is not used effectively; the **effectiveness factor** is defined as η = (actual rate)/(rate without diffusion limitation). Because of the mass balance, the rate

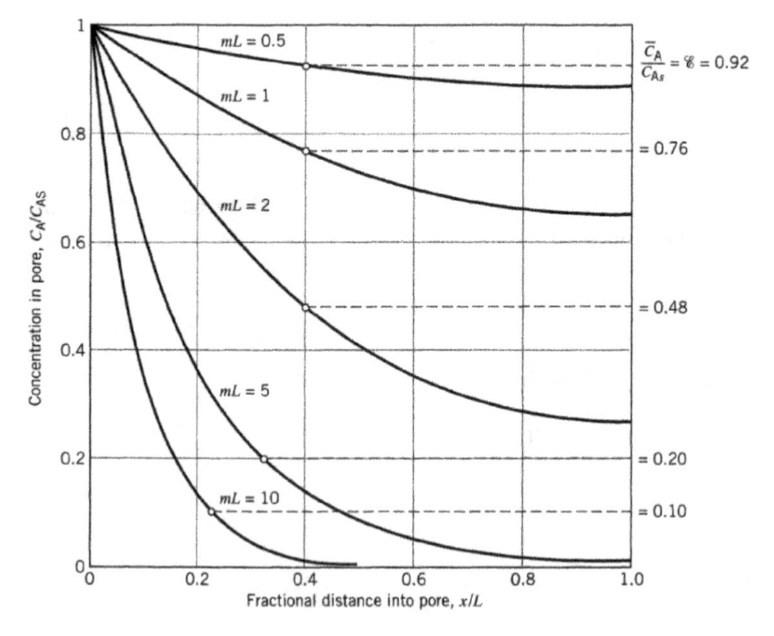

Fig. 4.5 Concentration of the reactant as a function of distance in a catalyst pore, depending on the values of the Thiele modulus ϕ = mL.

is equal to the rate of transport of the reactant through the pore mouth:

$$r = -\pi R^2 D(dc/dx)_{x=0} = \pi R^2 D c_0 (\phi/L) \cdot \tanh(\phi), \quad \text{with}$$

$$\tanh(\phi) = (e^{\phi} - e^{-\phi})/(e^{\phi} + e^{-\phi}).$$

In the absence of diffusion limitation, $c = c_0$ in the whole pore and

$$r_{\max} = 2\pi R \cdot L k_S c_0 \quad \text{and} \quad \eta = r/r_{\max} = (RD/2L^2 k_S) \cdot \phi \cdot \tanh(\phi)$$

$$\eta = \tanh(\phi)/\phi. \tag{4.19}$$

For a small Thiele modulus ($\phi < 0.3$), the effectiveness factor η is close to 1 (Fig. 4.6) and the catalyst will function effectively. For a large Thiele modulus ($\phi > 3$), $\tanh(\phi) \approx 1$ and $\eta \approx 1/\phi$. The concentration decreases strongly inside the pore and the reaction occurs almost exclusively near the pore mouth. The pores of the catalyst are most effectively used when the diffusion coefficient D is high, the reaction rate constant k is low, the pore radius is large and the pore length is small.

For geometries other than flat plate-like catalyst particles, the equations are slightly different, but the $\eta - \phi$ curve hardly changes. For spherical and cylindrical catalyst particles η deviates from 1 at higher ϕ than for the flat plate-like catalyst particles; this is a practical advantage. For reactions with order higher than 1, the $\eta - \phi$ curves are similar. However, the higher the order, the sooner η deviates from 1 and the sooner diffusion is problematic.

When diffusion is strongly hindered, $\eta = 1/\phi$ and $\tanh(\phi) = 1$, so that $r = \pi R^2 D c_0 (\phi/L) \cdot \tanh(\phi) = \pi R^2 c_0 \sqrt{D k_V}$. The apparent activation energy is not equal to the activation energy of the catalytic reaction, but $E_{\text{app}} = 0.5(E_0 + E_{\text{diff}})$, where E_0 is the activation energy of the catalytic

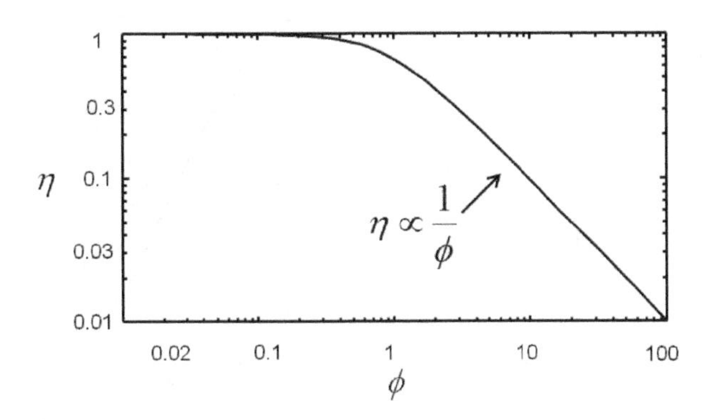

Fig. 4.6 Effectiveness factor as a function of the Thiele modulus.

reaction and $E_{\rm diff}$ the activation energy of diffusion. Because $E_{\rm diff}$ is usually much lower than E_0, the apparent activation energy is $E_{\rm app} \approx 0.5\, E_0$. Thus, when diffusion is limited, the apparent activation energy is lower than that of the chemical reaction. Diffusion limitation occurs at high rather than at low temperature, because, at high temperature, the chemical reaction is so fast that only the outer surface of the catalyst particles is involved in the reaction. The reaction rate is then determined by external diffusion from the gas phase to the gas film on the outer surface of the particles. Because the temperature coefficient of diffusion is small, the activation energy will be about zero at very high temperature. Figure 4.7 depicts the measured reaction rate of catalyst particles, plotted as a function of inverse temperature.

To determine whether diffusion constitutes a problem, one might measure the rate of the reaction as a function of $1/T$. However, $E_{\rm app}$ can be lower than E_o, not only because of diffusion limitation, but also because of strong adsorption (Fig. 4.2). A better method to check for diffusion limitation, therefore, is to decrease the size of the catalyst particles. The Thiele modulus is proportional to the length of the particles, and breaking the particles leads to an increase in the rate when diffusion is slow.

Diffusion limitation can occur not only in the pores of the catalyst particles, but also on their outer surface. A stagnant film forms on the

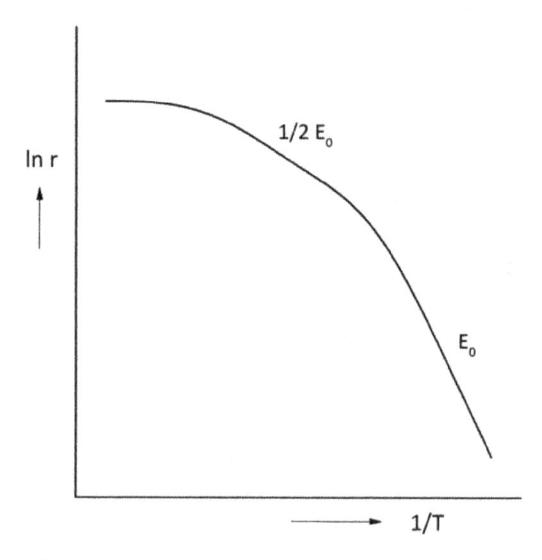

Fig. 4.7 Ln r as a function of reciprocal temperature in a catalyst particle.

surface and mass transport in this film is slower than in the gas or liquid between the catalyst particles. It is usually assumed that the molar flux in the film is proportional to the concentration difference over the film

$$F = k_f(c_0 - c_S), \tag{4.20}$$

where c_0 and c_S are the concentrations of the reactant between the catalyst particles and on the catalyst surface, respectively (Fig. 4.8), and k_f is the mass transfer coefficient ($k_f = D/\varepsilon$, where D is the diffusion coefficient and ε is the thickness of the film).

In the stationary state, the molar flux through the film is equal to the amount of material that is removed by the chemical reaction in the catalyst particle. Thus,

$$F = k_f(c_0 - c_S)S' \quad \text{and} \quad F = k_V c_S V \eta,$$

where S' is the outer surface of the catalyst particles and V the volume of the catalyst particles. This gives

$$c_0 - c_S = F/k_f S' \quad \text{and} \quad c_S = F/(k_V \eta V),$$

and thus

$$c_0 = F(1/k_f S' + 1/k_V \eta V).$$

When $a = S'/V$ (the specific surface) the volume specific rate $r = F/V$ is

$$r = \frac{c_0}{\frac{1}{k_f a} + \frac{1}{k_V \eta}}. \tag{4.21}$$

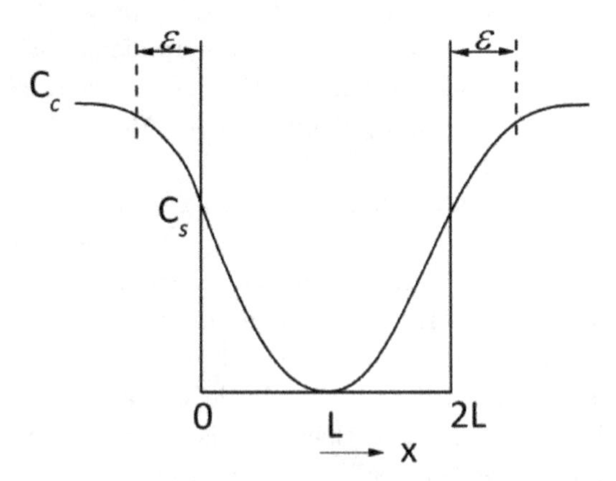

Fig. 4.8 Concentration profile inside and outside a catalyst pore due to limited diffusion in a stagnant film.

Equation (4.21) shows that film diffusion and pore diffusion behave like resistances in series in an electrical circuit. At low temperature the rate of the chemical reaction is slower than the rate of diffusion and is, thus, rate-determining. At high temperature the diffusion is faster, but the rate of the chemical reaction increases to a much greater extent. As a consequence, at high temperature, the diffusion through the film becomes rate-limiting. Then the rate is $r = k_f c_0 a$ and is determined solely by the outer surface of the catalyst particle. This extreme situation occurs in the oxidation of NH_3 with air to NO (Section 9.2.2). In industry this reaction is carried out in an adiabatic reactor, which contains nets of Pt-Rh wire. An ammonia/air mixture flows at a velocity of a few m/sec through the gauze (that has a temperature of 800 °C) and 98% of the NH_3 is converted to NO. The rate of the process is determined by mass transport to the Pt wire. Increasing the catalyst surface is not the solution; the outer surface of the Pt already has a high catalytic activity. The only way to bring about a change is to increase the gas velocity, which causes thinning of the stagnant gas film and, thus, increases the mass transfer coefficient (as when blowing on a cup of hot tea).

Problems with mass transfer do not usually occur in gas phase processes; only when the chemical reactions are very fast, as in the oxidation of ammonia, do they play a role. Film diffusion limitations occur more often in the liquid phase. A series of coupled diffusion limitations and chemical reaction occurs in the three-phase system gas–liquid–solid (G–L–S). This trickle-flow system is applied in industry for hydrogenation, for example for the hardening of vegetable oil to fat with the aid of a Ni/SiO_2 catalyst at 180 °C and 20 atm. Trickle flow operations are applied on a very large-scale in the hydrogenative desulfurization (HDS) of oil fractions, the catalyst being Co- or Ni-promoted MoS_2 supported on Al_2O_3 at 300–400 °C and 30–70 atm H_2 (Section 8.4).

Figure 4.9 represents a H_2 molecule moving from a gas bubble through the liquid phase to the surface of the catalyst particle and then into the catalyst pores. Along this path, diffusion limitation may occur in the film layers on the liquid side of the gas bubble and on the outer surface of the catalyst particle. Furthermore, diffusion limitation inside the catalyst pores may occur. Mass transfer through the film layers is described by the mass transfer coefficients k_{LG} and k_{LS}. Because k is inversely proportional to the thickness of the film layer, it is influenced by the stirring speed, as are the size of the gas bubbles and, thus, the specific area of the bubbles. Similar

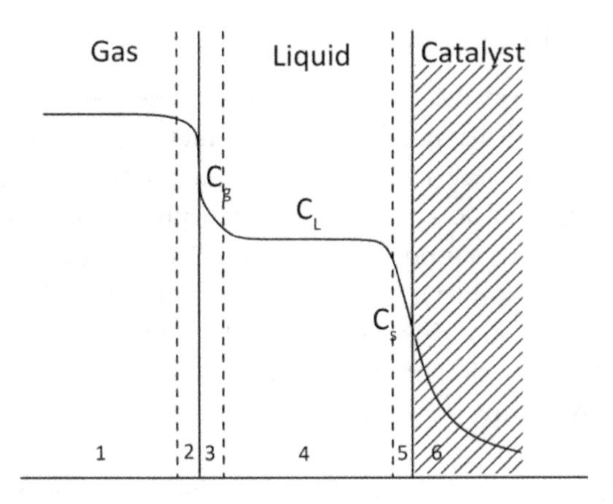

Fig. 4.9 Concentration profile of hydrogen in a G–L–S reactor. 1: gas phase, 2: gas film, 3: liquid film, 4: liquid phase, 5: layer around catalyst particle, 6: catalyst.

to the derivation of Eq. (4.21), one can derive (cf. [4.2]):

$$r = \frac{c_0}{\frac{1}{k_{LG}a_{LG}} + \frac{1}{k_{LS}a_{LS}} + \frac{1}{k\eta c_A^n a}}. \tag{4.22}$$

In Eq. (4.22), a_{LG}, a_{LS} and a are the specific surfaces of the bubbles, the outer catalyst surface and the internal catalyst surface, respectively. When diffusion is faster than the chemical reaction, the reaction rate is proportional to a and, thus, to the amount of catalyst. In that case the speed of stirring hardly influences the rate. In fast chemical reactions, the mass transfer from the gas bubbles to the liquid phase will be rate-limiting. The rate of hydrogen consumption is then a function of the stirring speed and does not depend on the amount of catalyst.

Diffusion limitations are also important in electrocatalysis, especially when reactions involving a gas occur at electrodes immersed in aqueous electrolytes (cf. Section 10.2.4). Typical examples of such reactions are the oxygen evolution reaction, suffering from bubble formation, and the CO_2 electroreduction, suffering by the low solubility and slow diffusion of CO_2 in the aqueous electrolyte. To eliminate these problems, one has either to employ stirring of the electrolyte solution or rotation of the electrode (for instance the rotating disc electrode technique is used to limit the diffusion layer thickness, Section 10.3.5) or to implement optimal electrode design, which can facilitate the transport of reaction species to/from the gas–liquid–solid interface [4.3] (cf. Section 10.4.1).

4.3 Bifunctional Catalysis

Some reactions are catalysed by more than one catalytic site [4.4]. For example, the isomerisation of hexane to methylpentane needs a metal catalyst as well as an acidic catalyst. The isomerisation is slow over Pt/SiO_2 (metal on neutral support) and over amorphous silica–alumina (ASA, $SiO_2{\cdot}Al_2O_3$, an acidic support, cf. Section 7.1.2), but fast over Pt/ASA (metal on an acidic support). Also the transformation of hexane to benzene by dehydrogenation and ring-closure needs a metal as well as an acidic catalyst. The isomerisation of alkanes and the dehydrogenative ring closure of alkanes are important steps in the improvement of the octane number of gasoline (Section 7.6).

In bifunctional catalysis, reactions take place consecutively on one type of site and on another type of site. How does transport of the intermediates between the sites S_1 and S_2 take place? Do the intermediates diffuse over the support surface from site S_1 to site S_2, or do they desorb from site S_1 and diffuse through the gas phase to site S_2? Consider a reaction of A to B to C, during which A reacts to B only on site S_1 and B reacts to C only on site S_2 (Fig. 4.10). The intermediate B forms on site S_1 but can only react to C on site S_2. Diffusion of B along the surface occurs with ease for neighbouring sites S_1 and S_2. If the sites are not neighbours, then the intermediate B must diffuse over the support surface from site S_1 to site S_2, which may take the same energy as desorption into the gas phase if the binding of the intermediate B to the support is weak.

Is it necessary for the diffusion distance to be microscopic, i.e. over molecular distances, or are macroscopic distances also feasible to obtain a high reaction rate? Consider the situation when all S_1 sites are in one reactor and all S_2 sites in a subsequent reactor (Fig. 4.11a). In the most extreme situation, the reaction of A to B on site S_1 may be limited by thermodynamics and give a low maximum conversion of A to B, x%. Since only B molecules react on sites S_2 in reactor 2, the maximum conversion to

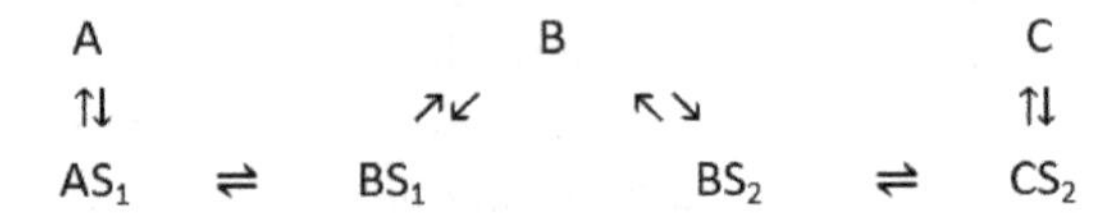

Fig. 4.10 Reaction of A to B to C. A can react only on catalyst site S_1 and B only on catalyst site S_2. B must diffuse through the gas or liquid phase from site S_1 to site S_2. Reprinted with permission from [4.4]. Copyright (1962) Elsevier.

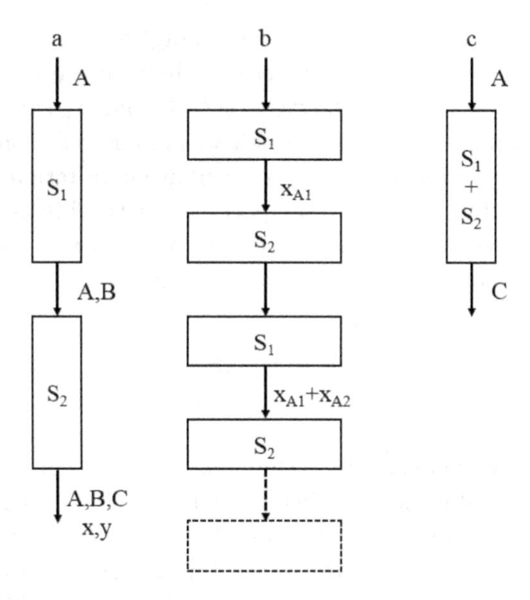

Fig. 4.11 Three possibilities to react A to B (catalysed by S_1) and B to C (catalysed by S_2). (a) one reactor filled with S_1 sites followed by a reactor filled with S_2 sites, (b) several reactors filled alternatingly with S_1 and S_2, and (c) one reactor with a homogeneous mixture of S_1 and S_2.

C would also be x%. This indicates that a sequence of several small reactors, filled alternatingly with S_1 and S_2 sites (Fig. 4.11b), will give a much higher conversion of A to C. In the second reactor only the B molecules produced in reactor 1 would react to C, but in reactor 3 new B molecules would be produced from the remaining A molecules, which would react to C in reactor 4, and so on. If the reaction of B to C is irreversible, full conversion of A to C would result, provided there is a sufficient number of small reactors. In the ultimate reactor configuration, the catalyst components S_1 and S_2 are homogeneously mixed in one reactor (Fig. 4.11c).

The next question is how well catalysts S_1 and S_2 must be mixed. Is it sufficient to mix the sites macroscopically by preparing a physical mixture of two finely ground catalysts, one containing sites S_1 and the other sites S_2, or must both sites be on the same support so that they are even closer together? Or should both sites make contact? If the two catalyst particles are in close contact, the effective diffusivity in the free gas space between the catalyst particles is much larger than within the high-surface-area catalyst solids, and the average interparticle distances are much smaller than the particle diameters. The problem of mass transport of the intermediate molecules between the S_1 and S_2 catalyst sites then becomes an intraparticle diffusion problem, similar to the one described in Section 4.3. Weisz derived an order-of-magnitude criterion that indicates how small the radius R of the two particles in the bifunctional catalyst must be so that no mass transport inhibition occurs in the bifunctional reaction. The requirement for the particle radius is [4.4]:

$$R^2 < 4.46 \times 10^{-5} \frac{273}{T} \frac{1}{\frac{dN}{dt}} D P_{B,\mathrm{eq}}, \tag{4.23}$$

where T is the temperature in Kelvin, dN/dt is the rate of the overall reaction, D is the diffusion coefficient of intermediate B and P_B is the pressure of intermediate B in the gas phase. For a typical gas diffusion constant of $D = 2 \cdot 10^{-3}\,\mathrm{cm^2 \cdot s^{-1}}$ and a typical reaction rate of $dN/dt = 10^{-6}\,\mathrm{mol \cdot s^{-1}\,cm^3}$, diffusion problems are negligible at 300 °C if

$$R(\mu m) < 2060\sqrt{P_B}, \quad \text{with } P_B \text{ in atm.} \tag{4.24}$$

This shows that if in a reaction A to B to C the equilibrium partial pressure of intermediate B at 300 °C is 10^{-4} atm, the radius of the catalyst particles must be smaller than $2060\sqrt{0.0001} = 21\,\mu\mathrm{m}$ for optimal cooperation between the two catalytic sites. The catalyst particle radius needed for a good cooperation is surprisingly large for "usual" reaction rates and small pressures of intermediates! A mechanical mixture of particles of one μm can still fully support a bifunctionally catalysed reaction as long as a partial pressure of 2×10^{-7} atm can be obtained for the intermediate. Multistep reactions may thus easily proceed with intermediates at concentrations below the detection limit!

Radii of tens of μm are no problem for mixtures of laboratory catalysts because such catalysts can be sieved to the right fraction. For mixtures of industrial catalysts, catalyst particle radii of tens of μm are a problem, however, because the dimensions of extrudates are in the order of a few cm and of spheres of a few mm. Industrial bifunctional catalysts thus must be prepared with both functions on the same support. Even when each catalyst component had a loading of only $10^{-4}\,\mathrm{mol \cdot g^{-1}}$ and a dispersion of 10% and the support had a surface area of $100\,\mathrm{m^2 \cdot g^{-1}}$, the average distance between the two functions would still be 4 nm, orders of magnitude smaller than needed for perfect cooperation between the two catalytic sites.

References

4.1 R. Prins, Eley–Rideal, the other mechanism, *Topics Catal.* 61, 714–721, 2018.

4.2 In steady state all the rates of mass transfer are equal and equal to the rate of the chemical reaction in the catalyst particle. Assuming that the hydrogenation reaction is first order in H_2 and n-th order in reactant A:

$$\begin{aligned}
r &= k_{\mathrm{LG}}(c_{\mathrm{O}} - c_L)a_{\mathrm{LG}} &\quad c_{\mathrm{O}} - c_L &= r/(k_{\mathrm{LG}}a_{\mathrm{LG}})\\
r &= k_{\mathrm{LS}}(c_L - c_S)a_{\mathrm{LS}} &\quad c_L - c_S &= r/(k_{\mathrm{LS}}a_{\mathrm{LS}})\\
r &= k c_S c_A^n \eta &\quad c_S &= r/(k c_A^n \eta a)
\end{aligned}$$

$$c_{\mathrm{O}} = r/(k_{\mathrm{LG}}a_{\mathrm{LG}}) + r/(k_{\mathrm{LS}}a_{\mathrm{LS}}) + r/(k c_A^n \eta a)$$

$$r = \frac{c_{\mathrm{O}}}{\frac{1}{k_{\mathrm{LG}}a_{\mathrm{LG}}} + \frac{1}{k_{\mathrm{LS}}a_{\mathrm{LS}}} + \frac{1}{k\eta c_A^n a}}.$$

4.3 Z. Xing, L. Hu, D. S. Ripatti, X. Hu, X. Feng, Enhancing carbon dioxide gas-diffusion electrolysis by creating a hydrophobic catalyst microenvironment, *Nature Commun.* 12, 136, 2021.

4.4 P. B. Weisz, Polyfunctional heterogeneous catalysis, *Adv. Catal.* 13, 137–190, 1962.

Questions

4.1 Derive the rate equation for a bimolecular reaction that follows the Eley–Rideal mechanism. What is the order of the reaction in the reactants A and B? Is it always possible to distinguish an Eley–Rideal mechanism from a Langmuir–Hinshelwood mechanism on the basis of kinetic results?

4.2 Why can Q_A in Eq. (4.9) be ignored?

4.3 Polymerisation grade ethene must be ultra-pure, because traces of acetylene cause crosslinking of polyethene. Traces of acetylene in ethene are removed in industry by hydrogenation over a Pd catalyst. When reacted separately, ethene reacts faster with hydrogen than acetylene does, but in a mixture of acetylene and ethene the acetylene reacts faster. How is this possible?

4.4 For reactions of an order higher than 1 the curves for the effectiveness factor as a function of the Thiele modulus are similar to the curve for order 1, but the higher the order, the quicker the effectiveness factor decreases from 1 and, thus, the sooner diffusion becomes a problem. Is this to be expected?

4.5 The rate r of the hydrogenation of nitrobenzene in an autoclave is measured as function of the catalyst mass m and the speed of stirring. Why is $1/r$ linear proportional to $1/m$ and why does $1/r$ decrease with increasing speed of stirring?

4.6 Formic acid (HCOOH) can be decomposed to CO_2 and H_2 on an Au catalyst. What will be the ratio of CO_2, H_2, D_2 and HD if HCOOD is decomposed on Au?

Chapter 5

Metal Surfaces

5.1 Surface Structures

Metals tend to crystallise in densely-packed crystal structures because the energy is lower when the atoms are close together and, thus, more tightly bound. The most important metal structures are: body-centred cubic (bcc) (e.g. Cr, V, Fe, Mo, W), face-centred cubic (fcc) (e.g. Al, Ni, Pt, Cu, Ag, Au) and hexagonal close-packed (hcp) (e.g. Mg, Zn, Co, Ru).

The bcc structure has three equivalent orthogonal main axes, atoms at the eight corners and one atom in the centre of the unit cell (Fig. 5.1a). The fcc structure (Fig. 5.1b) has three equivalent orthogonal main axes and atoms at the eight corners and six centres of the side planes (faces) of the unit cell, hence the term fcc structure (cf. Section 2.2.1). Planes are indicated by Miller indices, the reciprocals of the interceptions of the planes with the axes. The faces in the bcc and fcc structures are, therefore, the (100) planes. Atoms in the (100) faces of the fcc and bcc structures are surrounded by four atoms in the same plane; each atom in the (111) plane of the fcc structure is surrounded by six atoms in the close-packed plane. The bcc structure does not have a close-packed plane. The hcp structure (Fig. 5.1c) does not have orthogonal axes but has three axes at 120° in a close-packed plane and one axis perpendicular to this plane. Therefore, hcp planes are indicated by 4-digit Miller–Bravais indices. The third digit is the negative sum of the first two digits (e.g. $(11\bar{2}0)$) because the third vector is the negative sum of the other two vectors.

The fcc structure can be created not only by connecting cubic unit cells, but also by stacking layers of closely packed spheres (each sphere is

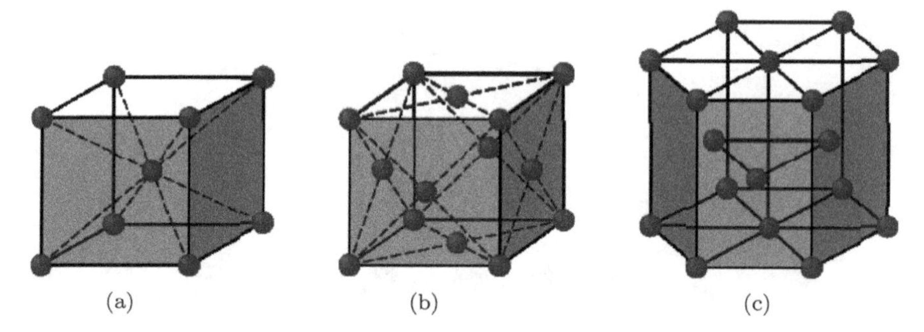

Fig. 5.1 Unit cells of (a) body-centred cubic, (b) face-centred cubic, and (c) hexagonal close-packed structures.

surrounded by six spheres) on top of each other (Section 2.2.1). This can be done in two ways, leading to the hcp and the fcc structures (Fig. 2.3). The position of the atoms in the fcc structure is repeated every three layers (in an ABCABC... sequence). In the hcp structure, the atoms of the third layer are above the atoms of the first layer, giving an ABABAB... sequence.

Heterogeneous catalysis takes place on the surface of catalyst particles, but surfaces are also important in corrosion, lubrication and in semiconductors. The surface is the interface between the solid phase and the gas or liquid phase and is the termination of the solid-state structure. Therefore, it is important to determine which planes form the surface of metal particles and what the structure of these planes is. Surface planes usually have a high density of atoms, which means that the surface atoms have many neighbouring atoms and low energy. Therefore, surface planes are often low-index planes such as (111), (100) and (110). Atoms in these planes have nine, eight and seven nearest neighbours, respectively (Fig. 5.2). An atom in the (111) plane has six neighbours in the plane and three in the plane below, while an atom in the (100) plane has four atoms in the plane and four below. The atoms in the (110) plane have two neighbours in the row of atoms, four neighbours in the plane below it and are above one atom in the third plane.

The lower the number of neighbouring atoms in a surface plane, the less stable the surface plane is. The (110) plane of many metals is already unstable and undergoes surface reconstruction. By removing the top rows of the (110) surface and depositing them above the rows in another (110) plane, a surface is created that consists of grooves that exhibit (111) planes (Fig. 5.3). In this way, the surface atoms have more neighbouring

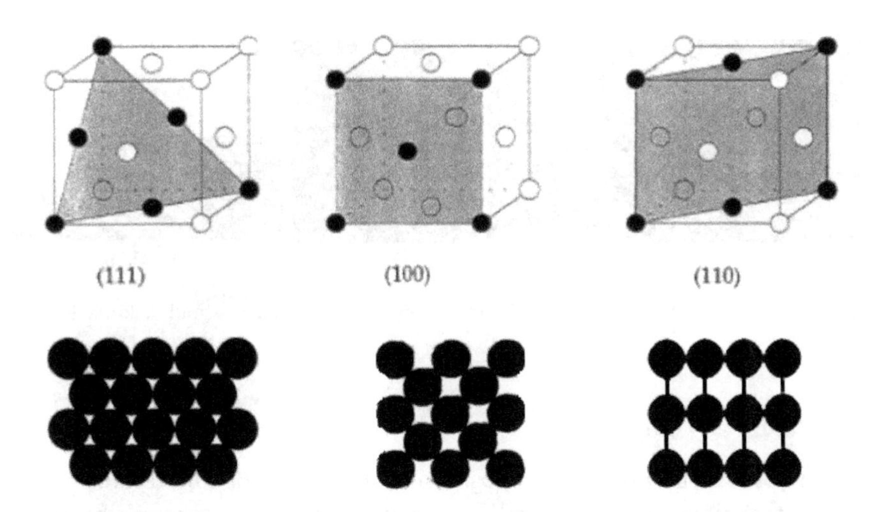

Fig. 5.2 Unit cell of the fcc structure with (111), (100) and (110) planes and the corresponding surface structures.

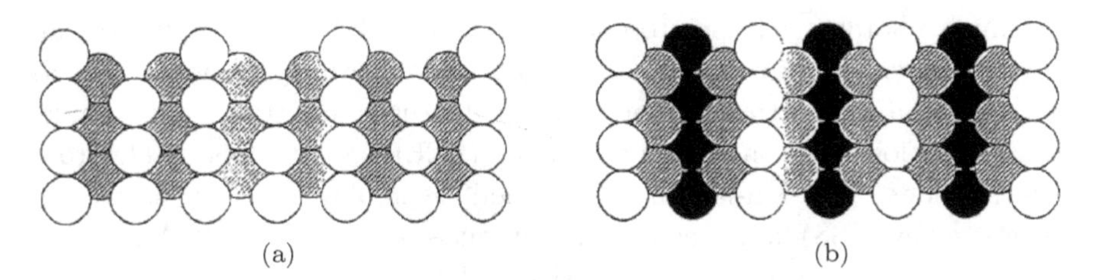

Fig. 5.3 Side (a) and top view (b) of the missing-row model of a reconstructed (110) surface.

atoms (nine instead of seven). On the other hand, the surface is no longer flat and is, thus, larger. Nevertheless, this so-called "missing-row" surface reconstruction occurs on (110) surfaces of many metals.

Stepped surfaces form when planes are placed on top of each other. When (111) layers are stacked at a regular distance, a (211) surface forms with edges, at which the atoms have only seven neighbours, four in the plane of the atom and three below it (Fig. 5.4). The terraces of the (211) planes have a (111) structure and the step has a (100)-type structure. If the edges make an angle with the rows of the atoms in the (111) plane, then an edge with so-called kink atoms is created with only six neighbouring atoms (Fig. 5.4).

Fig. 5.4 Flat (111) and (100) surfaces, a stepped (211) surface and a kinked (321) surface.

5.2 Surface Analysis

To study the structure and composition of the surface of a material, surface science techniques are employed, that aim particles (photons, electrons or ions) at a surface and detect the particles that leave it. When a surface is irradiated with photons and photons are detected (a photon-in photon-out experiment), the detected photons can have the same energy as the incoming photons (IR spectroscopy, XRD) or a different energy (Raman spectroscopy, X-ray fluorescence). X-ray photoelectron spectroscopy (XPS) is a photon-in electron-out method and Auger electron spectroscopy (AES) and low electron energy diffraction (LEED) are electron-in electron-out methods. The ion-in ion-out method is employed in ion scattering spectroscopy (ISS) and secondary ion mass spectroscopy (SIMS). Few of these techniques measure exclusively the atoms in the surface layer. A technique based on the detection of photons and electrons can never be completely surface-sensitive because photons and electrons can travel through matter. Only those techniques, which detect ions, are highly surface sensitive because ions have a high chance of being neutralised by electrons in the material. As a result, only ions originating from the surface layer will reach the detector. The principles of XPS and AES will be discussed below, as these techniques are generally used to analyse surfaces.

5.2.1 *X-ray Photoelectron Spectroscopy*

In XPS, a surface is irradiated by X-rays and the electrons that are created by ionisation of electrons from orbitals in surface and subsurface atoms are detected. If the ejected electron does not lose energy by interaction with atoms on its path to the surface, then its kinetic energy will be equal to $E_{\text{kin}} = h\nu - E_B$, where $h\nu$ is the energy of the X-ray photon and E_B is the binding energy of the electron to the atom. By measuring the kinetic energy of the ejected electron, its binding energy $E_B = h\nu - E_{\text{kin}}$ can be determined. The binding energy of an electron in an orbital of an atom is specific to the atom and, therefore, can be used for elemental analysis. The number of electrons that leave the surface with a specific kinetic energy is proportional to the number of corresponding atoms in the surface and is used for the quantitative analysis of the surface composition. The kinetic energy of the ejected electrons is measured by bending the path of the electrons in a magnetic or electric field. Figure 5.5 shows a hemispherical analyser, in which the electrons enter an electric field between two half spheres and are deflected by the electric field onto a detector. By varying the strength of the electric field between the half spheres, electrons of varying kinetic energy reach the detector.

Figure 5.6 shows the XPS spectrum of gold. The number of ejected electrons is plotted as a function of the binding energy of the ejected electrons ($E_B = h\nu - E_{\text{kin}}$). From left to right, in order of decreasing binding energy, the peaks are due to electrons ejected from the 4s, 4p, 4d, 4f, 5s, 5p and 5d orbitals. The spectrum was obtained with Al Kα X-ray radiation with an energy of 1,486 eV. Al Kα X-rays are created by bombarding an Al surface with electrons emitted by a hot metal wire and accelerated in an electric field. Electrons are ejected from the Al 1s atomic orbital and the

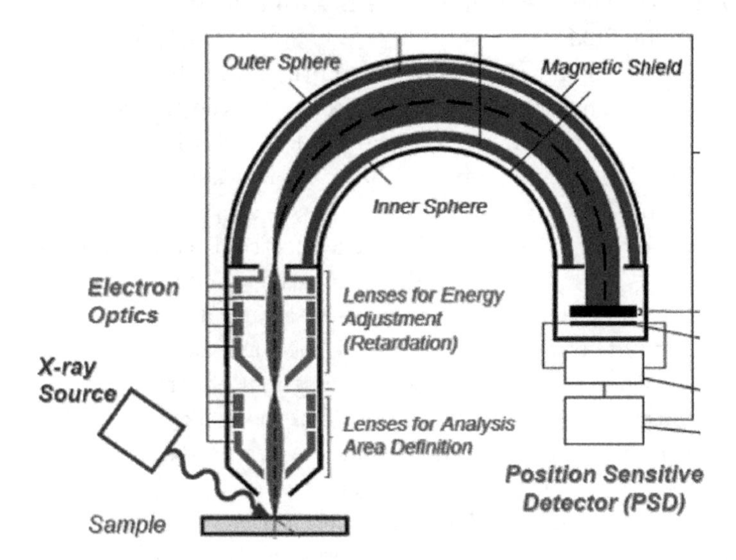

Fig. 5.5 Hemispherical electron analyser for measuring the kinetic energy of electrons.

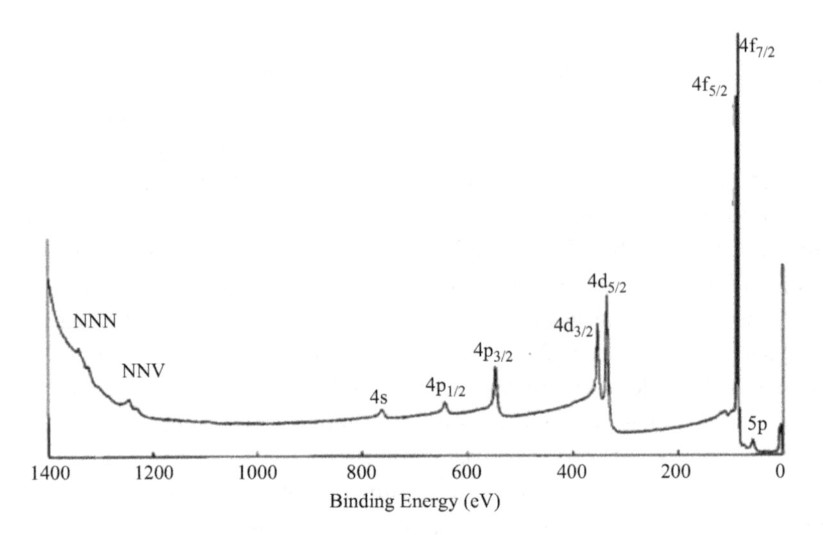

Fig. 5.6 XPS spectrum of Au measured with monochromatic Al Kα radiation. (http://mmrc.caltech.edu/SS_XPS/XPS_PPT/XPS_Slides.pdf).

resulting 1s hole (missing 1s electron) is filled by moving an electron from the 2p orbital to the 1s hole. The energy gain is used to create an X-ray photon with energy $h\nu = E_{2p} - E_{1s}$. This is analogous to the creation of Cu Kα X-rays when a hole in the Cu 1s shell is filled by a 2p electron (Cu 2p 1s) (Section 2.2.2).

The electrons, which are ejected from the atoms by the X-rays and do not lose energy in interaction with the electrons of the atoms in the material, are responsible for the sharp XPS peaks in Fig. 5.6. Electrons that undergo inelastic collisions lose energy and have a lower kinetic energy. Based on the energy conservation equation $h\nu = E_B + E_{\text{kin}}$, used to calculate E_B, these electrons are present in the XPS spectrum at an apparent higher binding energy E_B; therefore, XPS peaks have a tail to higher binding energy (Fig. 5.6). Electrons ejected from Au atoms at and directly under the surface contribute to the sharp XPS peaks because these electrons can reach the detector without interaction with atoms, but electrons from subsurface Au atoms lose energy, resulting in tailing of the peaks.

XPS peaks of electrons from p, d and f orbitals have a doublet structure (Fig. 5.6). The doublet splitting is larger for deeper lying levels (with higher binding energies): 96, 18 and 4 eV for the 4p, 4d and 4f levels, respectively. This so-called spin-orbit splitting is due to the fact that the electron spin can be aligned parallel or opposite to the orbital momentum of the hole in the p, d or f orbital, which is created upon electron ejection from this orbital. Since the s orbital has no orbital momentum, its XPS peak is single. XPS peaks of atomic levels with low binding energy are narrow (e.g. the 4f peaks), while peaks of deeper energy levels are broader (e.g. the 4p peaks). This is due to the uncertainty principle, which states that $\Delta E \cdot \Delta t = h$ is constant; ΔE is the uncertainty in the energy (thus the peak width) and Δt is the uncertainty in the lifetime of the excited state created by the ejection of the photoelectron. As the binding energy of a hole level increases, there are more levels at lower binding energy from which the hole can be filled. The filling of the hole becomes faster and Δt becomes smaller. Because $\Delta E \cdot \Delta t$ is constant, ΔE and, thus, the line width will be larger. Therefore, peaks at lower binding energy are narrower and are suitable for analytic purposes. For instance, the Au 4f lines are used to quantitatively measure Au.

The XPS peak position depends on the charge on the atom and on its surrounding atoms and is used to distinguish chemical structures. The shift of an XPS peak is about 1 eV for every oxidation step, from M^{n+} to $M^{(n+1)+}$, while the peak width is also about 1 eV. This enables the determination of the oxidation state of an element, for instance the oxidation state of iron (Fe^{2+} or Fe^{3+}) and oxygen (O^{2-} or OH^-). The O 1s spectrum of NiO, which has been transported through air, shows two peaks at about 529 and 531 eV (Fig. 5.7a). The 531 eV peak is in the same position as the O 1s peak of $Ni(OH)_2$ and indicates that the surface of NiO is covered with OH groups [5.1]. Figure 5.7b demonstrates that sulfide

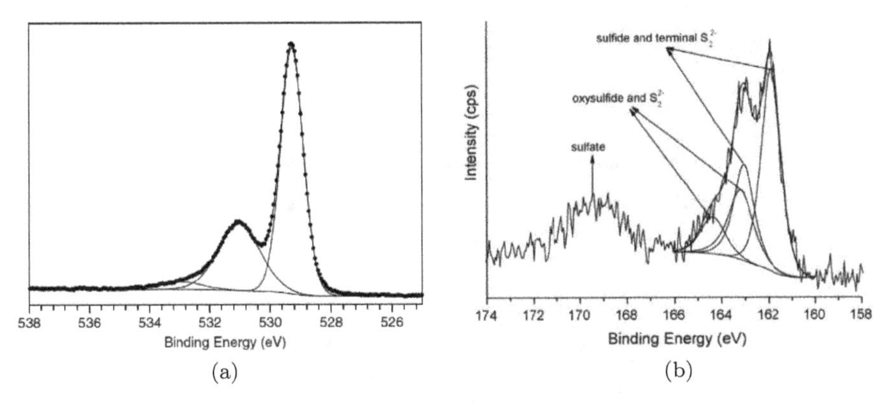

Fig. 5.7 (a) O 1s spectrum of NiO. Reprinted with permission from [5.1]. Copyright (2009) John Wiley & Sons, Inc. (b) S 2p spectrum of a sulfided Mo/Al_2O_3 catalyst. Reprinted with permission from [5.2]. Copyright (2010) Elsevier.

(S^{2-}) as well as oxysulfide, persulfide (S_2^{2-}) and sulfate (SO_4^{2-}) groups can be detected after sulfidation of an oxidic Mo/Al_2O_3 catalyst [5.2].

5.2.2 *Auger Electron Spectroscopy*

AES is another technique for the study of the surface of materials. As in XPS, with AES an electron is first removed from a core orbital (Fig. 5.8a). This can be done with X-rays (as in an XPS instrument) or, in special-purpose AES instruments, with high-energy electrons. Instead of measuring the kinetic energy of the ejected electron, as in XPS, one measures the kinetic energy of an electron that is ejected from the atom in an after effect (secondary effect). Once a hole is created by removing an electron from an inner orbital (Fig. 5.8a), the resulting hole is unstable because it is in an orbital with high binding energy. The cation becomes more stable when an electron from an orbital with lower binding energy is transferred to this core orbital (Fig. 5.8b). The energy that is freed in this electron transition can be emitted either as X-ray fluorescence (as in the production of Al X-rays, in the Al 2p to 1s transition) or can be used to eject another electron from the cation (Fig. 5.8c). The kinetic energy of this Auger electron depends on the energy of three atomic orbitals, $E_{\text{kin}} = E_2 + E_3 - E_1$, and is, thus, element-specific.

Auger electrons also appear in XPS spectra, as a result of the after effect of the primary creation of the XPS photoelectron. In Fig. 5.6, Auger NNN and NNV peaks are observed around 1,330 and 1,250 eV, respectively.

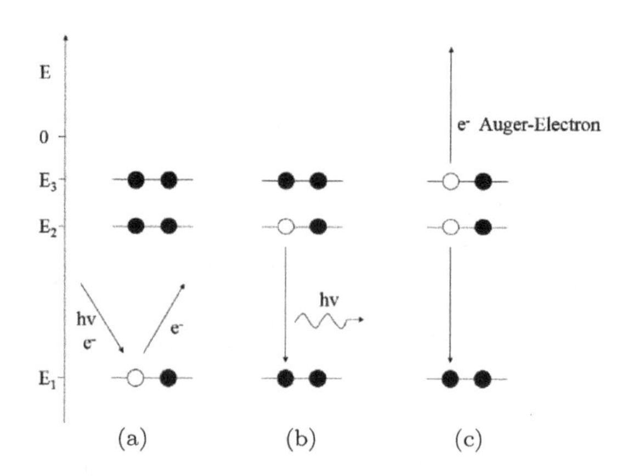

Fig. 5.8 Energy of the photon- or electron-induced ionisation of an atom followed by emission of an X-ray or an Auger electron.

The NNN peaks are due to transitions between $4d_{3/2}$, $4f_{5/2}$ and $4f_{5/2}$ or $4f_{7/2}$ levels, while the NNV peaks are due to transitions between $4d_{5/2}$, $4f_{5/2}$ or $4f_{7/2}$ and valence levels.

5.2.3 *Surface Sensitivity*

X-rays penetrate many layers of a material and, thus, create XPS electrons not only from atoms at the surface, but also from atoms under the surface. Most XPS electrons lose energy when travelling through the material because they interact inelastically with the electrons of the atoms that they pass on their way to the surface and on to the detector. The chance f that an electron will not lose energy and will still have the kinetic energy $h\nu - E_B$, obtained upon ejection from an atom by the X-ray photon, depends on the mean free path λ: $f = \exp(-r/\lambda)$. The longer the mean free path λ, the longer the distance r that the ejected electron can travel without losing energy. Electrons, which are created inside a material and have a high kinetic energy, move very fast and interact only weakly with the electrons that are still bonded to the atoms. Such fast electrons do not lose energy easily, and their mean free path is long (Fig. 5.9). Slow electrons interact stronger with the atoms; they will lose energy more quickly and have a shorter mean free path. Therefore, λ decreases with decreasing kinetic energy of the ejected electron until it reaches a minimum between 50 and 100 eV (Fig. 5.9). At low kinetic energy, λ increases again because electrons with a low kinetic energy cannot exchange energy with the electrons

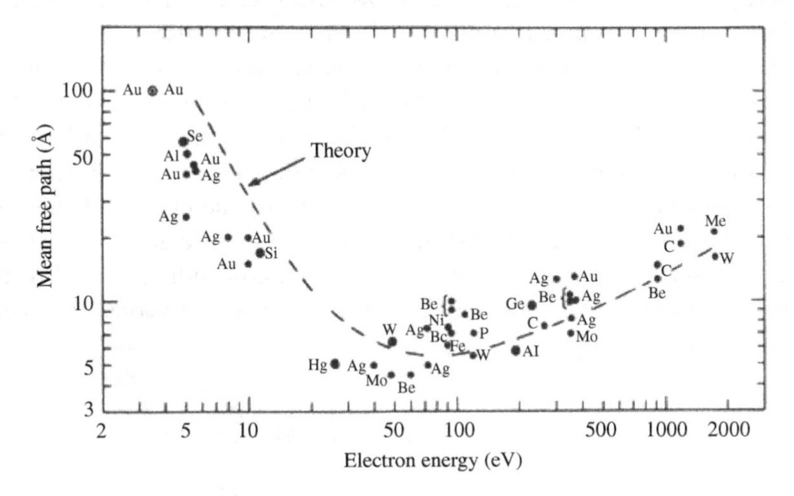

Fig. 5.9 Mean free path of electrons in metals as a function of their kinetic energy.

bonded to the atoms (excitation of bonded electrons requires at least a few eV). Therefore, the electrons do not lose energy and have a long mean free path.

The smallest mean free path is about 0.4 nm (Fig. 5.9). Hence electrons with a kinetic energy of about 70 eV, which are created 0.4 nm below the surface, have a chance of $\exp(-0.4/0.4) = 0.37$ of maintaining the original kinetic energy when travelling to the surface and on to the detector, while electrons that are created 1 nm under the surface have a chance of $\exp(-1/0.4) = 0.08$ to maintain the initial energy. This shows that XPS and AES not only measure electrons that originate from atoms in the surface layer but also from atoms in subsurface layers, although to a lesser extent.

5.3 Surface Enrichment

The composition of the surface of most materials is different from that of the bulk. This has significant consequences whenever surfaces play an important role. Surface enrichment occurs because atoms at the surface have different surroundings than in the bulk. Therefore, thermodynamics dictates that some atoms are at the surface and others in the bulk. Here, alloys will be used to illustrate surface enrichment, but the principles are also valid for other materials, such as metal oxides, metal sulfides and polymers.

Suppose an atom in the bulk of a binary alloy AB has p neighbouring atoms in the plane parallel to the surface and m neighbours perpendicular to that plane (m/2 above and m/2 below). As Fig. 2.3 shows, for the close packed layers of fcc and hcp metals $p = m = 6$. An atom at the surface will have the same number of neighbours in the surface plane (p) but only m/2 neighbours perpendicular to the plane because the neighbours above the plane are missing. The simplest model to treat the binding in metals is the **broken-bond model**, which is based on the assumption that only bonds between neighbouring atoms contribute to the energy of the alloy. A further simplification is to assume that surface enrichment occurs only in the top surface layer and that all other layers, including the subsurface layer, have the bulk composition. Based on these assumptions, the energy of atom A at the surface is

$$\begin{aligned} E_{As} &= p[x_s E_{AA} + (1 - x_s)E_{AB}] + \frac{m}{2}[x_b E_{AA} + (1 - x_b)E_{AB}] \\ &= \left(px_s + \frac{mx_b}{2}\right) E_{AA} + \left[p(1 - x_s) + \frac{m(1 - x_b)}{2}\right] E_{AB}, \end{aligned}$$

where x_s is the molar fraction of atom A at the surface, x_b the molar fraction of atom A in the bulk, E_{AA} the binding energy between two neighbouring A atoms and E_{AB} the binding energy between two neighbouring A and B atoms. The energy of an atom A in the bulk is $E_{Ab} = (p+m)[x_b E_{AA} + (1-x_b)E_{AB}]$. Similarly, the energies of atom B at the surface and in the bulk are E_{Bs} and E_{Bb}, respectively:

$$E_{Bs} = \left(px_s + \frac{mx_b}{2}\right) E_{AB} + \left[p(1-x_s) + \frac{m(1-x_b)}{2}\right] E_{BB} \quad \text{and}$$

$$E_{Bb} = (p+m)[x_b E_{AB} + (1-x_b)E_{BB}].$$

The energy difference ΔE between atom A at the surface and atom B in the bulk on the one hand and atom B at the surface and atom A in the bulk on the other hand (Fig. 5.10) is

$$\begin{aligned}\Delta E &= (E_{As} + E_{Bb}) - (E_{Bs} + E_{Ab}) = (E_{As} - E_{Ab}) - (E_{Bs} - E_{Bb}) \\ &= \left[p(x_s - x_b) - \frac{mx_b}{2}\right] E_{AA} - \left[2p(x_s - x_b) - mx_b + \frac{m}{2}\right] E_{AB} \\ &\quad + \left[p(x_s - x_b) - \frac{mx_b}{2} + \frac{m}{2}\right] E_{BB}.\end{aligned}$$

The binding energies E_{AA} and E_{BB} can be determined experimentally from the heats of sublimation of the pure metals A and B because in the broken-bond model the molar heat of sublimation of metal A is equal to $(p+m)E_{AA}/2$. The binding energy E_{AB} can be determined from the heat of mixing $\alpha = E_{AB} - (E_{AA} + E_{BB})/2$. Substitution of $E_{AB} \equiv (E_{AA} + E_{BB})/2 + \alpha$ ($\equiv$ means by definition) in ΔE gives

$$\Delta E = -\alpha \left[2p(x_s - x_b) - mx_b + \frac{m}{2}\right] + m(E_{BB} - E_{AA})/4.$$

The thermodynamic equilibrium composition of the alloy surface is reached when the free energy has a minimum as a function of x_s. Thus, x_s must be determined when $\delta(\Delta G)/\delta x_s = 0$, thus

$$\frac{\delta(\Delta H - T\Delta S)}{\delta x_S} = 0 \quad \text{or} \quad \frac{\delta(\Delta H)}{\delta x_S} = \frac{T\delta(\Delta S)}{\delta x_S}.$$

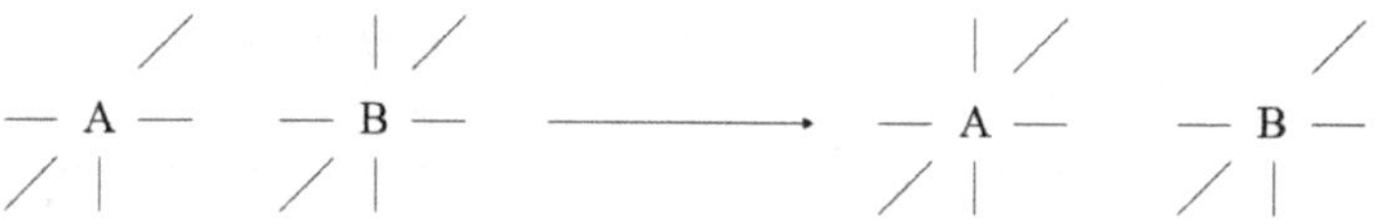

— A — — B — ⟶ — A — — B —

Fig. 5.10 Exchange between atom A at the surface and atom B in the bulk.

As explained in [5.3], this gives:

$$N_s \left\{ -\alpha \left[2p(x_s - x_b) - mx_b + \frac{m}{2} \right] + \frac{m(E_{BB} - E_{AA})}{4} \right\}$$
$$= kTN_s \ln \frac{x}{1-x} \frac{1-x_S}{x_S}$$

or $$\frac{m(E_{BB} - E_{AA})}{4} + \alpha \left[(2p+m)x_b - 2px_s - \frac{m}{2} \right] = -kT \ln \chi,$$

where χ is the overall, average concentration of A (in between x_s and x_b) and χ, the **surface enrichment factor**, is

$$\chi = \frac{x_S}{1-x_S} \frac{1-x}{x}. \tag{5.1}$$

When $\alpha \ll (E_{AA} + E_{BB})/2$ (which is the case for ideal solutions, for which $E_{AB} = (E_{AA} + E_{BB})/2$ and the heat of mixing is zero), then

$$\chi = \exp \left[\frac{-m(E_{BB} - E_{AA})}{4kT} \right]. \tag{5.2}$$

When BB bonds are stronger than AA bonds (i.e. B has the higher sublimation energy), $E_{BB} < E_{AA} < 0$ and $\chi > 1$. Thus, $x_S > x$ and the surface is enriched with A, the component with the lowest sublimation energy. Surface enrichment is more advantageous when the component with the highest binding energy has many neighbouring atoms; thus, in this case, B atoms tend to be in the bulk rather than at the surface. As a consequence, the other element A moves to the surface. In other words, it is more advantageous when the element with the lowest binding energy is located at the surface.

The molar fraction x_{As} of A at the surface is very sensitive to small differences in the sublimation energy of the metals A and B (Fig. 5.11). Already at small ΔE_{subl}, x_{As} is much larger than x_{Ab}. Surface enrichment can, thus, be strong. Surface enrichment also depends on m, the number of neighbouring atoms perpendicular to the surface. As m increases, the effect of the missing atoms above the surface plane is stronger, as is the enrichment. Rough surfaces (kinks and steps) have greater surface enrichment than smooth planes. The atoms of the element with the lower sublimation energy take up the energetically unfavourable kink and step positions.

The surface structure of endothermic and exothermic alloys is more difficult to predict than that of ideal solutions. Endothermic alloys have a negative heat of formation, and AA and BB bonds are stronger than two AB bonds. That is why endothermic alloys show segregation into two phases at low temperature when entropy no longer stabilises one homogeneous phase.

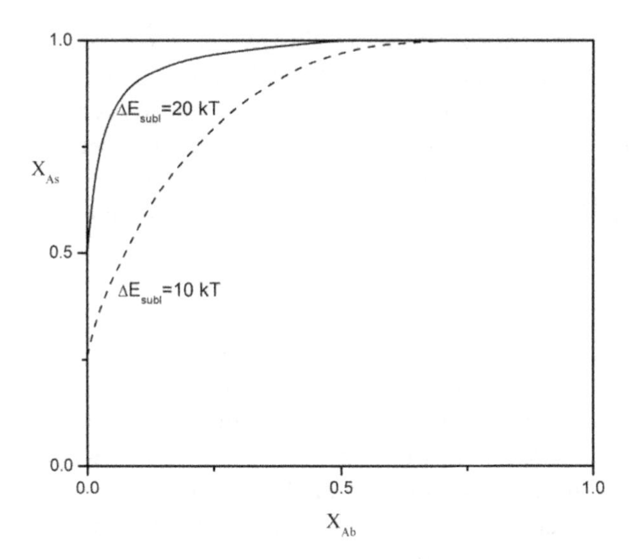

Fig. 5.11 Surface composition x_{As} as a function of bulk composition x_{Ab}.

Small metal particles consist of a kernel and an outer shell (cherry model). The outer shell has the lower surface energy and, in addition, is surface-enriched. For example, the kernel of a PtAu particle (1:1 molar ratio) contains 97% Pt, while the outer shell contains about 80% Au, the surface of which has even more than 80% Au because of the lower sublimation energy of Au than of Pt.

AB bonds of exothermic alloys are stronger than half the sum of AA and BB bonds. Some combinations of metals from the left and right side of the Periodic Table belong to this category (e.g. PdZr and Ni_3Al) as well as PtSn and Pt_3Sn. This has been ascribed to Lewis base (Pt, Pd, Ni)-Lewis acid (Al, Zr, Sn) type interactions. Also in exothermic alloys the surface is enriched in the component with the lowest sublimation energy, but because of the strong AB bonds this enrichment does not occur at the cost of the whole metal particle but mainly at the cost of the subsurface layer.

Factors other than surface energy may also cause surface enrichment. The **size of the atoms** explains why the surface of M(3d) M(4d) and M(3d) M(5d) alloys is often enriched in the 4d or 5d metal, even though the sublimation energies of 4d and 5d metals are larger than those of 3d metals (Section 5.4). For instance, the surface of Au–Cu particles is enriched in gold. The number of gold atoms that cover a surface is smaller than that of copper atoms, giving the gold-enriched surface a lower surface energy. The **gas phase** above the surface also influences the surface composition.

NiCu surfaces are enriched in nickel, instead of in copper, when the NiCu alloy is brought into contact with CO gas, because CO binds stronger to Ni atoms than to Cu atoms and pulls, as it were, the Ni atoms to the surface. For the same reason, the surface of an alloy in contact with air is enriched by the less noble component.

Temperature is another important factor because entropy is proportional to temperature. At higher temperature, mixing is very important (also of surface and bulk) and surface enrichment is lower. The driving force behind surface enrichment is the difference in energy; surface enrichment is, thus, stronger at lower temperature. However, the surface is enriched only when the metal atoms of one component diffuse from the bulk to the surface and the metal atoms of the other component diffuse in the opposite direction. At low temperature diffusion may be very slow, and the composition of the surface may differ from that predicted by thermodynamics. Therefore, a Pd layer deposited on top of Ag particles by electrodeposition is stable at room temperature.

5.4 Metal Binding

As explained in the foregoing section, the difference in the binding energy of metals A and B determines which metal is at the surface of an AB alloy. To predict surface enrichment, the binding energy and, thus, the heat of sublimation of metals should be known. These data are obtained from the **band model** of the bonding in metals, based on molecular orbital (MO) theory. The principles are apparent from the explanation of the bonding of two potassium atoms to a di-potassium K_2 molecule and the non-bonding of two calcium atoms to a di-calcium Ca_2 molecule. Similar as the 1s orbitals of two H atoms (Section 3.3.1, Fig. 3.11a), the 4s atomic orbitals (AOs) of two K atoms combine to a bonding MO with energy β and an antibonding MO with energy β' ($\beta' > \beta$) relative to the energy of the separate K 4s orbitals, where β is a measure of the interaction energy between the 4s orbitals on the neighbouring atoms. Each K atom has one electron in a 4s AO; in the K_2 molecule the two electrons can both occupy the bonding orbital (with opposite spin, Pauli principle). Thus, an energy gain of 2β is realised and K_2 is a stable diatomic molecule. The same energy diagram as used to explain the instability of the He_2 molecule (Fig. 3.11b), also applies to the Ca_2 molecule. Two Ca atoms can be combined to a Ca_2 molecule except that each Ca atom has two 4s electrons. Two 4s electrons occupy the bonding MO, which gives an energy gain of 2β, but the other two 4s

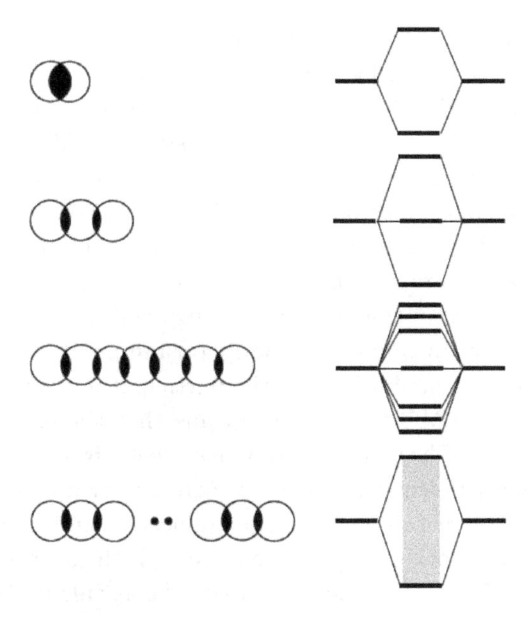

Fig. 5.12 Two, three and seven potassium 4s orbitals combining to two, three and seven MO energy levels and many potassium 4s orbitals in bulk potassium combining to a 4s energy band.

electrons must occupy the antibonding MO, leading to an energy loss of $2\beta'$, slightly larger than 2β. In all, there is an energy loss and the Ca_2 molecule is unstable, just like the He_2 molecule.

In a linear K_3 molecule, the three atomic K 4s AOs combine to three MOs: one bonding MO at energy $-\beta\sqrt{2}$, one nonbonding MO at energy 0 and one antibonding MO at energy $\beta\sqrt{2}$ (Fig. 5.12). Two electrons fill the bonding MO and one electron fills the nonbonding MO half. Thus, the energy of the K_3 molecule is $2\beta\sqrt{2}$ lower than that of three separate K atoms. When more than two K atoms are bonded to one central K atom, only bonding and nonbonding MOs are filled with electrons and such K_n molecules are stable. If each peripheral K atom is also surrounded by K atoms, then the situation is more complex. However, the total number of energy levels is equal to the total number N of the participating AOs (equal to the number of K atoms) and the lowest bonding level cannot be lower in energy than $-n\beta$, where n is the number of neighbouring atoms. Likewise, the highest antibonding energy level cannot be higher in energy than $n\beta'$, meaning that if there are N K atoms and each K atom is connected to n neighbours, then there are N levels between energy $-n\beta$ and $n\beta'$. For one mol of K, N is equal to the Avogadro number (6×10^{23}) and an extremely large number of levels will have an energy between $-n\beta$ and $n\beta'$. For fcc metals $n = 12$ and with $\beta = 2$ eV the average energy difference between adjacent energy levels will be in the order of 5×10^{-22} eV. As a consequence, whereas small molecules have distinct energy levels, the energy levels of solids form quasi-continuous bands (Fig. 5.12).

The energy difference between adjacent energy levels is much smaller than the thermal energy at room temperature ($2.4\,\text{kJ}\cdot\text{mol}^{-1} = 0.025$ eV) and, thus, electrons from the highest bonding orbitals are excited to empty orbitals with hardly any additional energy (or temperature). Since different MOs have different contributions by the participating AOs, the transfer of an electron from one MO to the next means that the electron density on the atoms changes. Electrons, thus, move freely from one atom to the next; metals conduct electricity. The prerequisite for a material to become metallic is that many atoms are interconnected and that each of the constituting atoms contains unpaired electrons. If the atoms do not contain 4s electrons, then it is still possible to create the 4s valence band, but there are no electrons in the band and electrons cannot flow from one atom to the next. On the other hand, if each constituting atom has the maximum number of electrons in its AO then the resulting valence band is completely filled and there is no bonding between the atoms.

This leads to the prediction that solid calcium (with $4s^2$ valence electron configuration), like the Ca_2 dimer, does not exist. However, the energy of the 3d level of Ca is just slightly higher than that of the 4s level. This has no consequence for the bonding in the Ca_2 dimer, but in solid calcium the effect is strong because the AOs form bands in solid Ca. Since s orbitals are larger than d orbitals, they interact stronger with each other and $\beta_{4s} > \beta_{3d}$. The width of the s band ($\sim 2n\beta_{4s}$) is, therefore, larger than that of the d band ($\sim 2n\beta_{3d}$) and, although the centre of the 3d band lies higher than that of the 4s band, the bands overlap. Electrons can flow from the filled 4s band into the empty 3d band, and both the 4s and the 3d band become partially filled, with an electron configuration of $(4s)^{2-x}(3d)^x$. Both bands contribute to electron conductivity, and calcium is a metal.

In contrast to the K atom, which has one unpaired electron in a 4s AO, transition metal atoms have more than one unpaired electron in their valence orbitals. For instance, the Ti atom has four electrons in the 3d and 4s AOs and the Ni atom has 10 electrons in the 3d and 4s AOs. While the maximum number of electrons in a 4s AO is only two, 10 electrons can occupy the 3d AOs because there are five d AOs, each of which can contain two electrons (Pauli principle). The first five electrons can, therefore, occupy each of these five d orbitals with parallel spins, while the next five electrons pair with the first five electrons. For Ti this means that the maximum number of unpaired electrons is four, either three in the 3d orbitals and one in the 4s orbital (d^3s^1) or all four in four 3d orbitals (d^4). For Ni, the 10 electrons can be distributed in three ways over the d and s orbitals: d^8s^2, d^9s^1 or $d^{10}s^0$. The configuration d^8s^2 has two unpaired d electrons, the configuration d^9s^1 has one unpaired d electron and one unpaired s electron and the configuration $d^{10}s^0$ has only paired electrons. The configuration with the lowest energy depends on the energy difference between the d and s orbitals as well as on the spin state. The energy of the d orbital is lower than that of the s orbital, but a configuration with unpaired electrons has lower energy than a configuration with paired electrons. Thus, a Ti atom has the d^4 configuration and a Ni atom the d^8s^2 configuration. Both the d and the s AOs on the metal atoms combine to valence bands. As for solid Ca, the s band is broader than the d band and the bands overlap in energy (Fig. 5.13). The difference between the transition metals and Ca is that the centre of the s band for the transition metals lies above that of the d band.

When an AO does not contain electrons, the valence band of the constituting AOs is empty too, and the binding energy between the atoms

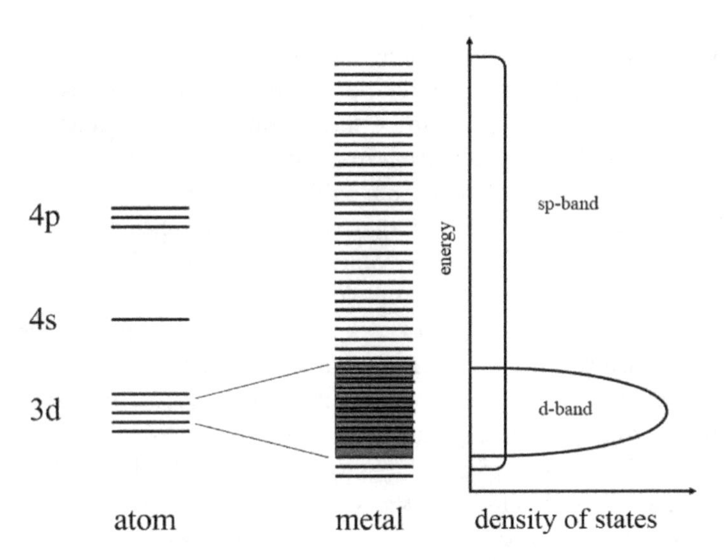

Fig. 5.13 Density of states of metals with broad sp band and narrow d band.

is zero. This is the case for elements on the left side of the Periodic Table. Elements on the right side have filled AOs and the valence d and s bands of the corresponding solids will also be filled, and the binding energy between the atoms will be zero. Going from the left side to the middle of the Periodic Table, the number of electrons in the d and s orbitals of the atoms increases and the electrons will fill the bonding MOs in the valence band of the solid. As the number of electrons increases, more bonding MOs are filled and the binding between the metal atoms increases. In the middle of the Periodic Table the d and s atomic orbitals will be half filled, giving the d^5s^1 configuration. This configuration has maximum bonding because all bonding orbitals of the metal valence band are filled. Towards the right side of the Periodic Table, the electrons will also fill antibonding orbitals, causing a decrease in the total energy of the solid. As a consequence, the binding energy (and, thus, the sublimation energy) of metals increases from the left side to the middle of the Periodic Table, after which it decreases again (Fig. 5.14). The resulting curve is a parabola. The sublimation energy of the 5d elements is higher than that of the 4d elements, which in turn is higher than that of the 3d elements, because 5d orbitals are larger than 4d orbitals than 3d orbitals. As a consequence, the overlap and, thus, the interaction energy β increase in the order $\beta_{3d} < \beta_{4d} < \beta_{5d}$, as do the widths of the 3d, 4d and 5d bands (width $\approx 2n\beta$). Therefore, tungsten is the strongest metal (Fig. 5.14) and is used to coat the surface of tools.

The 3d elements behave anomalously between V and Cu, with a minimum rather than a maximum for Mn (Fig. 5.14): Cr and Mn are antiferromagnetic and Fe, Co and Ni are ferromagnetic. Because $\beta_{3d} < \beta_{4d} < \beta_{5d}$, the magnetic energy is more important for the 3d than for the 4d and 5d elements. Another deviation from the parabola occurs in the middle of the 4d series, where the maximum sublimation energy is not reached for Mo with six electrons in the d and s AOs but is reached for Nb with five electrons. The reason is that the configuration of the Mo atoms with the lowest energy is d^4s^2, not d^5s^1, while for Nb it is d^4s^1.

Based on the data in Fig. 5.14 it is possible to make qualitative predictions about surface enrichment. For instance, the sublimation energy of Cu is lower than that of Ni because the sublimation energy decreases in the right half of the 3d parabola and Cu is one position to the right of Ni in the Periodic Table. Therefore, the surface of a CuNi alloy under vacuum is enriched in Cu and, similarly, the surfaces of AgPd and AuPt alloys are enriched in Ag and Au, respectively. The surface of PtPd is enriched in Pd because Pt and Pd are in the same column of the Periodic Table and the 5d

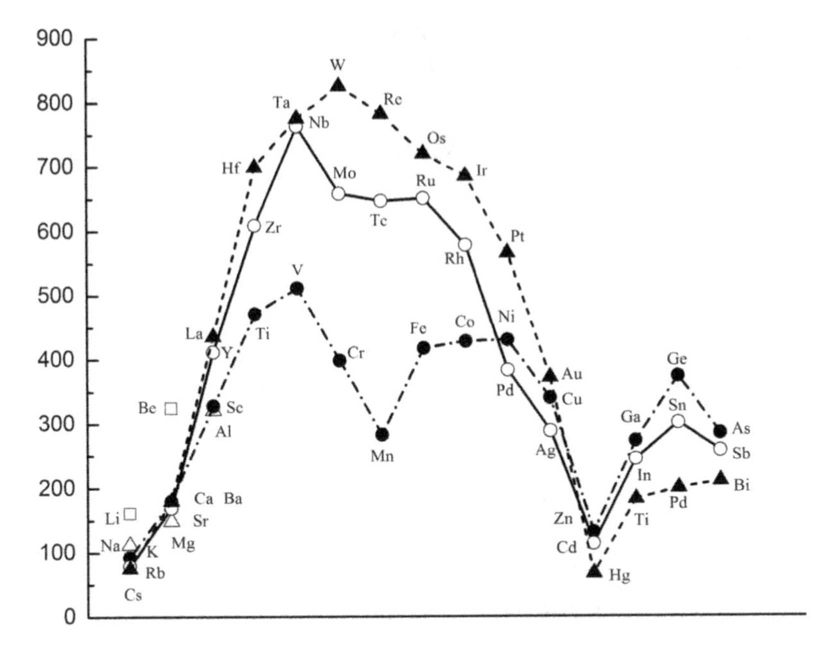

Fig. 5.14 Sublimation energy of metals depending on their position in the periodic table.

parabola is above the 4d parabola. Figure 5.14 also enables the prediction of the surface composition of sp metals (e.g. Ga, Ge, As).

References

5.1 M. C. Biesinger, B. P. Payne, L. W. M. Lau, A. Gerson, R. St. C. Smart, X-ray photoelectron spectroscopic chemical state quantification of mixed nickel metal, oxide and hydroxide systems, *Surf. Interface Anal.* 41, 324–332, 2009.

5.2 L. Qiu, G. Xu, Peak overlaps and corresponding solutions in the X-ray photoelectron spectroscopic study of hydrodesulfurisation catalysts, *Appl. Surf. Sci.* 256, 3413–3417, 2010.

5.3 $\frac{\delta(\Delta H)}{\delta x_S} = \frac{\delta[N_S \int(\Delta E)\delta x_S]}{\delta x_S} = \Delta E \cdot \mathrm{N}_s = \mathrm{N}_s - \alpha[2p(x_s - x_b) - mx_b + m/2] + m(E_{BB} - E_{AA})/4\}$.

The calculation of $\frac{\delta(\Delta S)}{\delta x_S}$ requires knowledge of the entropy of mixing. According to statistical thermodynamics, the entropy of mixing of a homogeneous mixture of N atoms of two components with molar fraction x is $S = -kN[x \ln x + (1 - x) \ln(1 - x)]$. The entropy of mixing of two elements to a surface of composition x_s and a bulk of composition x_b thus is

$$S_1 = -kN_s[x_s \ln x_s + (1 - x_s) \ln(1 - x_s)] - kN_b[x_b \ln x_b + (1 - x_b) \ln(1 - x_b)].$$

Because of the surface enrichment of A in the surface, the bulk concentration x_b of A is lower than x_s and the overall, average concentration x of A is in between:

$$x \equiv x_s - \lambda \equiv x_b + \Lambda, \text{ with } N_{Ab}\Lambda = N_{As}\lambda \ (\equiv \text{ means by definition}).$$

Replacing x_b by $x - \Lambda$ gives

$$x_b \ln x_b = (x-\Lambda)\ln(x-\Lambda) = (x-\Lambda)\ln x(1-\Lambda/x) = (x-\Lambda)[\ln x + \ln(1-\Lambda/x)]$$

and if $\Lambda/x \ll 1$,

$$\ln(1-\Lambda/x) = -\Lambda/x \quad \text{and} \quad (x-\Lambda)\ln(1-\Lambda/x) = -\Lambda.$$

This gives

$$x_b \ln x_b = x\ln x - \Lambda \ln x - \Lambda,$$

and similarly

$$(1-x_b)\ln(1-x_b) = (1-x)\ln(1-x) + \Lambda\ln(1-x) + \Lambda.$$

Substitution in the equation for the entropy S_1 gives

$$\begin{aligned} S_1 = & -kN_s[x_s\ln x_s + (1-x_s)\ln(1-x_s)] \\ & -kN_b[x\ln x - \Lambda\ln x - \Lambda + (1-x)\ln(1-x) + \Lambda\ln(1-x) + \Lambda], \end{aligned}$$

and with

$$S_0 = -k(N_s + N_b)[x\ln x + (1-x)\ln(1-x)],$$

one obtains

$$\begin{aligned} \Delta S = S_1 - S_0 &= kN_s[x\ln x + (1-x)\ln(1-x) - x_s\ln x_s \\ &\quad -(1-x_s)\ln(1-x_s)] + kN_b\Lambda\ln x/(1-x) \\ &= kN_s[x\ln x + (1-x)\ln(1-x) \\ &\quad -x_s\ln x_s - (1-x_s)\ln(1-x_s) + \lambda\ln x/(1-x)]. \end{aligned}$$

Because

$$\begin{aligned} \delta\lambda/\delta x_s &= \delta(x_s - x)/\delta x_s = 1, \\ \delta(\Delta S)/\delta x_s &= kN_s[-\ln x_s - 1 + \ln(1-x_s) + 1 + \ln x/(1-x)] \\ &= kN_s\ln[x/(1-x)\cdot(1-x_s)/x_s]. \end{aligned}$$

Questions

5.1 Why are XPS and AES spectroscopy measurements always carried out under vacuum?
5.2 Auger spectra often have several lines. For instance, when an electron is removed from the 1s level of the oxygen anion ($1s^22s^22p^6$) in MgO, the Auger spectrum has three lines. Why?
5.3 The surface sensitivity of XPS is greatest when the kinetic energy of the ejected electron is about 50 eV. This cannot be realised for all elements by XPS in the laboratory. Why not? How could XPS be carried out in order to obtain the best surface sensitivity for many elements?
5.4 The XPS spectrum of a piece of gold shows sharp 4f and 4d peaks, each followed by a rather intense tail at the high binding-energy side (Fig. 5.6). How does the 4f and 4d spectrum of a monoatomic layer of gold on another metal, e.g. platinum, look like?
5.5 Is it possible to distinguish between the N atoms of NH_4NO_3 with XPS?
5.6 Why is the surface enrichment of rough surfaces (kinks and steps) greater than the surface enrichment of smooth planes?
5.7 Why is a Pd layer on top of Ag unstable?
5.8 What is the surface composition of an In-Sn (1:1) alloy at room temperature and at 500 °C?
5.9 Why is zinc a metal? Why does it have a low melting point?

Chapter 6

Metal Catalysis

Metal catalysts are used in many reactions because they have the ability to dissociatively adsorb molecules, let the fragments react at their surfaces and allow product molecules to desorb (Sabatier, Noble Prize in Chemistry 1912, together with Grignard). Thus, in hydrogenation and oxidation reactions metals play an important role as catalyst. For instance, the initial steps in the reaction of H_2 and CO to hydrocarbons (the Fischer–Tropsch reaction) do not take place between molecules but between atoms. H_2 is not reactive, but H atoms are radicals and are very reactive. The CO bond is strong and must be broken before the C and O atoms can react with the H atoms. Fe and Co do this and are, therefore, used in the Fischer–Tropsch process (Section 6.4.1.3). Fe is also used as a catalyst in the Haber–Bosch process, the hydrogenation of N_2 to ammonia (Haber, Nobel Prize in Chemistry in 1918 and Bosch in 1931, together with Bergius, for high pressure chemistry). This reaction is one of the most important catalytic reactions and is of fundamental importance in the fertiliser industry, sustaining about one third of the world population (Section 6.5). Ni, Pd and Pt are used in the hydrogenation of unsaturated bonds (Section 6.2). Ni hydrogenates vegetable oils to fats and CO to methane (Section 6.4.1.2); in the reverse reaction, Ni reforms methane with steam to synthesis gas (Section 6.3). In the refinery, Pd is involved in the hydrogenation of ethyne (acetylene) to ethene, and Pt improves the octane number of gasoline in the catalytic reforming process (Section 7.5.3). Cu is the catalyst in the formation of methanol from synthesis gas (Section 6.4.2). In oxidation, Pt and Rh catalyse the oxidation of ammonia to NO_x, which is further converted to nitric acid (Section 9.2.2). Ag catalyses the oxidation of methanol to formaldehyde (Section 9.3.1) and Pt and Rh are the catalysts for treating exhaust gas, by reacting NO to N_2 and CO to CO_2 (Section 9.1.2).

P. Sabatier

V. Grignard

F. Haber

C. Bosch

F. Bergius

6.1 Dissociation of H_2

Under usual conditions, H_2 molecules do not react with unsaturated molecules in the absence of a catalyst. In the presence of a catalyst that dissociates H_2 to H atoms, hydrogenation reactions are already fast below room temperature because H atoms react very fast with unsaturated molecules. Metals are ideal catalysts for hydrogenation because they adsorb H_2 dissociatively. MO theory explains why a metal surface splits H_2 in two H atoms, even though the binding energy of H_2 is $436\,\text{kJ}\cdot\text{mol}^{-1}$.

As discussed in Section 3.3.1, the two 1s AOs of the H atoms of a H_2 molecule combine to a bonding $\sigma(H_2)$ MO occupied by two electrons and an empty $\sigma^*(H_2)$ antibonding MO. When a H_2 molecule approaches the surface of a transition metal with the H–H axis parallel to the surface, the $\sigma(H_2)$ MO interacts with the s, p_z, and d_z^2 AOs of a single metal atom or with suitable combinations of the s, d_z^2 and d_{xz} orbitals of two metal atoms (Figs. 6.1a and 6.1b).

The filled $\sigma(H_2)$ MO and the empty symmetric d(Acc) metal AO combine to a bonding $[\sigma(H_2) - d]$ and antibonding $[\sigma(H_2) - d]^*$ MO pair (Fig. 6.1c). The energy of the bonding $[\sigma(H_2) - d]$ combination is lower than the energy of the $\sigma(H_2)$ orbital; the occupation of this MO with two electrons results in an attractive interaction between the hydrogen molecule and the metal surface. The new orbital is a mixture of the H_2 orbital and the metal orbital. Thus, electron transfer from H_2 to the metal takes place, and the strength of the H–H bond decreases and that of the M–H bond increases.

The $\sigma^*(H_2)$ orbital (antisymmetric with respect to the z axis) interacts with a d_{xz} or d_{yz} AO on one metal atom or with antisymmetric

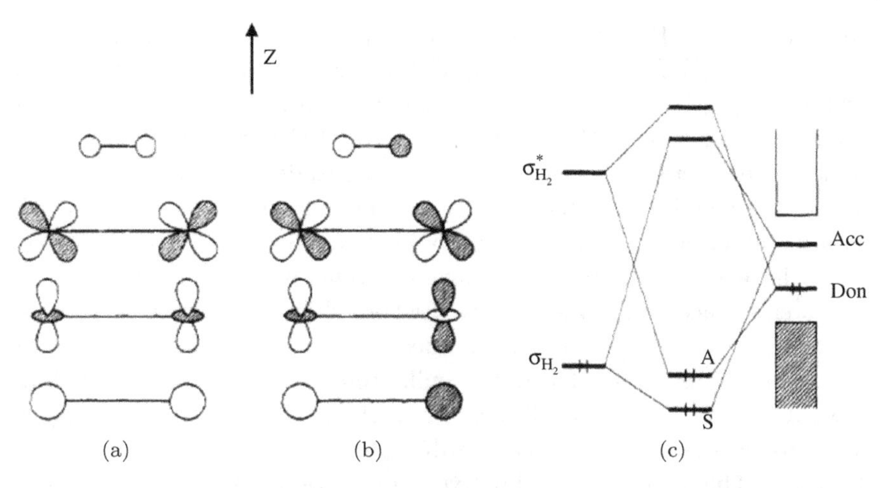

Fig. 6.1 Interaction between $H_2\sigma$ and σ^* MOs and metal *s*, *p* and *d* AOs. Reprinted with permission from [6.1]. Copyright (1984) the American Chemical Society.

combinations of *s*, d_z^2 and d_{xz} orbitals of two metal atoms (Fig. 6.1b) to give an MO pair $[\sigma^*(H_2) - d]$ and $[\sigma^*(H_2) - d]^*$ (Fig. 6.1c). Occupation of the bonding $[\sigma^*(H_2) - d]$ MO (denoted A in Fig. 6.1c) with electrons results in an attractive interaction between the H_2 molecule and the metal surface. The energy of the $[\sigma^*(H_2) - d]$ MO A is lower than the energy of the d(Don) orbital (Fig. 6.1c) and not all of the electron density in the new MO A is on the metal. As a consequence, there is a back-donative interaction with an electron flow from metal orbitals to the σ^* antibonding H_2 MO. The H–H bond strength decreases and the M–H bond strength increases. Both interactions lead to weakening of the H–H bond and formation of the M–H bond, even though they occur by charge transfer in opposite directions [6.1].

As the H_2 molecule approaches the metal surface, the donation and back-donation interactions become stronger and increase the H–H bond distance until the H–H bond breaks. The $[\sigma(H_2) - d]$ and $[\sigma^*(H_2) - d]$ orbitals as well as the $[\sigma(H_2) - d]^*$ and $[\sigma^*(H_2) - d]^*$ orbitals then have the same energy; the difference in the energy of the σ and σ^* orbitals disappears. At the same time, combinations of two electrons from the metal with two electrons from the H_2 molecule are used to form the MH bonds.

The relative position of the Fermi level ε_d of the metal (the highest energy at which metal orbitals are filled) determines the strength of the back-donating interaction. In Section 3.3.1 we saw that the interaction is proportional to $(\mathrm{H}_{AB})^2/\Delta E_{AB}$, where H_{AB} is the exchange interaction $\mathrm{H}_{AB} = \int A\left(\frac{e}{r}\right) B\ dx$ and ΔE_{AB} is the energy difference between the orbitals. As ε_d increases in energy (moves upward), the energy difference between the d and $\sigma^*(H_2)$ level decreases and the strength of the back-donation interaction increases. Because ε_d depends on the coordination of the surface metal atoms, the chemisorption energy of an atom or molecule on a metal surface depends on the topology of the metal surface. Surface atoms have fewer nearest neighbours (n) than bulk atoms and, thus, a narrower surface band (width $\sim 2n\beta$, where β is the interaction between neighbouring atoms, cf. Section 5.4). Because the Fermi level of the surface atoms must be equal to that of the bulk atoms ε_d, the surface band shifts upward. There is also a small increase in the occupation of the d valence electrons on the surface as the coordination number of the surface atom decreases. The increased coordinative unsaturation of the surface atoms is reflected by an upward shift in the ε_d and in a greater contribution of the back-donating interaction; H_2 molecules interact more strongly with rough surfaces and step edges.

The Fermi energy ε_d decreases from left to right in the Periodic Table. As a consequence, the H(1s)-M(d) antibonding states for Cu and Au are lower than ε_d; the antibonding states are filled and cause repulsion. This leads to a higher activation energy for H_2 adsorption on Cu and Au surfaces and explains why the adsorption of H_2 on Cu, Ag and Au is activated, while adsorption on group 8 metals has no activation energy. On Au the dissociative adsorption of H_2 is even endothermic (Section 3.3.2, Fig. 3.16). Bonding on metal atoms with a lower coordination number is stronger than bonding on nine-coordinated metal atoms, as in a (111) plane. Therefore, dissociative H_2 chemisorption on stepped and kinked Au surfaces and on small Au particles is weakly exothermic. Cu and Au catalysts can, therefore, be used for the selective hydrogenation of molecules that contain different unsaturated groups, such as acrolein ($CH_2{=}CH{-}CH{=}O$) and cinnamaldehyde ($C_6H_5{-}CH{=}CH{-}CH{=}O$) [6.2].

Within the accuracy of the DFT method, the heat of adsorption of H_2 on the Pt(111) surface ($Q_{\mathrm{ads}} = E(\mathrm{H_2}) + E(\mathrm{Pt}) - 2E(\mathrm{H{-}Pt})$) is $80\,\mathrm{kJ{\cdot}mol^{-1}}$ for all surface sites (H atoms adsorbed on atop, bridge or threefold hollow sites, Fig. 6.2). The activation energy for hopping (diffusion) of a hydrogen atom between these sites is 2 to $13\,\mathrm{kJ{\cdot}mol^{-1}}$ and indicates that the potential energy surface is quite flat [6.3]. Because the binding energy of H_2 is $436\,\mathrm{kJ{\cdot}mol^{-1}}$, the binding energy of a hydrogen atom on the Pt(111) surface is $(-436 - 80)/2 = -258\,\mathrm{kJ{\cdot}mol^{-1}}$ H.

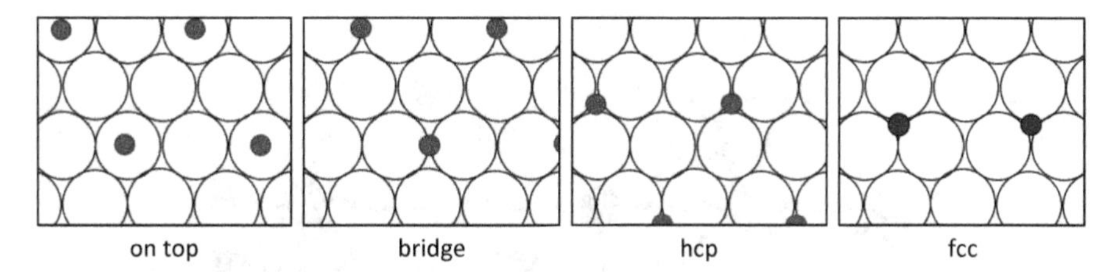

Fig. 6.2 Atop, bridge and threefold hollow (hcp and fcc) sites on a flat metal surface.

6.2 Hydrogenation of Ethene

Having answered the question how a H_2 molecule dissociates to H atoms on a metal surface, the next question is how the H atoms react with unsaturated molecules. Catalytic hydrogenation by metals started at the end of the 19th century with small unsaturated hydrocarbons such as ethyne (acetylene) and ethene (Sabatier) and is still important today. Scientists are still fascinated by this topic, because ethene and ethyne are simple molecules, industrialists remain interested because of many possible applications. Ethene chemisorbs on a transition metal surface in π or di-σ mode [6.4]. In the π mode, ethene sits atop a single metal atom, while in the di-σ mode it binds parallel to a bridge metal–metal bond, forming two σ metal–carbon bonds (Fig. 6.3). The Dewar–Chatt–Duncanson model explains the π interaction of ethene with the surface (Section 3.3.1, Fig. 3.15). Electron donation takes place from the π orbitals of ethene to the empty d states on the metal surface, and back donation of electron density occurs from the occupied metal d band into the antibonding π^* orbital. This leads to elongation of the C–C bond from 1.30 Å in the gas phase to 1.395 Å on the Pd(111) surface and a binding energy of ethene to the Pd(111) surface of $-27\,\text{kJ}\cdot\text{mol}^{-1}$. The binding of ethene on Pd(111) is stronger in the di-σ mode ($-62\,\text{kJ}\cdot\text{mol}^{-1}$) [6.6]. In the di-$\sigma$ mode, there is substantial back donation from the surface and the C=C bond elongates to 1.45 Å, thus becoming more like a C—C single bond. Furthermore, the H atoms are no longer in the plane of the carbon atoms but are bent away from the metal surface as in sp^3 hybridisation of the carbon atoms (Fig. 6.3). The favoured type of adsorption depends on the electronic and geometric structure of the metal surface as well as on the reaction conditions; on Ni(111), Pd(111) and Pt(111) it is the di-σ bridge mode.

The hydrogenation of ethene occurs according to the **Horiuti–Polanyi mechanism** by consecutive surface-catalysed additions of hydrogen

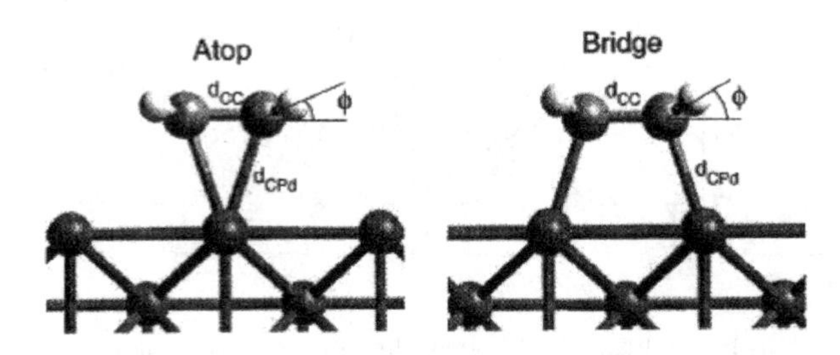

Fig. 6.3 Ethene chemisorbed on atop and bridge sites. Reprinted with permission from [6.5]. Copyright (2002) Elsevier.

atoms [6.7],

$$H* + C_2H_4 \rightarrow C_2H_5*, \tag{6.1}$$

and

$$H* + C_2H_5* \rightarrow C_2H_6 + 2*, \tag{6.2}$$

where the asterisk $*$ denotes an adsorption site. DFT calculations indicate that, at low surface coverage (when adsorbed species are far apart), π-bonded ethene easily diffuses over the surface to a site where it can form the energetically more stable di-σ surface complex, without undergoing hydrogenation. Hydrogenation from the di-σ adsorption state is, therefore, the only reaction path at low coverage. The first reaction step is to add a H atom to adsorbed ethene to form an ethyl surface intermediate (Eq. (6.1)). The reaction energy profile calculated by DFT for this reaction step on Pd(111) is shown in Fig. 6.4a [6.6]. The apparent activation energy is only $26\,\text{kJ}\cdot\text{mol}^{-1}$ because, for a Langmuir–Hinshelwood mechanism at low surface coverage of ethene, the heat of adsorption lowers the intrinsic activation energy on the surface to an apparent activation energy of $E_{\text{app}} = E_0 - Q_A = 88 - (30 + 32) = 26\,\text{kJ}\cdot\text{mol}^{-1}$ (Eq. (4.5)).

In the past, reactions at surfaces were studied in surface-science equipment at low pressure, where the coverage of the surface is low, and theoretical calculations were performed in the same coverage regime. Lately, however, interest in *in situ* (operando) conditions has grown and studies at higher surface coverages have become standard. On the experimental side this was stimulated by new techniques that became available and, on the theoretical side, progress in computational techniques made it possible to calculate reaction of molecules at surfaces in the presence of other species. Such experimental and theoretical studies have shown that reactions at low and high surface coverage can be totally different.

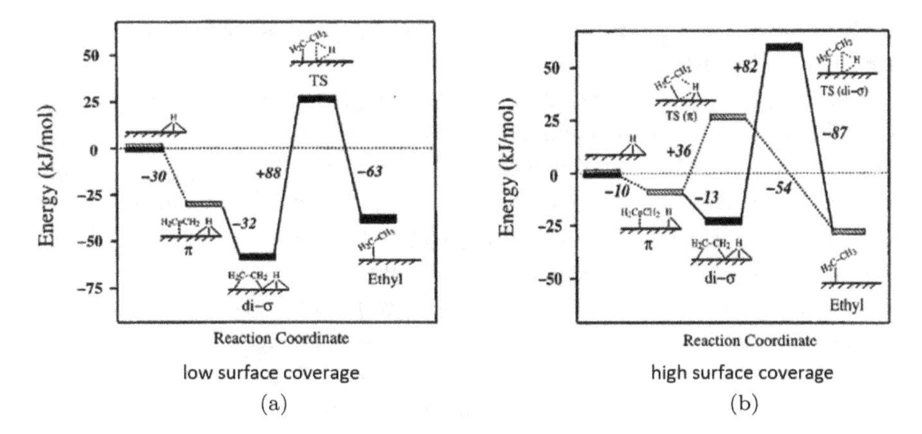

Fig. 6.4 Reaction profiles for the reaction of ethene to an ethyl fragment on Pd(111) at low (a) and high (b) surface coverage. Reprinted with permission from [6.6]. Copyright (2000) the American Chemical Society.

For instance, as we will see in the following for the hydrogenation of ethene, at lower coverage, weakly π-bonded species desorb or are converted into more stable di-σ surface intermediates, which are responsible for reaction. At higher coverage, the relatively weakly π-bonded species can hardly convert to stable intermediates and are responsible for reaction. Another example of the influence of surface coverage, the dissociation of CO on a metal surface, will be discussed in Section 6.4.1.1.

The reaction of ethene on Pt was studied by infrared-visible sum frequency generation spectroscopy (SFG). In this technique, a tunable infrared laser and a visible light laser with constant frequency are directed at the interface between two media and an output beam is generated at a frequency which is the sum of the two input beams. Only vibrations of molecules at the interface are observed, not from molecules in the gas or liquid phase above the catalyst surface. The C–H stretching frequencies of di-σ and π-bonded ethene and ethylidyne (CH_3–C$*$, a decomposition product of ethene) were observed with IR-visible SFG *in situ* during ethene hydrogenation on Pt(111) at atmospheric pressure and room temperature [6.8]. Simultaneous measurements of the reaction rate with a gas chromatograph allowed to correlate the reaction rate and surface adsorbate concentration. Higher surface coverage of ethylidyne decreased the coverage of di-σ-bonded ethene, but hardly influenced the surface coverage of π-bonded ethene and the ethene hydrogenation rate. Therefore, it was suggested that π-bonded ethene and not di-σ-bonded ethene is the key intermediate for ethene hydrogenation, although the di-σ-bonded ethene binds stronger to Pt(111) than π-bonded ethene.

The explanation of this, at first sight, surprising result is the presence of other adsorbates, which change the adsorption energy, activation energy and reaction selectivity of ethene [6.6]. Through-space repulsive interactions dominate when ethene shares metal atoms with a carbon atom, but when ethene and the carbon atom are separated by a single metal bond, the through-surface interactions increase the stability of ethene in the π-bonded mode. In the di-σ mode, the interactions are predominantly repulsive and weaken ethene σ adsorption. These interactions change not only the stability of the intermediates but also the kinetics. Ethylidyne and other CH_x surface intermediates cover most of the surface of Pd and Pt under reaction conditions but are only spectator species (do not react) [6.4]. The concentration of di-σ ethene is lower in the presence of ethylidyne but the surface coverage of the π-bonded ethylene intermediate is hardly influenced [6.8]. The more dominant reaction pathway at 1 atm is, therefore, the hydrogenation of π-bonded ethene [6.9]. DFT calculations of the hydrogenation of ethene on Pd(111) at high coverage of H atoms showed that lateral repulsive interactions between di-σ-bonded ethene and a H atom weakened the di-σ binding energy of ethene from -62 to $-23\,\text{kJ}\cdot\text{mol}^{-1}$ (cf. Figs. 6.4a and 6.4b) [6.6]. At high surface coverage, the activation energy of the pathway by the π-bonded intermediate (Fig. 6.4b, dashed lines) is lower than that of the pathway by di-σ-bonded ethene (solid lines).

Repulsive interactions also occur for π-adsorption of ethene. The π-bonded species, however, require only a single metal atom adsorption site, while the di-σ-intermediate require two metal atom sites. At high coverage, the repulsive surface interactions force the H atom from the threefold hollow into the bridge site, and the H and π-bonded ethene do not share metal atoms and, thus, there are no lateral repulsive interactions. Co-adsorbed di-σ-bonded ethene and a bridge-bonded (or threefold bonded) H, on the other hand, always share surface metal atoms; the lateral interactions are repulsive and both adsorbates are less strongly chemisorbed.

At high coverage, ethene is π-adsorbed between hydrogen and other surface intermediates and cannot, therefore, move to a neighbouring site and convert to the di-σ-intermediate. As the bridge-bonded H atom approaches the π-bonded ethene, it undergoes weak repulsive interactions, not strong enough to remove ethene from the surface. During reaction ethene slides upward to form a "five-centre" type of transition state among the two carbon atoms, the H atom, and two adjacent Pd atoms (Fig. 6.4b).

Hydrogenation of ethene from this π-bonded state at high coverage on Pd(111) occurs at greater ease, with an intrinsic activation energy of only $36\,kJ{\cdot}mol^{-1}$ (Fig. 6.4b), significantly lower than the activation energy for hydrogenation of di-σ-bonded ethene ($82\,kJ{\cdot}mol^{-1}$). At higher coverage of ethene, the intrinsic activation energy is equivalent to the apparent activation energy since the reference state is then the adsorbed state (Eq. (4.7)). This pathway is accessible only at high surface coverage, requires at least three metal atom centres and is, therefore, unsuitable for homogeneous catalysts with a single metal atom.

Strong repulsive interactions between the adsorbates at high coverage weaken the metal-C and metal-H bonds, enable the facile insertion of H into the metal-C bond and, thus, lower the activation energy. A subsurface H atom has the same effect; it weakens the interaction of adsorbates with the metal surface, thus lowering the intrinsic activation energy for hydrogenation. In general, a weaker metal-adsorbate bond lowers the intrinsic activation energy for hydrogenation as well as other coupling reactions. The reverse is true for bond dissociation where a strong adsorbate-metal bond generally lowers the activation energy. This is evident from the reverse activation energy for C—H bond breaking of ethyl to form di-σ-ethene. At lower coverage the activation energy is only $63\,kJ{\cdot}mol^{-1}$, whereas at higher surface coverage it is $87\,kJ{\cdot}mol^{-1}$ (Fig. 6.4).

At normal pressure the hydrogenation of ethene occurs by means of π-bonded ethene. The relatively weakly bonded π-bonded ethene is reactive only at higher coverage because at lower coverage it desorbs or is converted to more stable di-σ-bonded ethene before it reacts. This illustrates the importance of surface coverage effects on the intrinsic activation energies and, hence, on the controlling reaction mechanism. It also indicates that caution must be exercised in creating catalytic reaction pathways based on information obtained only for the most stable adsorption intermediates; *in situ* experiments should also be performed.

6.3 Synthesis of CO and H_2

Gasoline- and diesel-powered vehicles and kerosene-powered airplanes rely on fuels, which consist of a mixture of hydrocarbons and differ in boiling point range and degree of branching. Gasoline has a boiling point range of 40–200 °C and contains branched hydrocarbons to reach a high octane number, kerosene has a boiling point range of 150–270 °C and diesel has a boiling point range of 250–350 °C and contains unbranched hydrocarbons to reach a high cetane number. Fuels are produced from oil by distillation and further upgrading, but alternative sources are attracting increasing attention because sources of clean oil are scarce, whereas coal and natural gas are abundant. Biomass can also be a source for fuels. However, the whole fuel-energy picture may change dramatically in the coming one or two decades if electricity-driven cars take over from fuel-driven cars.

The H:C ratio of coal is about 1, while the H:C ratio of hydrocarbons is between 1 (aromatics) and slightly above 2 (alkanes). Transformation of coal to fuels, thus, requires the addition of hydrogen or the removal of carbon. The addition of hydrogen is the preferred route. In the direct addition of hydrogen, coal is heated with hydrogen in the presence of a catalyst. This **Bergius process** was developed at the beginning of the 20th century. Because the impurities in coal end up on the catalyst, the catalyst is not recycled and must, therefore, be inexpensive (e.g. iron sulfide). In another version of the Bergius process a hydrogen transfer agent, such as tetralin (tetrahydronaphthalene), is heated with the coal. The agent transfers hydrogen atoms to coal molecules and tetralin becomes naphthalene. The naphthalene is distilled out of the resulting hydrogenated coal liquid and is fed into a separate reactor where it is hydrogenated with H_2 over a catalyst. This catalyst can be regenerated rather easily because it does not come into contact with the coal and is not contaminated by the impurities. The catalyst can be more expensive and more active (MoS_2). A Bergius plant in China uses an Fe catalyst and produces 1.1 MT/y liquid products and two other plants are under construction.

F. Fischer H. Tropsch

Coal can also be converted indirectly to fuel. The coal is first gasified with steam to synthesis gas ($CO+H_2$) (Section 1.3), and the synthesis gas is used in the **Fischer–Tropsch (FT) reaction** to produce alkanes, alkenes and alcohols (Section 6.4.1.3). Synthesis gas can be obtained by gasification of coal as well as of natural gas (methane), oil fractions and biomass. Just as the Bergius process, the FT process for the production of fuel was developed in Germany at the beginning of the 20th century. Further development of the FT process took place in South Africa, which, like Germany, has coal reserves but no oil. The increasing demand for transportation fuel and access to large natural gas reserves have increased interest in FT synthesis for the production of "clean" fuels, especially diesel (no sulfur, no aromatics), by the conversion of synthesis gas. Malaysia (1993) and Qatar (2011), with large reserves of natural gas, have introduced the FT process to convert natural gas via synthesis gas to liquid fuels in a gas-to-liquids (GTL) process.

The production of synthesis gas has been discussed in Section 1.3 and the use of synthesis gas for the production of methane (methanation) and higher hydrocarbons (FT reaction) will be discussed in Section 6.4. In both reactions, CO and H_2 dissociate on a metal catalyst surface and the resulting C and H atoms react to hydrocarbon molecules. Therefore, before discussing the methanation and FT reactions, the dissociation of CO will be discussed in Section 6.4.1.1.

6.4 Hydrogenation of CO

6.4.1 *CO Hydrogenation to Hydrocarbons*

6.4.1.1 *CO dissociation*

The dissociation of CO is the first elementary reaction step in the FT reaction and in the methanation reaction. Combined catalytic activity measurements on supported catalysts, ultra-high vacuum (UHV) experiments on well-defined single crystals and DFT calculations of the dissociation mechanism of CO on Ni surfaces demonstrated that the hydrogenation of CO over Ni is highly structure sensitive [6.10]. **Structure sensitivity** has been defined as: "The turnover frequency (TOF) of structure-sensitive reactions under fixed conditions depends strongly on surface crystalline anisotropy as expressed on clusters of varying size or on single crystals exposing different faces" [6.11]. The CO methanation rate at 250 °C, 0.01 bar CO and 0.99 bar H_2 could be fitted with an exponent of 2.6 as a function of 1/d, the inverse of the Ni particle diameter [6.10]. Catalytic activity depends on the number of active sites, and the activity is proportional to 1/d when all the surface atoms of the catalyst particles are active. When only the edge atoms are active the activity will be proportional to $1/d^2$ and when only corner or kink sites are active it will be proportional to $1/d^3$.

The exponent of 2.6 indicates that terraces are inactive for methanation, whereas low-coordinated surface sites are active, proving that methanation is a highly structure-sensitive reaction, in accordance with the observation that steam reforming is structure-sensitive.

The DFT calculations [6.10] showed that the surface coverages under UHV and industrial conditions are very different. Under UHV conditions (at low CO coverage and in the absence of hydrogen), dissociation of CO is strongly favoured at defects with a low coordination number on the Ni(111) surface: The activation energy on the (111) terrace was 275 kJ·mol^{-1}, but was only 165–185 kJ·mol^{-1} at defects. The activation energy for hydrogen-assisted CO dissociation at low CO coverage was even lower ($\sim$120 kJ·mol^{-1}), but the free activation energy was high, due to the large loss of entropy at the very low hydrogen pressure. Therefore, dissociation of CO at low coverage in the absence and presence of hydrogen is difficult. Under methanation conditions the situation is different than under UHV conditions (Fig. 6.5). The most favourable reaction site for CO dissociation on Ni at high CO coverage in the absence of hydrogen is a double-step site with a high activation energy of 185 kJ·mol^{-1} [6.10]. The presence of

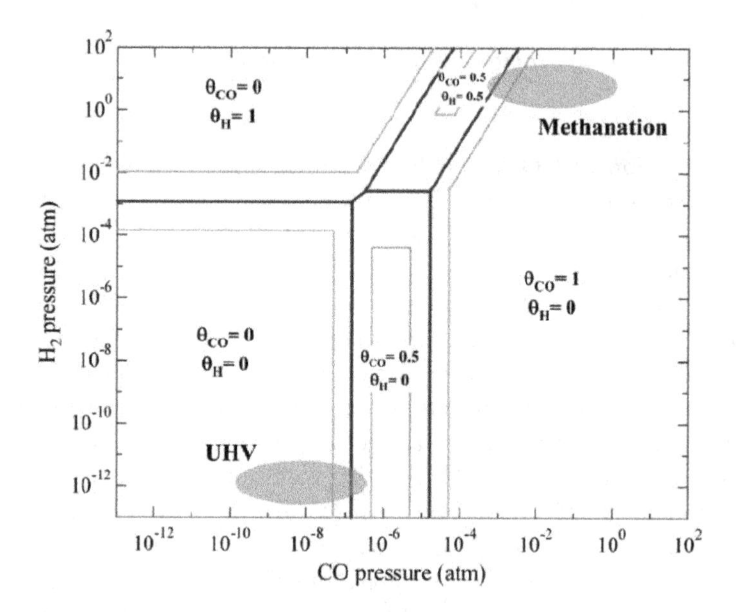

Fig. 6.5 Calculated surface coverage of CO and H with the lowest free energy as a function of H_2 and CO pressure at 230 °C. UHV and methanation conditions are shaded in grey. Reprinted with permission from [6.10]. Copyright (2008) Elsevier.

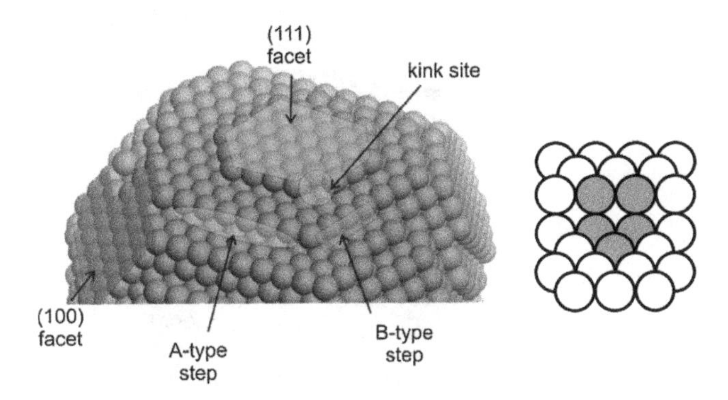

Fig. 6.6 Left: Metal particle terminated by (111) and (100) facets and A- and B-type steps. Reprinted with permission from [6.12]. Copyright (2014) Elsevier. Right: (210) surface with B5 site composed of a threefold hollow site on the (111) plane and a bridge site on the (100) step plane.

hydrogen (as in industrial methanation), significantly reduces the activation energy by causing C—O bond stretching as the CO—H intermediate forms. The resulting effective activation energy is only 104 kJ·mol^{-1} at a double-stepped Ni surface. Such steps serve as a model for edges as well as corners on a supported catalyst (Fig. 6.6).

The easier dissociation of CO on steps is a general phenomenon. Surface metal atoms have fewer neighbour atoms than bulk metal atoms and hence have a narrower d band (Section 6.1). To match the Fermi energies of the bands of surface and bulk metal atoms, the narrower surface band is shifted up in energy relative to the bulk band and this causes a stronger back-bonding interaction with antibonding MOs of adsorbing molecules. Edge surface atoms have less neighbours than terrace surface atoms and, thus, interact more strongly with the reactants than terrace sites. Hence, molecule fragments adsorbed on edge sites will have a lower energy than adsorbed on terrace sites (Fig. 6.7). When the potential energy surfaces of the two adsorption states do not differ much, then also the transition state for edge adsorption will be lower in energy than that for terrace adsorption. This is the case for many reactions on metals because they have late transition states, meaning that the geometries of the adsorbed molecular fragments and metal surface in the transition state resemble the final state more than the initial state [6.13]. The interaction between the adsorbed fragments and the surface will then be similar in transition and final states. As a consequence, the activation energy from initial state to transition state is lower for edge site adsorption than for terrace-site adsorption (Fig. 6.7).

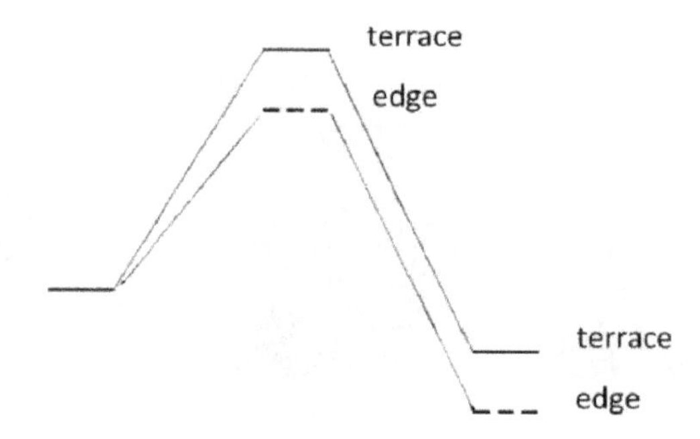

Fig. 6.7 Energy diagram for the dissociation reaction AB → A + B through adsorption on an edge or terrace site.

Fig. 6.8 CO adsorption on terrace site (left) and on a B5 step edge site (right). Reprinted with permission from [6.13]. Copyright (2003) the American Chemical Society.

In addition to the attraction between molecular fragments A and B and surface metal atoms, there will be repulsion when A and B share bonding with surface atoms. The repulsion is stronger when atoms A and B are adsorbed at two neighbouring threefold hollow sites (so that A and B share a surface atom) than in a situation in which the A and B atoms are further apart. For instance, in the transition state for dissociation of CO on a metal surface, the C and O atoms share bonding to one metal atom on a terrace site. On a stepped (211) surface, the C atom can adsorb on a threefold hollow site at the bottom of the step, while the O atom can adsorb on a bridge site at the top of the step (Fig. 6.8). On this B_5 site the C and O atoms have no metal atoms in common, and there is no repulsion [6.13]. As a result, the transition state for adsorption of CO on a terrace site will move up more in energy than the transition state for adsorption of CO on an edge site (cf. Fig. 6.7) and it will be even more likely that CO dissociation occurs on edge metal sites and not on terrace metal sites.

6.4.1.2 *Methanation*

The C and O atoms that form by dissociation of CO on the metal surface react with adsorbed H atoms. Oxygen atoms are removed from the surface as water; DFT calculations of the removal of O atoms from a stepped Co(0001) surface indicate that the removal occurs in elementary reaction steps from O to OH to H_2O at surface steps [6.14]. The activation energy for the removal of an O atom in the presence of O or OH groups on neighbouring metal atoms was 31 kJ·mol^{-1} for the first step, 106 kJ·mol^{-1} for the second step and 62 kJ·mol^{-1} for the third step

$$\mathrm{O*} + \mathrm{O*} + 3\,\mathrm{H*} \rightarrow \mathrm{O*} + \mathrm{OH*} + 2\,\mathrm{H*} + \;* \qquad (6.3)$$

$$\mathrm{O*} + \mathrm{OH*} + 2\,\mathrm{H*} \rightarrow 2\,\mathrm{OH*} + \mathrm{H*} + \;* \qquad (6.4)$$

$$2\,\mathrm{OH*} + \mathrm{H*} \rightarrow \mathrm{H_2O*} + \mathrm{OH*} + \;*, \qquad (6.5)$$

while the lowest activation energy for O removal of isolated O and H atoms (low surface coverage) was 154 kJ·mol^{-1}. Therefore, the DFT results suggest that the formation of water in the hydrogenation of CO occurs readily only at high surface coverage and that the addition of water facilitates the removal of oxygen.

The simplest reaction that C atoms can undergo is the methanation reaction. The hydrogenation of CO follows a Horiuti–Polanyi reaction scheme, with dissociative adsorption of CO and H_2 on the metal surface, followed by step-by-step reaction to the adsorbed products and desorption. The energy diagram (Fig. 6.9) gives the results of a DFT calculation for a stepped Ni(211) surface [6.15]. H atoms and CH_3 groups tend to be located atop a metal atom, while reacting CH_x groups ($x = 0 - 2$) are found on threefold hollow sites on the flat surface and a fourfold hollow site on the stepped surface. The activation energies of each elementary step on both flat and stepped surfaces are small and similar, as are the energy profiles. Therefore, CH_x hydrogenation is not structure-sensitive.

More important than energy is free energy because it determines the thermodynamics and kinetics. Figure 6.10 shows the free energy calculated with DFT of the reactants, intermediates and transition states for hydrogenation of CO at high CO coverage over a Ni(211) surface [6.10]. Adsorbed C atoms form by hydrogen-assisted dissociation of CO at edge sites, and hydrogenation of CH_x intermediates occurs on the terraces because the edge sites are blocked by CO. The dissociation of a COH intermediate on a step site is rate-determining below 500 °C. A model in which CO dissociation was assisted by hydrogen was consistent with the kinetic results of the

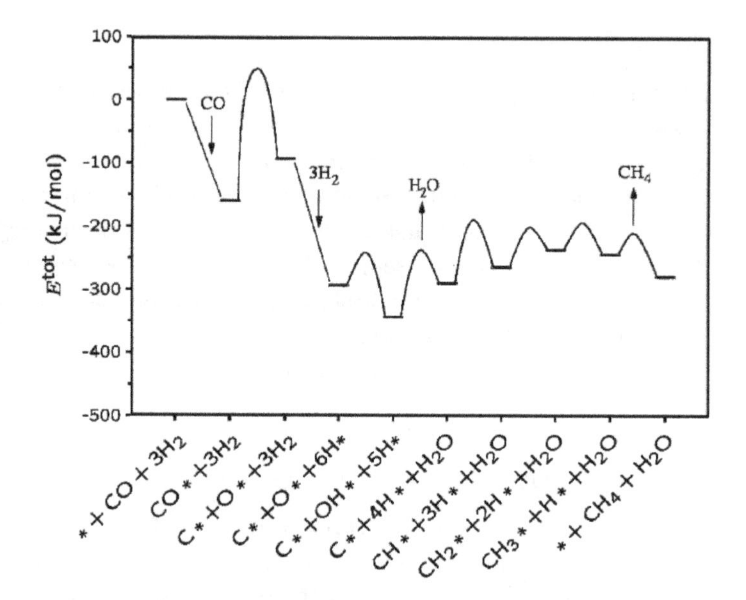

Fig. 6.9 Calculated potential energy diagrams for CO hydrogenation over stepped Ni. Reprinted with permission from [6.15]. Copyright (2004) Elsevier.

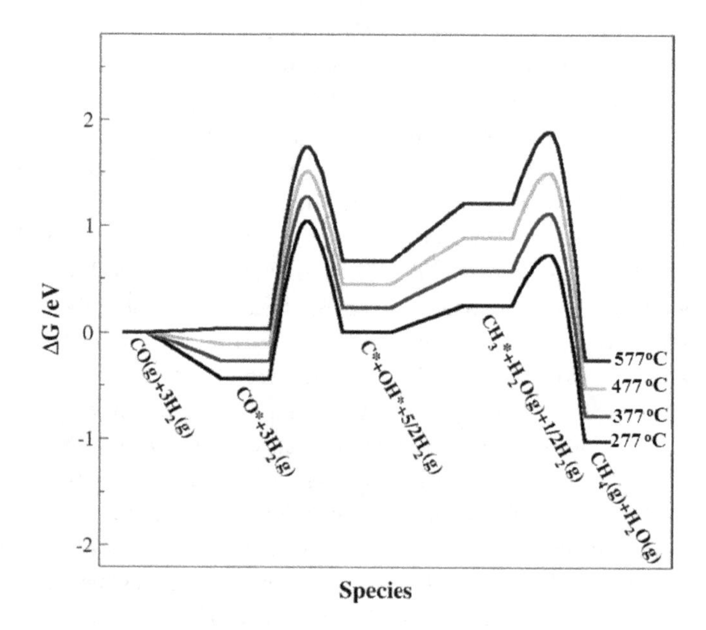

Fig. 6.10 Free energy of the methanation reaction on Ni(211) at high CO coverage. CO dissociation via COH is rate-limiting until 500 °C. Reprinted with permission from [6.10]. Copyright (2008) Elsevier.

methanation of CO: the experimentally observed activation energy, the slightly negative order in CO partial pressure (−0.3–0.5) and the positive order in hydrogen pressure (0.6–0.8).

6.4.1.3 *Fischer–Tropsch reaction*

CO can not only be hydrogenated to methane, but also to higher alkanes, alkenes and alcohols:

$$n\,\mathrm{CO} + (2n+1)\mathrm{H_2} \rightarrow \mathrm{C}_n\mathrm{H}_{2n+2} + n\,\mathrm{H_2O} \tag{6.6}$$

$$n\,\mathrm{CO} + 2n\,\mathrm{H_2} \rightarrow \mathrm{C}_n\mathrm{H}_{2n} + n\,\mathrm{H_2O} \tag{6.7}$$

$$n\,\mathrm{CO} + 2n\,\mathrm{H_2} \rightarrow \mathrm{C}_n\mathrm{H}_{2n+1}\mathrm{OH} + (n-1)\mathrm{H_2O}. \tag{6.8}$$

The product is determined by the catalyst and the conditions (CO/H_2 ratio, temperature). For example, on iron more alkenes and alcohols form than on cobalt. The Fischer–Tropsch synthesis (FTS) of alkanes consists of several reaction steps:

- initiation by CO dissociation and CH_x formation,
- propagation via CH_x–CH_x coupling reactions,
- termination and desorption of the hydrocarbons,
- removal of the oxygen atoms.

The FTS of alkanes proceeds through CO adsorption, C–O bond cleavage and incorporation of the formed C_1 species into a growing chain:

$$\begin{array}{ccccccccccc} & k_0 & & k_p & & k_p & & k_p & & k_p & \\ \mathrm{H_2}+\mathrm{CO} & \rightarrow & C_1* & \rightarrow & C_2* & \rightarrow & C_3* & \rightarrow & C_4* & \rightarrow & \cdots \\ & & \downarrow k_t & & \downarrow k_t & & \downarrow k_t & & \downarrow k_t & & \\ & & C_1 & & C_2 & & C_3 & & C_4 & & \end{array}$$

The FTS can be treated as a polymerisation reaction with a stationary state, in which the number of active sites is constant and the rates of the propagation reaction (r_p) and the termination reaction (r_t) are constant. In that case, the FTS can be modelled with the Anderson–Schulz–Flory model for polymerisation reactions. For a chain of i carbon atoms, the chance of chain growth α_i is equal to the ratio between the rate of propagation and the sum of the rates of propagation and termination

$$\alpha_i = \frac{r_{p,i}}{r_{p,i} + r_{t,i}},$$

while the chance that the chain is terminated by desorption as an alkene or hydrogenation to an alkane is equal to $1 - \alpha_i$. If the rates of propagation $r_{p,i}$ and termination $(r_{t,i})$ of chain fragments C_i* at the catalyst surface are independent of the chain length, then α is independent of i. In that case,

$$c_{i+1}/c_i = \alpha \quad \text{and} \quad c_i = \alpha c_{i-1} = \alpha^2 c_{i-2} = \cdots = \alpha^{i-1} c_1,$$

and

$$n_i = \frac{N_i}{\sum_i N_i} = \frac{c_i}{\sum_i c_i} = \frac{\alpha^{i-1} c_1}{\sum_i \alpha^{i-1} c_1} = \alpha^{i-1}(1-\alpha), \tag{6.9}$$

where N_i is the number of molecules with chain length i and $\Sigma_i N_i$ is the total number of molecules. The resulting number distribution n_i is referred to as the Flory distribution (P. Flory, Nobel Prize in Chemistry, 1974). Ignoring the beginning and end of the polymer chain, the molar weight is equal to the weight m_0 of the monomer times a whole number and the weight distribution function w_i becomes

$$w_i = \frac{W_i}{\sum_i W_i} = \frac{i m_0 N_i}{\sum_i i m_0 N_i} = \frac{i N_i}{\sum_i i N_i} = \frac{i\alpha^{i-1}}{\sum_i i\alpha^{i-1}} = i\alpha^{i-1}(1-\alpha)^2. \tag{6.10}$$

The number distribution n_i decreases continuously with i, but the weight distribution function w_i goes through a maximum (Fig. 6.11).

Reactions that follow the Anderson–Schulz–Flory mechanism always give a broad distribution of products; when a molecule with i monomer units forms, a molecule with $i + 1$ units can also form. FT reactions with high α

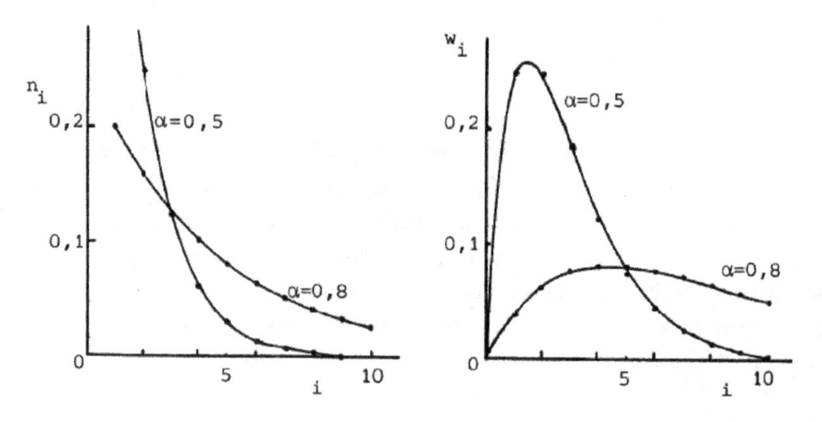

Fig. 6.11 Anderson–Schulz–Flory functions of number and weight distribution.

have a very broad product distribution (Fig. 6.11, right, $\alpha = 0.8$), while FT reactions with low α have a narrow product distribution. The only product that can be made with 100% selectivity is methane because then $r_p = 0$ and $\alpha = r_p/(r_p + r_t) = 0$. Experimental distribution curves deviate from the Anderson–Schulz–Flory distribution because the chain growth factor α is about constant for long chains but decreases for shorter chains from $\sim C_{10}$ to C_3. Furthermore, the first two hydrocarbons do not follow the Anderson–Schulz–Flory distribution: ethane falls below and methane above the Anderson–Schulz–Flory distribution curve. The chain growth factor α depends on temperature. At high temperature the FT process gives shorter products (Sasol, Secunda, South Africa) than at low temperature (wax) or medium temperature (wax, diesel).

If methanation is required, then the formation of higher hydrocarbons and alcohols must be suppressed; chain growth should be slow, C_1 hydrogenation fast (low α) and CO dissociation slow. CO dissociation is often rate-determining because it generally has a higher activation energy than the CH_x—CH_y coupling, and the reaction order in CO will be close to -1. The preferred metal is Ni, with a weak M—C bond and high activation energy for CO dissociation (higher than on Fe, Co and Ru). If higher hydrocarbons (gasoline, kerosene, diesel) are required, then termination should be rate-limiting and CO dissociation and C_1 formation should be fast. If CO dissociation is fast relative to chain termination, long chains will form and the reaction order in CO is close to zero.

Fe, Co and Ru are highly selective catalysts for FTS; Fe and Co are used industrially. The iron catalyst is prepared by reducing iron oxide in synthesis gas. Under process conditions the catalyst becomes a mixture of iron oxide, iron and iron carbide phases, of which the Hägg carbide χ-Fe_5C_2 is regarded to be the active phase. The Co catalyst is used in supported form and consists of metallic Co particles supported on silica. It is prepared by reducing an oxidic Co/SiO_2 precursor in hydrogen. The H_2/CO ratio of synthesis gas from coal is lower (≤ 1) than required for the FT reaction. Because iron oxide is a water–gas-shift catalyst (Section 1.3.3) and can increase the H_2/CO ratio, iron-based catalysts are used when coal is used as the source of synthesis gas. Cobalt is used when methane (natural gas) is the feedstock because the synthesis gas made from methane has a higher H_2/CO ratio (~ 3).

The high-temperature FT process (330–350 °C) is carried out with a Fe catalyst that yields C_1–C_{15} hydrocarbons and a large fraction of α-olefins and oxygenates. Co or Fe catalysts are used in low-temperature

FT technology (200–250 °C) to produce linear long-chain hydrocarbons; the main product is (ultra-clean) diesel and jet fuel. In the FT process of Shell in Qatar (140,000 barrels/day) very long hydrocarbons are produced from natural gas by FTS over Co/SiO_2 and these are hydrocracked (Section 8.1.3) to gasoline, kerosene and diesel. In the Synfuels process in Ningxia (China, 100,000 barrels/day) coal is gasified to synthesis gas, which then undergoes FTS over an Fe catalyst at medium temperature (~275 °C). Efficient removal of heat from the reactor is very important since FT reactions are highly exothermic. In the reactors of the Shell FT process in Qatar, the heat is removed in multi-tubular fixed-bed reactors. These tubes contain the catalyst and are surrounded by boiling water which removes the heat of reaction. Sasol uses a slurry reactor in their process in Qatar, with internal cooling coils. The synthesis gas is bubbled through the hydrocarbon products and the finely-dispersed catalyst, which is suspended in the liquid medium.

FTS is technologically well-developed but the reaction mechanism, in which the simple molecules hydrogen and CO are converted into a mixture of many hydrocarbons, is very complex and still under study: How are the CO molecules dissociated, which hydrocarbon fragments play a role in the chain growth, does the chain growth includes oxygen-containing fragments, which sites on the catalyst surface play a role, what are the rate determining steps? DFT calculations showed that CO dissociation on flat metal terraces requires an activation energy higher than 200 $kJ{\cdot}mol^{-1}$, higher than experimentally observed in FTS. A lower activation energy of 100–150 $kJ{\cdot}mol^{-1}$ was calculated for CO dissociation on step sites on Co and even lower on Ru, or for CO dissociation assisted by H atoms on terraces. An older proposal that CO dissociation takes place after insertion of CO into the growing chain has become less popular. Today CO dissociation on Co and Ru is generally believed to take place on step edge sites [6.16].

A second subject of debate is the mechanism of chain growth. Which surface structure is optimal for chain growth, what is the chemical nature of the inserting monomer and the coupling intermediate? A number of proposals involving insertion of species such as $CH_x (x = 0, 1, 2)$ or CO into growing chains of various types (e.g. alkyls, alkenyls) have been made. In a mechanism involving CH_x species, C atoms are formed on stepped sites and hydrogenated to methylidyne species, CH, which are involved in a reversible chain growth process [6.17]. The CH fragments can migrate over the surface to a neighbouring site on a terrace. Whether chain growth takes place on edge or terrace sites is still under discussion [6.18–6.21]. Mechanisms

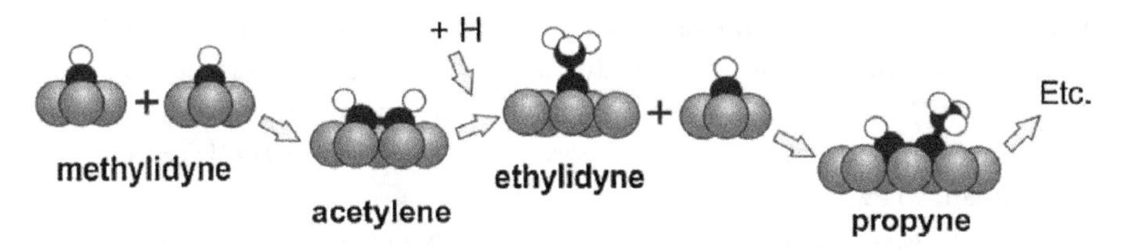

Fig. 6.12 Mechanism for FT chain growth: After CO and H_2 dissociation and hydrogenation of C atoms to methylidyne, CH, methylidynes dimerise to chemisorbed acetylene, CH + CH → HC≡CH, and this is hydrogenated to ethylidyne HC≡CH + H → C–CH_3. Ethylidyne reacts with methylidyne C≡CH_3 + CH → HC–C–CH_3. The chain continues by alternating hydrogenations and reactions with methylidyne. Reprinted with permission from [6.21]. Copyright (2016) Elsevier.

for chain growth are based on acetylene-type structures because DFT calculations indicate that such unsaturated structures are more stable than hydrogenated alkyl or alkylene structures [6.18, 6.20, 6.21]. After formation of methylidyne, CH, by CO dissociation and hydrogenation of the C atom, methylidyne groups can dimerise to acetylene, HC≡CH, which is hydrogenated to ethylidyne, C–CH_3. Ethylidyne then couples with CH to form propyne, HC≡C–CH_3, and so on (Fig. 6.12). A fairly large number of surface sites might be involved in the growth of a single chain. Step sites produce CH_x monomers which spillover onto a terrace, where several chains grow at the same time. Diffusion of surface species is essential in the overall reaction sequence.

The availability of a synchronised pathway for CO dissociation and C–C coupling with low activation energy supports the proposal that B_5 sites are the unique reaction centres for a high yield of chain growth FT products. Only on such sites does CO dissociate with activation energies lower than those typical of chain termination (80–100 kJ·mol^{-1} for Co), so that the relative rate of CO dissociation is fast enough to supply the C_1 building blocks required for the chain growth reaction.

The increased selectivity of methane formation and the decrease in the reaction rate with decreasing Co particle size are consistent with the proposal that CO dissociates on a B_5 site. When metal particles are too small, B_5 sites are no longer stable, leading to an increase in the CO dissociation activation energy and competition between CO dissociation and hydrocarbon removal from the surface; the overall rate of the reaction decreases and more short-chain hydrocarbons are produced.

Another mechanistic point that is under discussion is the rate of removal of chemisorbed oxygen from the metal surface. Some scientists believe that oxygen removal is fast, others think it might be slow. The apparent activation energy for water formation from Co surfaces is about 130 kJ·mol^{-1}. This high value indicates that oxygen removal via water formation cannot *a priori* be considered as a fast step in the FTS reaction mechanism [6.22].

DFT studies of the surface of χ-Fe_5C_2 (which is supposed to be the active part of an iron FT catalyst) and of reactions taking place at its surface have been carried out. One study explored a mechanism starting with hydrogenation of surface carbon [6.23]. On a surface with partial C coverage, direct and H-assisted CO activations are feasible, the latter being the preferred path. The catalytically active surface is dynamic: carbon atoms of the carbide surface participate in the surface reactions and CO dissociation on vacant sites leads to restoration of the carbide structure. Another study identified several feasible CO dissociation pathways over the Hägg carbide [6.24]. The CO dissociation rate depended on the presence and topology of interstitial C atoms close to the active site. CO dissociation proceeded preferentially via direct C−O dissociation on the stepped sites on the χ-Fe_5C_2 surface. H-assisted CO dissociation was preferred when the surface did not have a stepped character.

6.4.2 *Hydrogenation of CO and CO_2 to Methanol*

6.4.2.1 *CO hydrogenation to methanol*

CO can not only be hydrogenated to hydrocarbons but also to oxygenates. For instance, on Rh catalysts a mixture of hydrocarbons, alcohols and aldehydes is formed. The formation of higher alcohols and aldehydes suggests that also the formation of oxygenates follows a polymerisation mechanism. However, while there is a large market for a mixture of hydrocarbons (fuel), there is no market for mixtures of oxygenates. As a result, industrially only the production of methanol has been realised: $CO + 2\,H_2 \rightarrow CH_3OH$ (about 40 MT/y). Methanol is used to produce formaldehyde (for phenol and urea resins), acetic acid, methyl-t-butyl ether (MTBE), methylamines and methylated molecules. The formation of methanol is exothermic ($\Delta H = -90.7$ kJ·mol^{-1}) and, thus, requires elevated pressure; industrial conditions are 200–300 °C and 5–10 MPa. The industrial $Cu/ZnO/Al_2O_3$ catalyst contains Cu and ZnO nanoparticles and about 10% Al_2O_3 as a **structural promoter** [6.25].

Although in-depth research has been carried out on the mechanism of the methanol synthesis, several aspects, such as the nature of the catalytically active sites and the role of copper and zinc oxide are still research topics. There is general agreement that metallic copper is the active component, either as pure Cu sites or as Cu–Zn alloy sites. The substantial increase in catalytic activity of Cu by ZnO can be explained in several ways: ZnO may influence the dispersion and morphology of the Cu particles and may stabilise Cu particles with more active defect sites. It may also provide Zn atoms for the substitution of Zn in the Cu steps, which strengthens the binding of the intermediates and increases the activity of the catalyst [6.26]. Furthermore, it was proposed that the catalyst is bifunctional and that ZnO provides basic sites [6.27]. Hydrogen atoms, created by chemisorption of H_2 on Cu particles, would migrate to the ZnO and hydrogenate HCOO–Zn formate species (formed from CO and Zn–OH) to CH_3O–Zn methoxide species. Hydrolysis would then free methanol.

Methanol synthesis is based on associative (non-dissociative) chemisorption of CO. Originally, it was assumed that methanol forms from CO, but isotope labelling experiments under industrial conditions showed that CO_2 is actually the main carbon source [6.25, 6.28]. Because water forms in the production of methanol from CO_2, H_2O can react with CO in the water–gas-shift (WGS) reaction to reform CO_2. The WGS reaction is fast on Cu catalysts and, therefore, only a small amount of CO_2 or H_2O in the feed is necessary to start methanol production:

$$\mathrm{CO_2 + 3\,H_2 \rightarrow CH_3OH + H_2O} \tag{6.11}$$

$$\underline{\mathrm{CO + H_2O \rightarrow CO_2 + H_2}} \tag{6.12}$$

$$\mathrm{CO + 2\,H_2 \rightarrow CH_3OH.} \tag{6.13}$$

A DFT study indicated that both CO and CO_2 may act as a source of carbon atoms in the methanol reaction. At low conversion, the hydrogenation of CO_2 had a higher thermodynamic driving force in the industrial temperature range than the hydrogenation of CO [6.29], and the WGS reaction proceeded in the reverse direction ($\Delta G > 0$), converting some CO_2 to CO. At higher conversion, the ΔG of the hydrogenation of CO was more negative than that of CO_2, and the WGS reaction run in the forward direction ($\Delta G < 0$). The thermodynamics indicate that reaction conditions and conversion levels can significantly influence the competitive CO and CO_2 hydrogenation and may even reverse the direction of the WGS reaction. Below 200 °C the source of C in the methanol product shifted from CO_2 to CO [6.30]. These studies show that caution must be exercised when drawing conclusions about the main carbon source in CH_3OH.

It is assumed that the reaction from CO_2 to CH_3OH goes through adsorbed formate (HCO_2^*) that may form from adsorbed bicarbonate (HCO_3^*):

$$2\,CO_2 + H{*} + 3{*} \rightarrow 2\,CO_2{*} + H{*} + {*} \rightarrow CO{*} + O{*} + CO_2{*} + H{*}$$
$$\rightarrow CO{*} + CO_3{*} + H{*} + {*} \rightarrow CO{*} + HCO_3{*} + 2{*}$$
$$\rightarrow CO{*} + HCO_2{*} + O{*} + {*}\,.$$

A DFT calculation indicated that the hydrogenation reactions $HCO_2{*} \rightarrow H_2CO_2{*} \rightarrow H_2CO{*} \rightarrow H_3CO{*} \rightarrow H_3COH{*} \rightarrow H_3COH$ (Fig. 6.13) are probably the elementary steps in the hydrogenation of formate to methanol on a low-coordinated site on a 29-atom copper cluster [6.31]. The same reactions probably occur on stepped Cu surfaces. The activation energy for the rate-limiting step of the methanol synthesis reaction on the Cu cluster is lower than on Cu(111) (Fig. 6.13) and corresponds to a much faster reaction rate at 300 °C. The activity enhancement is due to the presence of active low-coordinated Cu sites in the nanoparticle, which significantly stabilise the intermediates, the transition states and, therefore, lower the activation energy for the rate-limiting hydrogenation step.

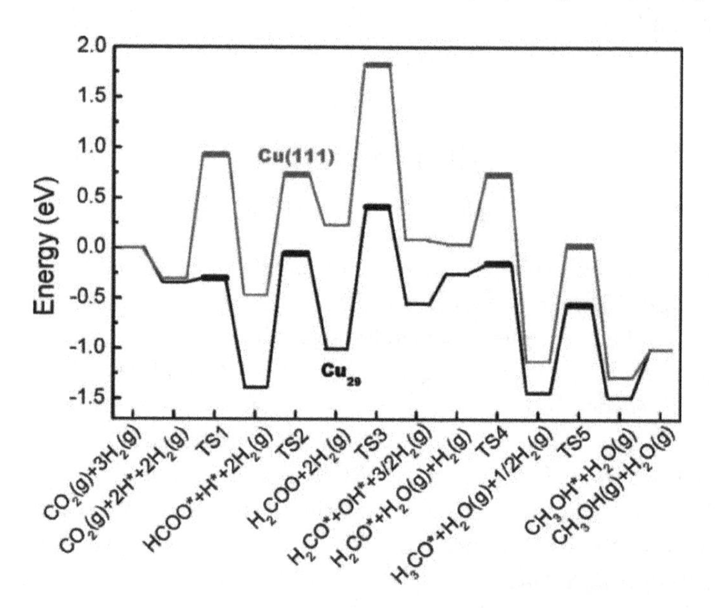

Fig. 6.13 Potential energy diagram for methanol synthesis on a Cu(111) surface (upper curve) and a Cu_{29} nanoparticle (lower curve). Reprinted with permission from [6.31]. Copyright (2010) the PCCP Owner Societies.

On the other hand, experiments conducted at lower temperature and pressure than used in industry suggest that carboxyl (COOH) is the crucial intermediate for methanol synthesis and that formate plays only a co-adsorbate spectator role [6.29]. The proposed elementary hydrogenation steps are $CO_2 \rightarrow COOH* \rightarrow C(OH)_2* \rightarrow COH* \rightarrow HCOH \rightarrow H_2COH \rightarrow CH_3OH$ [6.32].

DFT calculations of a stepped Co surface enabled a comparison of CO hydrogenation to methanol via the carbon atom (first giving HCO, then H_2CO and finally H_3CO), hydrogenation of CO via the oxygen atom (giving COH, then HCOH and H_2COH) and dissociative routes through coupling of $CH_2 + O$ to CH_2O and of $CH_3 + OH$ to CH_3OH [6.33]. All three pathways were favoured at step sites, and the preferred mechanism is $CO + 4H \rightarrow HCO + 3\,H \rightarrow H_2CO + 2\,H \rightarrow H_3CO + H \rightarrow CH_3OH$. CO and HCO bind to the surface through the C atom, H_2CO lies flat and binds through its C and O atoms and H_3CO binds through the O atom. In this mechanism, hydrogenation of H_3CO has the highest activation energy ($123\,kJ{\cdot}mol^{-1}$).

Another mechanism, based on experiments that showed that Cu crystallites with more lattice strain have a higher catalytic activity [6.27], suggests that strain leads to steps at the surface and that steps are more active for methanol formation. The effect of steps was confirmed by DFT calculations, which showed for the hydrogenation of CO_2 and CO that the energy of the intermediates and the transition-state was lower for the (211) surface than for the (111) surface, rendering the steps more active than the terraces. Alloying of Zn into the Cu step further increased the adsorption strength of HCO, H_2CO and H_3CO and decreased the activation energies. The most active surface was therefore a stepped Cu surface with Zn alloyed into the step.

Thus, although the mechanism of the hydrogenation of CO to methanol has not yet been firmly established, there is general agreement that sites on stepped surfaces are more active than sites on flat surfaces because, as for the hydrogenation of CO to methane, on B_5 sites adsorbates do not share metal atoms with other adsorbates.

6.4.2.2 *CO_2 hydrogenation to methanol*

Not only CO can be used as source for methanol, but also CO_2. CO_2 might play an important role in future energy scenarios. Today fossil fuels provide the main part of the world energy and, although use of renewable energy sources is increasing, this will continue in the short and medium term.

This massive use of fossil fuels in industry and transport produces large amounts of CO_2, which is a major contributor to climate change. CO_2 emissions reached 35 GT in 2020 but must decline to 10 GT by 2050. To decrease CO_2 emissions, one could hydrogenate CO_2 to valuable chemicals but even urea, the biggest consumer of CO_2, only has an annual consumption of 0.1 GT/y [6.34]. This is not surprising because the chemicals market is roughly an order of magnitude smaller than the energy market. Therefore, to consume CO_2, a solution in the energy domain must be found. Methanol and ammonia are possible solutions because they can be used as hydrogen-storage agents.

Methanol is an important industrial bulk chemical used to produce several industrial chemicals and fuels. For instance, the transformation of methanol to hydrocarbons is an emerging sector (Section 7.5.2). Methanol can also be used for the generation of energy. Wind and photovoltaics are forms of energy which are more sustainable than fossil fuels. Their drawback is that they depend on time of day and season of the year. Therefore, there is a need to store such energy at peak production times for use in times of low production. One way to do this is to convert such energy into chemical energy, e.g. in hydrogen. Hydrogen may be obtained by water electrolysis and, in the future, by photocatalytic water splitting. This hydrogen can, in turn, be stored in an energy dense liquid form such as methanol. The advantage of liquid methanol is that existing infrastructures for fuel transportation and distribution can be used and form the basis of the so-called "methanol economy", a concept proposed in 2005 by Olah *et al.* [6.35]. (Olah received the Nobel Prize in Chemistry in 1994). They claimed that the use of methanol could close the carbon-fuel-cycle. A 5,000 T/y methanol plant, based on CO_2 and called "George Olah renewable methanol plant", started in 2012 in Iceland, where a geothermal power station supplies the CO_2 as well as the energy (5 MW/y) needed to split water to H_2.

A reviewer of the book by Olah *et al.* [6.35] pointed out, however, that the methanol economy is not without difficulties [6.36]. He calculated that a world oil consumption of 4 MT is equivalent to 8.5 GT methanol and would need the electrolysis of 15 GT water to generate the hydrogen necessary for the production of this methanol. This would require 5,000 1-GW nuclear reactors or 1000,000 5-MW wind turbines to produce the electricity for electrolysing the water. This is about 18 times the present capacity installed in China, the country that has by far the largest wind power capacity (280 GW, yearly adding 70 GW) [6.37].

Currently, CH_3OH is produced from CO and mixtures of CO and CO_2 on copper-based catalysts, as discussed in the preceding Section 6.4.2.

The most used catalyst for CO hydrogenation is $Cu/ZnO/Al_2O_3$, but it suffers from low activity, selectivity and stability in CO_2 hydrogenation. The exothermic formation of methanol from CO_2 (Eq. (6.11)), in which the number of molecules decreases, demands a low temperature and high pressure for a high conversion. However, CO_2 is not very reactive and a temperature above 240 °C is necessary for acceptable CO_2 conversion. At the same time, selectivity is also a problem because the accompanying reverse water–gas shift reaction ($CO_2 + H_2 \rightarrow CO + H_2O$) is undesirable as it consumes hydrogen and lowers the methanol yield. The large amount of water produced by both reactions has an inhibitory effect on the catalyst because H_2O adsorbs stronger than CO_2 on the catalyst. The CAMERE process [6.38] circumvents the inhibition by first transforming the CO_2 to CO and H_2O in the reverse WGS reaction, then removing the water and finally hydrogenating CO to methanol in the classic way (Section 6.4.2.1). The catalyst used in the CAMERE process is $Cu/ZnO/ZrO_2/Ga_2O_3$.

Most of the catalysts employed in CO_2 hydrogenation are based on copper, often a modified form of the $Cu/ZnO/Al_2O_3$ catalyst developed for CO hydrogenation. The idea behind this choice is that in the hydrogenation of CO, the actual molecule that is hydrogenated is CO_2, as discussed in Section 6.4.2. Although a lot of research is being carried out on the CO_2 to methanol reaction, from an industrialist point of view, a totally satisfactory catalyst is at present not yet available [6.34].

6.5 Hydrogenation of N_2 to Ammonia

6.5.1 *Fe Catalyst*

Ammonia is produced on a very large-scale in all industrialised countries. It is required for the manufacture of fertilisers (urea, ammonium nitrate, ammonium phosphate and ammonium sulfate) and nitric acid. An estimated one third of the world's food production is based on these artificial fertilisers. Ammonia is made from N_2, produced from air, and H_2, which is produced from natural gas, oil or coal. The production of N_2, H_2 and ammonia and the role of catalysis at most stages of ammonia production was discussed in Section 1.3. In this section the focus is on the catalyst and the mechanism of ammonia formation.

Ammonia was first synthesised in industry in 1913, and until 1992 only unsupported iron was used as a catalyst. Fe is a robust industrial catalyst and is easy to prepare by reduction of magnetite (Fe_3O_4), which is inexpensive and widely available. It was discovered by chance in the

BASF laboratory at the beginning of the 20th century that impurities in the magnetite substantially increase the activity of the Fe catalyst. Al_2O_3, K_2O and CaO turned out to promote the catalyst activity. Substances which do not catalyse but increase the rate of a catalysed reaction are called **promoters.** XPS, AES and SEM showed that the promoters are strongly enriched at the surface [6.39]. High temperature (400 °C) treatment of magnetite in a N_2/H_2 mixture, corresponding to the actual reaction conditions of ammonia synthesis, reduces the iron oxide to the metallic state, while the valence state of the promoter cations is unaffected. Reduction of the Fe_3O_4 particles occurs without an appreciable change in the external volume. As a consequence, a porous structure (pore diameter 10–50 nm) with high surface area results, whereby Al_2O_3 acts as a **structural promoter** that prevents the small Fe particles from sintering; CaO plays a similar role. Upon reduction, K and O atoms tend to spread uniformly over the iron particles. The catalytically active phase is, thus, metallic Fe covered with 20–50% of a monolayer of a K + O composite. This adlayer acts as an **electronic promoter** and affects the specific activity of the iron surface.

Kinetic measurements of the rate of ammonia synthesis at high pressure on Fe single crystal surfaces showed that the Fe catalyst is **structure-sensitive** and that the order of the activity of the crystal faces was (111) > (100) > (110). Dissociative nitrogen adsorption proceeds by $N_2 \rightleftharpoons N_2{*} \rightarrow 2N{*}$ (Fig. 6.14). The initial activation energy for dissociative nitrogen adsorption is fairly low (almost zero for Fe(111)). Nevertheless, dissociative nitrogen chemisorption is a very slow process, caused by the low sticking

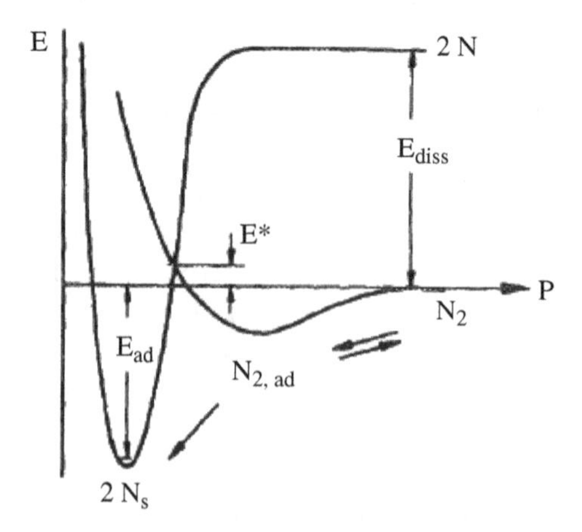

Fig. 6.14 Potential energy of dissociative adsorption of N_2 on Fe. Reprinted with permission from [6.40]. Copyright (1983) American Vacuum Society.

coefficient of the molecular state to the surface (10^{-2}) and a low probability of dissociation of physisorbed N_2 (10^{-4}) relative to desorbing again. Nitrogen adsorption is, thus, the rate-determining step, in agreement with the observation that nitrogen adsorption and ammonia formation proceed at about the same rate on actual catalysts.

Adsorption of H_2 on single crystal surfaces of Fe is dissociative with a fairly high initial sticking coefficient (~ 0.1) and an adsorption energy between about 65 and 110 kJ·mol^{-1} [6.40]. Under the conditions of ammonia synthesis, the adsorbed H atoms are very mobile on the surface, and their concentration depends essentially on their adsorption–desorption equilibrium. H coverage is fairly low at $\sim 400\,°C$, even at high pressures, and the inhibition of nitrogen adsorption plays a minor role.

The addition of alkali metals improves the activity of an ammonia synthesis catalyst. Alkali metals are **electronic promoters**, which act primarily by electrostatic interaction between the dipole moment (μ_{ads}), induced when the adsorbate is bonded to the surface, and the electrostatic field ε_{alkali}, induced by the adsorbed alkali atom [6.41]. The change in the adsorption energy induced by the promoter is $\Delta E_{int} = -\varepsilon_{alkali} \cdot \mu_{ads}$ ($\varepsilon_{Na} \sim -1V/Å$). The presence of alkali metal atoms will, therefore, stabilise molecules with a negative dipole moment on the metal surface. Because the dipole moment of N_2 on a metal surface is negative, while that of N, NH and NH_2 intermediates is positive, the transition state for the dissociation of the N–N bond will be lower in the presence of alkali metal atoms, and the transition states of the N, NH and NH_2 intermediates will increase in energy. In accordance with experimental studies, DFT calculations indicate that alkali atoms promote NH_3 synthesis, by lowering the N_2 dissociation barrier and the coverage of other intermediates along the reaction path [6.42].

The ammonia synthesis reaction follows a Horiuti–Polanyi mechanism, with dissociative adsorption of the reactants on the metal surface, followed by a step-by-step reaction to the adsorbed products and desorption.

$$N_2 + 2* \rightarrow 2N*$$

$$H_2 + 2* \rightleftharpoons 2H*$$

$$N* + H* \rightleftharpoons NH* + *$$

$$NH* + H* \rightleftharpoons NH_2* + *$$

$$NH_2* + H* \rightleftharpoons NH_3* + *$$

$$NH_3* \rightleftharpoons NH_3 + * .$$

In these reactions, $*$ is an empty surface site.

6.5.2 *Ru Catalyst*

Since 1992, Ru supported on graphitic carbon and promoted with BaO and Cs_2O has found industrial application in the Kellogg Advanced Ammonia Process (KAAP). Ru is an order of magnitude more active than promoted Fe [6.43] and ammonia synthesis can, thus, proceed at lower temperature and, as consequence, at lower pressure (cf. Section 1.3.4). However, Ru is much more expensive than Fe and is not widely available, so it is used in supported form. The support must be inert and have a high surface area for high dispersion of the Ru particles. Graphitic carbon fulfils these requirements and does not gasify to methane. Despite its higher activity, Ru catalysts are used in only 10 plants worldwide.

The ammonia synthesis reaction follows the same Horiuti–Polanyi mechanism on a Ru catalyst as it does on a Fe catalyst, with dissociative adsorption of the reactants on the metal surface followed by step-by-step reaction to the adsorbed products and desorption. The reaction $N_2 + 3H_2 \rightarrow 2\ NH_3$ on a Ru(0001) surface is exothermic by $174\,\text{kJ}\cdot\text{mol}^{-1}$ (1.8 eV), as calculated by DFT [6.15, 6.42]. The activation energy for the rate-determining dissociation of N_2 decreased from $183\,\text{kJ}\cdot\text{mol}^{-1}$ (1.9 eV) for the flat Ru(0001) surface to $39\,\text{kJ}\cdot\text{mol}^{-1}$ (0.4 eV) for the stepped Ru surface, while the activation energy of the hydrogenation steps changed only slightly (Fig. 6.15). Step sites are much more reactive than terrace sites and, on real catalysts, the reaction mainly occurs on step sites. The experimental value for the activation energy of N_2 dissociation on a Ru(0001) surface is $36\,\text{kJ}\cdot\text{mol}^{-1}$ but the value calculated by DFT is $169\,\text{kJ}\cdot\text{mol}^{-1}$. This discrepancy was explained by the presence of a small number of steps on the flat (0001) surface. Metal surfaces cannot be prepared with 100% precision and a (0001) surface may have an average step density of about 1%. To confirm this suggestion, small amounts of Au were deposited on a Ru(0001) surface [6.44]. The gold decoration substantially suppressed the nitrogen coverage and N_2 dissociation rate. This indicates that the low activation barrier, found for the (assumed ideal) Ru(0001) surface, was dominated by fewer than 1% steps, which are blocked by Au deposition (cf. Section 5.3 for the explanation why gold decorates step sites).

On the flat as well as on the stepped surface, one N atom of the N_2 molecule is at the threefold hollow site, while the other N atom is at a bridge position. The step site is strongly favoured because the two N atoms do not share Ru atoms as nearest neighbours and the transition state configuration avoids indirect repulsive interactions that are responsible for the high activation energy on the terrace. The difference in activation energy is,

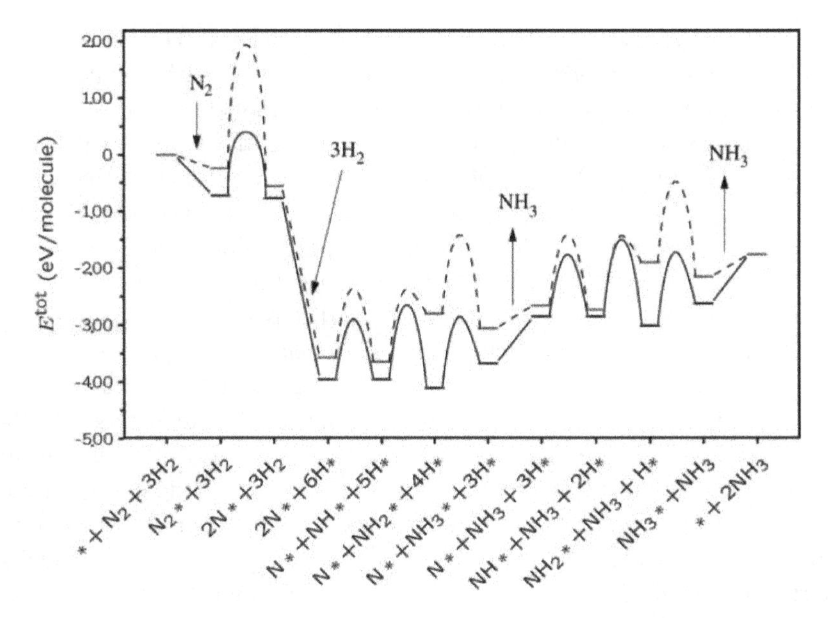

Fig. 6.15 Calculated energy diagram for NH_3 synthesis over a flat (dashed lines) and stepped (solid lines) Ru(0001) surface. Reprinted with permission from [6.42]. Copyright (2003) Elsevier.

therefore, partly due to the fact that a B_5 site with five Ru atoms, rather than a site with four Ru atoms on the terrace, is associated with the transition-state complex. For the same reason, the dissociation of CO (which is iso-electronic with N_2) takes place on step sites (Section 6.4.1.1). Particle models predict that metal crystallites smaller than ca. 1.5 nm do not possess B_5 sites and that the probability for B_5 sites has a maximum for particles of 2–2.5 nm and decreases for larger particles [6.45]. Results obtained for Ru catalysts on different supports are in accordance with this **structure sensitivity**.

6.6 Volcano Curves

Whereas all metals, except gold, can dissociate H_2 and act as hydrogenation catalyst, not all metals are good catalysts for dissociating N_2 and for reacting H and N atoms to ammonia. The best metal can be established by measuring the activities of all metal catalysts, as was done at the beginning of the 20th century when developing a catalyst for ammonia synthesis. Today, this is achieved much quicker by high-throughput experimentation in many parallel reactors. Another possibility is to perform DFT calculations of all reactants, intermediates, products and their transition states for all metals of interest. However, even with the aid of modern computers this is a formidable task. Therefore, scientists have looked for ways of calculating this for one metal by DFT and determining the parameters for other metals by empirical relations. If a linear relationship existed between the energy of a reaction intermediate and the adsorption energy of one of its constituting atoms, then it would be possible to estimate the adsorption energy of adsorbed species. Such relationships will be discussed qualitatively below.

The centre of gravity of the d band decreases in energy when going from left to right in the transition series — e.g. from Co to Cu, from Rh to Ag and from Ir to Au. This is a consequence of the ineffective shielding of the nucleus by the outer electrons. In the Periodic Table, the nuclear charge of neighbouring elements differs by one unit, as does the number of electrons. If the additional electron were to be in an orbital outside the orbitals of the other electrons, then it would be in the field of a nuclear charge of $Z - (Z - 1) = +1$. In reality, however, the orbital of the added electron is partly located inside the orbitals of the other electrons and will experience an effective nuclear charge greater than one. Therefore, the electron is attracted more strongly by the nucleus, and the energy of the d level decreases, as does the centre of gravity of the d band.

Although the band filling increases from left to right in the transition series, the decrease in the band centre is larger, and the Fermi level decreases on the right side of the transition series [6.46]. The decrease in the centre of the d band leads to a weaker interaction between the $2\pi*$ CO orbitals and the metal d orbitals, to a weaker metal to CO bond and to a stronger C≡O bond of adsorbed CO from left to right in the transition series [6.47]. Whereas CO adsorbs strongly on metals on the left side of the transition series, adsorption is weaker on metals on the right side. The general trend of weakening of metal-adsorbate bonding from left to right in the transition series is not only valid for associative chemisorption but also for dissociative adsorption. Dissociation of CO occurs with ease on metals on the left side, while only associative chemisorption takes place on metals on the right side. The final state for CO dissociation, with a C atom and an O atom on the surface of a metal on the left side in the transition series, has a low energy because the metal surface binds the C and O atoms strongly. Metals on the right side in the transition series bind C and O atoms weakly and the final state will be higher in energy. Because the final state energy is higher, also the transition energy of CO dissociation on a metal on the right side in

Table 6.1 Activation energies (in kJ·mol^{-1}) for CO dissociation on M(111).

Fe	166	Co	251	Ni	355	Cu	517
Ru	227	Rh	315	Pd	424	Ag	592
Os	227	Ir	336	Pt	419	Au	581

the transition series will be higher than for a metal on the left side. This is analogous to the chemisorption of CO on edge and terrace sites (Fig. 6.7).

Metals on the left side bind the C and O atoms very strongly and the activation energy for desorption of the products will be very high. Metals on the right side bind the adsorbate and product less strongly. Adsorption and desorption are no longer a problem, but the activation energy for the surface reaction is high. Metals in the middle bind adsorbate and product neither too strongly, nor too weakly. The activation energy for CO dissociation at the surfaces of Fe, Ru, Os and Co is significantly lower than at the surfaces of other group 8 metals (Table 6.1). The high reactivity of Fe leads to the formation of bulk iron carbides during the FT process. Os is not used in industry because OsO_4 (a gas, which would form during regeneration of an Os catalyst) is very toxic. The selectivity of Rh for lower oxygenates is attributed to the relatively low rate of CO dissociation, which results in a high probability of termination of the growing alkyl or alkenyl chain through reaction with CO.

The catalytic activity of several metals for CO dissociation as a function of the adsorption energy shows that metals to the left and right in the transition series are not as efficient as metals in between, such as Ru and Co (Fig. 6.16) [6.48]. Re, Pt and Pd are weak catalysts because either they bind C and O atoms too strongly (Re) or bind CO too weakly (Pt, Pd). Fe, Ni and Rh are in the middle. Note that the activity axis is logarithmic. Nevertheless, Ni is the catalyst of choice in industry because of cost (Ni : Ru : Co = 1 : 200 : 1.7) and availability (Ni : Ru : Co = 1 : 10^{-5}: 0.03). The curve in Fig. 6.16 is a **volcano curve**. The explanation that bonding is too strong on one side and too weak on the other was originally given by Sabatier (**Sabatier principle**). In Russia the volcano curve is referred to as the **Balandin curve**.

Also, the difference in selectivity in the hydrogenation of CO can be explained by a volcano curve. At the right side of the transition series, Cu has a low adsorption energy for carbon and is inactive. To the left in the transition series, the surface changes from a low activity surface,

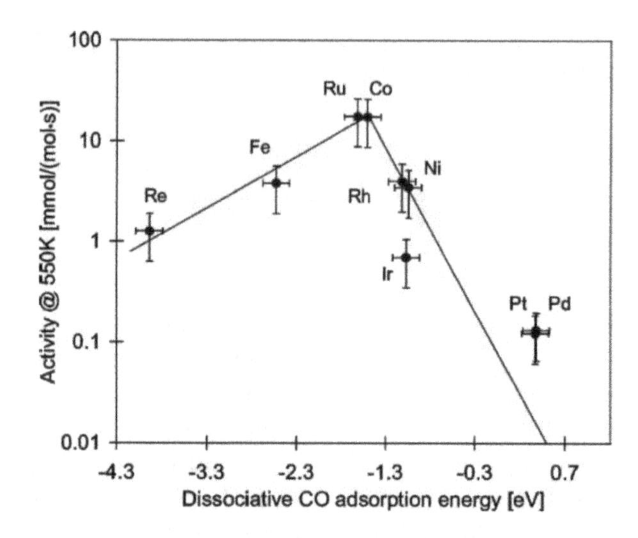

Fig. 6.16 Activity of different supported transition metals as a function of the reaction energy for dissociative CO chemisorption. Reprinted with permission from [6.15]. Copyright (2004) Elsevier.

on which hardly any CO dissociates but CH_x intermediates are rapidly hydrogenated (Ni), to a surface that can dissociate CO, has a slightly lower rate for CH_x hydrogenation and a higher α and CO conversion. Further left in the transition series, surface reactivity increases further, the rate of the chain termination is controlling and the chain-growth parameter α is high.

The reverse of the hydrogenation of CO to methane is the **steam reforming** of methane to synthesis gas ($CH_4 + H_2O \rightarrow CO + 3\,H_2$). Because the forward and backward reaction of a catalytic reaction follow the same path (Section 1.3.2), knowledge gained for methanation can be applied to steam reforming and vice versa. The energy of the reactants, intermediates and final products in steam reforming was calculated according to DFT (Fig. 6.17) [6.49]. The overall steam reforming reaction is energetically uphill (endothermic), but entropy increases ($\Delta n > 0$) and the change in free energy can, thus, be negative when the reaction is run at high temperature. Experiments showed that metals are the catalysts of choice and that activity decreases in the order Ru $\sim$ Rh $>$ Ni $\sim$ Ir $\sim$ Pt $\sim$ Pd. The thermodynamic data (Fig. 6.17) explain this order in the reactivity of metals in steam reforming (and in CO hydrogenation) qualitatively. The energy from $CH_4 + H_2O(g)$ to $C* + O*$ strongly increases for the noble metals Au, Ag and Cu because C and O atoms bind only weakly to these

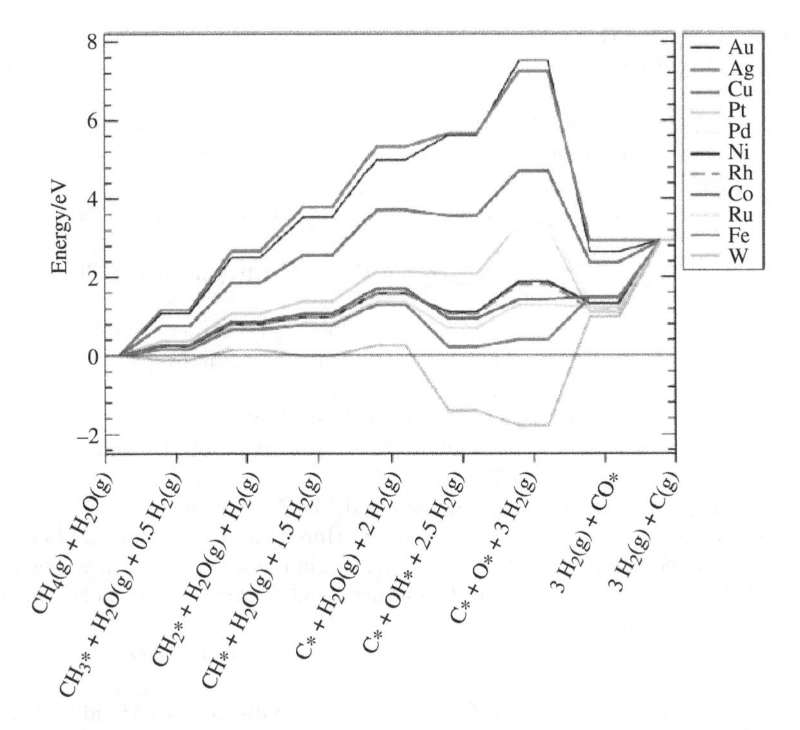

Fig. 6.17 Potential energy for steam reforming, determined from the linear-scaling relationships over metals ranging from noble to highly reactive. Reprinted with permission from [6.49]. Copyright (2008) Elsevier.

metal surfaces. Thus, these metals are unsuitable for steam reforming. The other extreme is W, which binds the C and O atoms so strongly that further reaction to CO and H_2O is slow. W is oxidised during steam reforming! Pt, Pd, Ni, Rh, Co and Ru have a more favourable energy profile, with Ni, Rh, Co and Ru having the best profiles. DFT calculations indicate that Pt and Pd have high activation energy for the adsorption and dissociation of water molecules and are, therefore, less efficient than Ni, Rh, Ir, Co and Ru.

References

6.1 J. Y. Saillard, R. Hoffmann, C–H and H–H activation in transition metal complexes and on surfaces, *J. Am. Chem. Soc.* 106, 2006–2026, 1984.

6.2 P. Claus, Heterogeneously catalysed hydrogenation using gold catalysts, *Appl. Catal. A* 291, 222–229, 2005.

6.3 G. Papoian, J. K. Nørskov, R. Hoffmann, A comparative theoretical study of the hydrogen, methyl, and ethyl chemisorption on the Pt(111) surface, *J. Am. Chem. Soc.* 122, 4129–4144, 2000.

6.4 F. Zaera, On the mechanism for the hydrogenation of olefins on transition-metal surfaces: The chemistry of ethylene on Pt(111), *Langmuir* 12, 88–94, 1996.
6.5 Q. Ge, M. Neurock, Correlation of adsorption energy with surface structure: Ethylene adsorption on Pd surfaces, *Chem. Phys. Lett.* 358, 377–382, 2002.
6.6 M. Neurock, V. Pallassana, R. A. van Santen, The importance of transient states at higher coverages in catalytic reactions, *J. Am. Chem. Soc.* 122, 1150–1153, 2000.
6.7 J. Horiuti, M. Polanyi, Exchange reactions of hydrogen on metallic catalysts, *Trans. Faraday Soc.* 30, 1164–1172, 1934.
6.8 P. S. Cremer, X. Su, Y. R. Shen, G. A. Somorjai, Ethylene hydrogenation on Pt(111) monitored *in situ* at high pressures using sum frequency generation, *J. Am. Chem. Soc.* 118, 2942–2949, 1996.
6.9 F. Zaera, T. V. W. Janssens, H. Oeffner, Reflection absorption infrared spectroscopy and kinetic studies of the reactivity of ethylene on Pt(111) surfaces, *Surf. Sci.* 368, 371–376, 1996.
6.10 M. P. Andersson, F. Abild-Pedersen, I. N. Remediakis, T. Bligaard, G. Jones, J. Engbæk, O. Lytken, S. Horch, J. H. Nielsen, J. Sehested, J. R. Rostrup-Nielsen, J. K. Nørskov, I. Chorkendorff, Structure sensitivity of the methanation reaction: H_2-induced CO dissociation on nickel surfaces, *J. Catal.* 255, 6–19, 2008.
6.11 M. Boudart, Turnover rates in heterogeneous catalysis, *Chem. Rev.* 95, 661–666, 1995.
6.12 C. J. Weststrate, I. M. Ciobîcă, A. M. Saib, D. J. Moodley, J. W. Niemantsverdriet, Fundamental issues on practical Fischer–Tropsch catalysts: How surface science can help, *Catal. Today* 228, 106–112, 2014.
6.13 Z.-P. Liu, P. Hu, General rules for predicting where a catalytic reaction should occur on metal surfaces: A Density Functional Theory study of C–H and C–O bond breaking/making on flat, stepped, and kinked metal surfaces, *J. Am. Chem. Soc.* 125, 1958–1967, 2003.
6.14 X.-Q. Gong, R. Raval, P. Hu, A density functional theory study on the water formation at high coverages and the water effect in the Fischer–Tropsch synthesis, *Mol. Phys.* 102, 993–1000, 2004.
6.15 T. Bligaard, J. K. Nørskov, S. Dahl, J. Matthiesen, C. H. Christensen, J. Sehested, The Brønsted–Evans–Polanyi relation and the volcano curve in heterogeneous catalysis, *J. Catal.* 224, 206–217, 2004.
6.16 W. Chen, B. Zijlstra, I. A. W. Filot, R. Pestman, E. J. M. Hensen, Mechanism of CO dissociation on a cobalt Fischer—Tropsch catalyst, *ChemCatChem.* 10, 136–140, 2018.
6.17 Z.-P. Liu, P. Hu, A new insight into Fischer–Tropsch synthesis, *J. Am. Chem. Soc.* 124, 11568–11569, 2002.
6.18 W. Chen, I. A. W. Filot, R. Pestman, E. J. M. Hensen, Mechanism of cobalt-catalyzed CO hydrogenation: 2. Fischer–Tropsch synthesis, *ACS Catal.* 7, 8061–8071, 2017.

6.19 R. Pestman, W. Chen, E. J. M. Hensen, Insight into the rate-determining step and active sites in the Fischer–Tropsch reaction over cobalt catalysts, *ACS Catal.* 9, 4189–4195, 2019.
6.20 J. Cheng, P. Hu, P. Ellis, S. French, G. Kelly, C. M. Lok, Chain growth mechanism in Fischer–Tropsch synthesis: A DFT study of C–C coupling over Ru, Fe, Rh, and Re surfaces, *J. Phys. Chem. C* 112, 6082–6086, 2008.
6.21 C. J. Weststrate, P. van Helden, J. W. Niemantsverdriet, Reflections on the Fischer–Tropsch synthesis: Mechanistic issues from a surface science perspective, *Catal. Today* 275, 100–110, 2016.
6.22 A. C. Kizilkaya, J. W. Niemantsverdriet, C. J. Weststrate, Oxygen adsorption and water formation on Co(0001), *J. Phys. Chem. C* 120, 4833–4842, 2016.
6.23 M. O. Ozbek, J. W. Niemantsverdriet, Elementary reactions of CO and H_2 on C-terminated χ-Fe_5C_2 (001) surfaces, *J. Catal.* 317, 158–166, 2014.
6.24 R. J. P. Broos, B. Zijlstra, I. A. W. Filot, E. J. M. Hensen, Quantum-chemical DFT study of direct and H- and C-assisted CO dissociation on the χ-Fe_5C_2 Hägg carbide, *J. Phys. Chem. C* 122, 9929–9938, 2018.
6.25 K. C. Waugh, Methanol synthesis, *Catal. Lett.* 142, 1153–1166, 2012.
6.26 M. Behrens, F. Studt, I. Kasatkin, S. Kühl, M. Hävecker, F. Abild-Pedersen, S. Zander, F. Girgsdies, P. Kurr, B.-L. Kniep, M. Tovar, R. W. Fischer, J. K. Nørskov, R. Schlögl, The active site of methanol synthesis over Cu/ZnO/Al_2O_3 industrial catalysts, *Science* 336, 893–897, 2012.
6.27 R. Burch, S. E. Golunski, M. S. Spencer, The role of copper and zinc oxide in methanol synthesis catalysis, *J. Chem. Soc. Faraday Trans.* 86, 2683–2691, 1990.
6.28 G. C. Chinchen, P. J. Denny, D. G. Parker, M. S. Spencer, D. A. Whan, Mechanism of methanol synthesis from CO_2/CO/H_2 mixtures over copper/zinc oxide/alumina catalysts: Use of ^{14}C-labelled reactants, *Appl. Catal.* 30, 333–338, 1987.
6.29 L. C. Grabow, M. Mavrikakis, Mechanism of methanol synthesis on Cu through CO_2 and CO hydrogenation, *ACS Catal.* 1, 365–384, 2011.
6.30 Y. Yang, C. A. Mims, D. H. Mei, C. H. F. Peden, C. T. Campbell, Mechanistic studies of methanol synthesis over Cu from CO/CO_2/H_2/H_2O mixtures: The source of C in methanol and the role of water, *J. Catal.* 298, 10–17, 2013
6.31 Y. Yang, J. Evans, J. A. Rodriguez, M. G. White, P. Liu, Fundamental studies of methanol synthesis from CO_2 hydrogenation on Cu(111), Cu clusters, and Cu/ZnO(000-1), *Phys. Chem. Chem. Phys.* 12, 9909–9917, 2010.
6.32 Y.-F. Zhao, Y. Yang, C. Mims, C. H. F. Peden, J. Li, D. H. Mei, Insight into methanol synthesis from CO_2 hydrogenation on Cu(111): Complex reaction network and the effects of H_2O, *J. Catal.* 281, 199–211, 2011.
6.33 J. Cheng, P. Hu, P. Ellis, S. French, G. Kelly, C. M. Lok, Some understanding of Fischer–Tropsch synthesis from density functional theory calculations, *Top. Catal.* 53, 326–337, 2010.

6.34 R. Guil-López, N. Mota, J. Llorente, E. Millán , B. Pawelec , J. L. G. Fierro, R. M. Navarro, *Materials* 12, 3902, 2019.

6.35 G. A. Olah, A. Goeppert, G. K. Surya Prakash, *Beyond Oil and Gas: The Methanol Economy*, Wiley-VCH, Weinheim, 2006.

6.36 J. O. Metzger, Book review of beyond oil and gas: The methanol economy, in *Angew. Chem. Int. Ed.*, Eds. G. A. Olah, A. Goeppert, G. K. Surya Prakash, Wiley-VCH, Weinheim, 45, pp. 5045–5047, 2006.

6.37 Wind power in China. https://en.wikipedia.org/wiki/Wind_power_in_China.

6.38 O. S. Joo, K. D. Jung, I. Moon, A. Y. Rozovskii, G. I. Lin, S. H. Han, S. J. Uhm, Carbon dioxide hydrogenation to form methanol via a reverse-water-gas- shift reaction (the Camere process), *Ind. Eng. Chem. Res.* 38, 1808–1812, 1999.

6.39 G. Ertl, Surface science and catalysis studies on the mechanism of ammonia synthesis, *Catal. Rev.-Sci. Eng.* 21, 201–223, 1980.

6.40 G. Ertl, Primary steps in catalytic synthesis of ammonia, *J. Vac. Sci. Technol. A* 1, 1247–1253, 1983.

6.41 J. K. Nørskov, S. Holloway, N. D. Lang, Microscopic model for the poisoning and promotion of adsorption rates by electronegative and electropositive atoms, *Surf. Sc.* 137, 65–78, 1984.

6.42 A. Logadottir, J. K. Nørskov, Ammonia synthesis over a Ru(0001) surface studied by density functional calculations, *J. Catal.* 220, 273–279, 2003.

6.43 D. E. Brown, T. Edmonds, R. W. Joyner, J. J. McCarroll, S. R. Tennison, The genesis and development of the commercial BP doubly promoted catalyst for ammonia synthesis, *Catal. Lett.* 144, 545–552, 2014.

6.44 S. Dahl, E. Törnqvist, I. Chorkendorff, Dissociative adsorption of N_2 on Ru(0001): A surface reaction totally determined by steps, *J. Catal.* 192, 381–390, 2000.

6.45 C. J. H. Jacobsen, S. Dahl, P. L. Hansen, E. Törnqvist, L. Jensen, H. Topsøe, D. V. Prip, P. B. Møenshaug, I. Chorkendorff, Structure sensitivity of supported ruthenium catalysts for ammonia synthesis, *J. Mol. Catal. A* 163, 19–26, 2000.

6.46 R. Hoffmann, A chemical and theoretical way to look at bonding on surfaces, *Rev. Mod. Phys.* 60, 601–628, 1988.

6.47 B. Hammer, J. K. Nørskov, Why gold is the noblest of all the metals, *Nature* 376, 238–240, 1995.

6.48 M. P. Andersson, T. Bligaard, A. Kustov, K. E. Larsen, J. Greeley, T. Johannessen, C. H. Christensen, J. K. Nørskov, Toward computational screening in heterogeneous catalysis: Pareto-optimal methanation catalysts, *J. Catal.* 239, 501–506, 2006.

6.49 G. Jones, J. Geest Jakobsen, S. S. Shim, J. Kleis, M. P. Andersson, J. Rossmeisl, F. Abild-Pedersen, T. Bligaard, S. Helveg, B. Hinnemann, J. R. Rostrup-Nielsen, I. Chorkendorff, J. Sehested, J. K. Nørskov, First principles calculations and experimental insight into methane steam reforming over transition metal catalysts, *J. Catal.* 259, 147–160, 2008.

Questions

6.1 Explain why the chain growth factor α in the Fischer–Tropsch reaction is relatively low for smaller hydrocarbons but high for $n = 2$.

6.2 When ethene adsorbs on a Pd(111) surface with a binding energy of $-27\,\text{kJ}\cdot\text{mol}^{-1}$, the C–C bond becomes longer. Do the metal–metal surface bonds remain unperturbed?

6.3 The hydrogenation of ethene occurs by a Horiuti–Polanyi mechanism, with consecutive surface-catalysed addition of hydrogen. Would another mechanism be possible?

6.4 The apparent activation energy of the reaction of ethene to adsorbed ethyl is only $+26\,\text{kJ}\cdot\text{mol}^{-1}$, although the intrinsic activation energy from the surface is $+88\,\text{kJ}\cdot\text{mol}^{-1}$ (Fig. 6.6). Explain this based on the information in Section 4.1.1.1.

6.5 At high ethene coverage, the apparent activation energy of the reaction of ethene to ethyl is equivalent to the actual intrinsic activation energy. Explain this with the aid of Section 4.1.1.1.

6.6 The formation of methane from synthesis gas seems to be structure-insensitive, but the reverse reaction, steam reforming, is structure-sensitive. Can both conclusions be true, or do they contradict each other?

6.7 How can a double labelling experiment prove that methanol is formed by non-dissociative chemisorption of CO?

6.8 Isotope labelling experiments under industrial conditions showed that the main carbon source of methanol is not CO but CO_2. How would you conduct this experiment?

6.9 Explain why, in principle, synthesis gas with a H_2/CO ratio as low as 0.5 could be used as feed for the Fischer–Tropsch reaction over an iron catalyst.

6.10 Aromatics have a high octane number (good for gasoline) and linear alkanes a high cetane number (good for diesel). Is the Fischer–Tropsch product better suited for gasoline or for diesel?

6.11 Why does Au preferentially adsorb on the defect sites of a Ru(0001) surface?

6.12 A Cu/ZnO catalyst is used in the low temperature WGS reaction to produce H_2 ($CO + H_2O \rightarrow CO_2 + H_2$). Give the two chemical steps that explain how the catalyst works.

6.13 The purge gas in the NH_3 synthesis is relatively rich in argon. Where does the argon come from?

6.14 How is urea made? What is it used for?

Chapter 7

Catalysis by Solid Acids

7.1 Solid Acid Catalysts

In this chapter we will discuss reactions that are catalysed by solid acids and that are important in refinery processes and in the manufacture of organic base chemicals. Base chemicals are alkenes and aromatics. In the petrochemical industry, small alkenes (ethene and propene) are made by thermal cracking of naphtha or diesel and aromatics (benzene, toluene and xylenes) are obtained by separation of oil fractions. Base chemicals are used to manufacture other chemicals, often first to so-called intermediate chemicals that are used for preparing final products. For instance, polyethene is made from ethene, polypropene from propene, phenol and acetone from cumene (which is made from propene and benzene), styrene from ethene and benzene, polyethylene terephthalate from glycol (made from ethene oxide, which is made from ethene and oxygen, cf. Section 9.3.1) and terephthalic acid (made from *p*-xylene, Section 9.3.3.2). Before discussing acid-catalysed reactions, we will first introduce the two main classes of solid acids: zeolites and amorphous silica–alumina.

7.1.1 *Zeolites*

Zeolites are crystalline aluminosilicates with microporous structures and they can be transformed into solid acids with quite high acid strength. Their frameworks consist of Si, Al and O atoms and they contain cations and water within their pores: $M^{n+}_{x/n}[(\mathrm{AlO_2})_x(\mathrm{SiO_2})_y]^{x-}\cdot z\mathrm{H_2O}$. More than 40 zeolites occur naturally as minerals, mined in many parts of the world, but even more zeolites (more than 160) are made synthetically. Zeolites are used in a variety of applications, with a global market of several million tons per annum. The major application is in detergents, where zeolites replace sodium tripolyphosphate as builder, to improve the performance of surfactants and other detergent additives by softening the water through ion exchange of calcium, magnesium and heavy metal cations. Other applications are in petrochemical cracking, in separation and removal of gases and solvents, agriculture, animal husbandry and construction. They are also called **molecular sieves**.

Zeolite frameworks may be considered as being built up of tetrahedra of oxygen atoms, with a silicon or aluminium atom in the centre (Fig. 7.1, left). The formal charge of a tetrahedron of oxygen atoms is $4 \times (-2/2) = -4$ for the four O atoms ($-2/2$, because each O atom is shared by two tetrahedra) and $+4$ for the Si^{4+} cation and $+3$ for the Al^{3+} cation. Because of the different charges of the Al^{3+} and Si^{4+} cations, SiO_4 tetrahedra have no formal charge while AlO_4 tetrahedra have a charge -1. Therefore, a zeolite must contain as many cations as needed to compensate the formal charge of -1 for every Al ion in the zeolite lattice. This explains the formula $M_x^+[(AlO_2)_x(SiO_2)_y]^{x-}{\cdot}zH_2O$ for monovalent cations M. The cations are not inside the lattice, but are positioned close to the Al ions, against the walls of the inner zeolite surface.

The SiO_4 and AlO_4 tetrahedra are linked together by their corners and form various n-membered rings in three dimensions (Fig. 7.1), where n is the number of tetrahedra in the ring. The zeolite faujasite (also called zeolite NaY) is built up of four-membered rings (4MRs), six-membered rings (6MRs) and 12-membered rings (12MRs) (Fig. 7.2, left). Each ring consists of an equal number of O atoms and Si plus Al atoms and the ring number

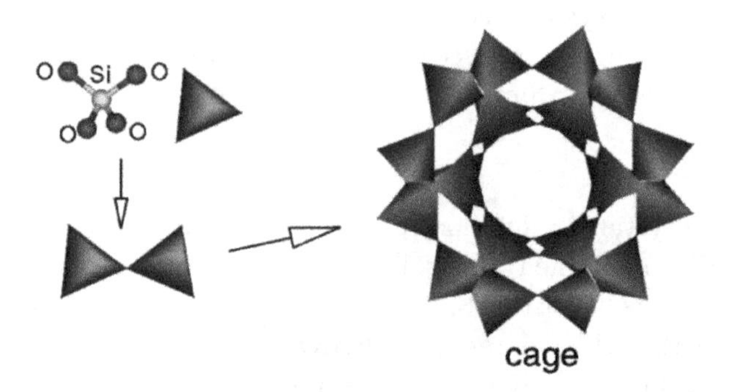

Fig. 7.1 Schematic picture of the formation of a zeolite cage by condensation of silicate ions.

is equal to the number of O atoms (or to the sum of the Si and Al atoms). In FAU the 4MRs and 6MRs combine to so-called sodalite cages, which are connected through double 6MRs (D6MRs) and thus give a structure with tetrahedral symmetry. Six sodalite cages enclose a 12MR, with a diameter of 0.74 nm. These large pore openings give access to supercages with a diameter of 1.2 nm. Each supercage has four 12MR, with the centres of the rings oriented as the corners of a tetrahedron, which give access to neighbouring supercages. The inner space of zeolite FAU thus is easily accessible for organic molecules. The "pores" of FAU are not linear, but extend in the four directions of a tetrahedron.

When sodalite cages are connected by 4MRs instead of 6MRs, zeolite LTA (Linde Type A) is obtained (Fig. 7.2, right). The largest ring in the LTA structure is an 8MR, enclosed by four sodalite cages and four D4MRs. The parent LTA material, zeolite 4A, has Na^+ cations as counter ions to the Al ions (Fig. 7.3). LTA has a Si/Al = 1 ratio and the 8MRs contain four Al atoms in alternating positions because, according to **Loewenstein's rule**, two neighbouring MO_4 tetrahedra cannot both contain an Al atom in the centre (an Al atom in the zeolite framework cannot have Al as next-nearest neighbouring atoms). AlO_4 tetrahedra have a negative charge and, thus, neighbouring AlO_4 tetrahedra would repel each other. As a consequence, there are four Na^+ cations near the four Al atoms in the 8MR pore opening, leaving a diameter of 0.4 nm accessible to molecules. When 75% of the Na^+ ions are replaced by larger K^+ ions (giving zeolite K–A), the pore opening is reduced to 0.3 nm. In zeolite Ca–A the Ca^{2+} cations are shifted away from the plane of the 8MR pore opening, and the opening is therefore

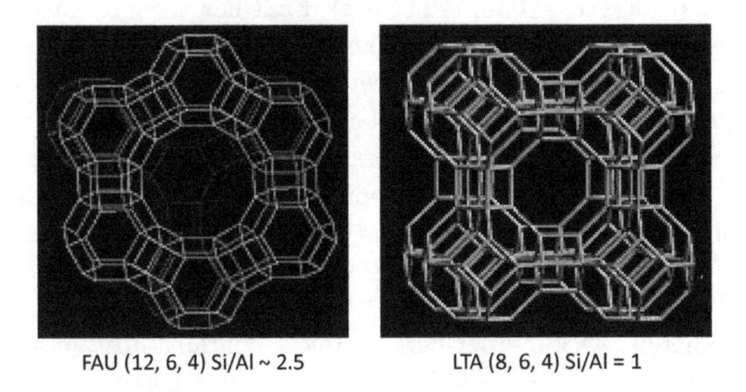

Fig. 7.2 Structures of zeolite faujasite (FAU) with 12, 6 and 4 rings and Si/Al ~2.5 and zeolite A (LTA) with 8, 6 and 4 rings and Si/Al = 1.

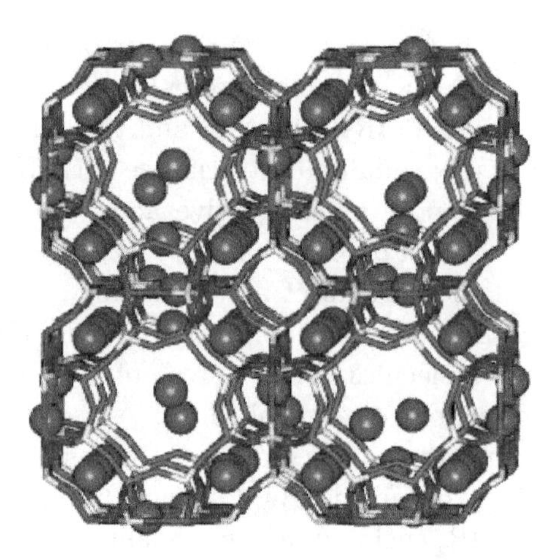

Fig. 7.3 Sodium zeolite A (zeolite 4A), with the Si and Al atoms of the zeolite framework indicated in red and yellow and the Na cations in purple.

larger in size (0.5 nm). LTA has straight pores in the x, y and z directions perpendicular to the 8MRs. LTA zeolites are used as adsorbents.

Other zeolites do not contain cages, but all zeolites contain pores. Mordenite (MOR) and zeolite beta (BEA) have 12MRs as the largest rings, like FAU, and the pores are straight with ellipsoidal pore openings of 0.70×0.65 nm and 0.76×0.64 nm, respectively (Fig. 7.4). MOR has a one-dimensional pore system and BEA a three-dimensional system. ZSM-5 (Zeolite Socony–Mobil Five, MFI) and ferrierite (FER) belong to the family of 10MR zeolites. ZSM-5 has a three-dimensional pore system with straight 10MR pores in one direction, connected by sinusoidal 10MR pores in a perpendicular direction (0.54×0.56 nm). FER has straight parallel 10MR pores, connected by 8MR pores. When the pore size is sufficiently large, molecules can enter the pores. Depending on the size of the molecule, limiting pore sizes are roughly between 0.3 and 1 nm in diameter.

Because the extra-framework metal cations (Fig. 7.3) are not part of the zeolite lattice, but are loosely bound to the inner and outer zeolite surface, they can be exchanged for other metal cations when in aqueous solution. This is exploited in water softening, where the "hard" Ca^{2+} and Mg^{2+} cations in water replace the Na^+ or K^+ cations in the zeolite. Many commercial washing powders therefore contain substantial amounts of zeolite; this is the largest application of zeolites. Waste water, containing heavy metals, and nuclear effluents with radioactive isotopes, can also be cleaned this way. For example, after the Fukushima Daiichi nuclear disaster

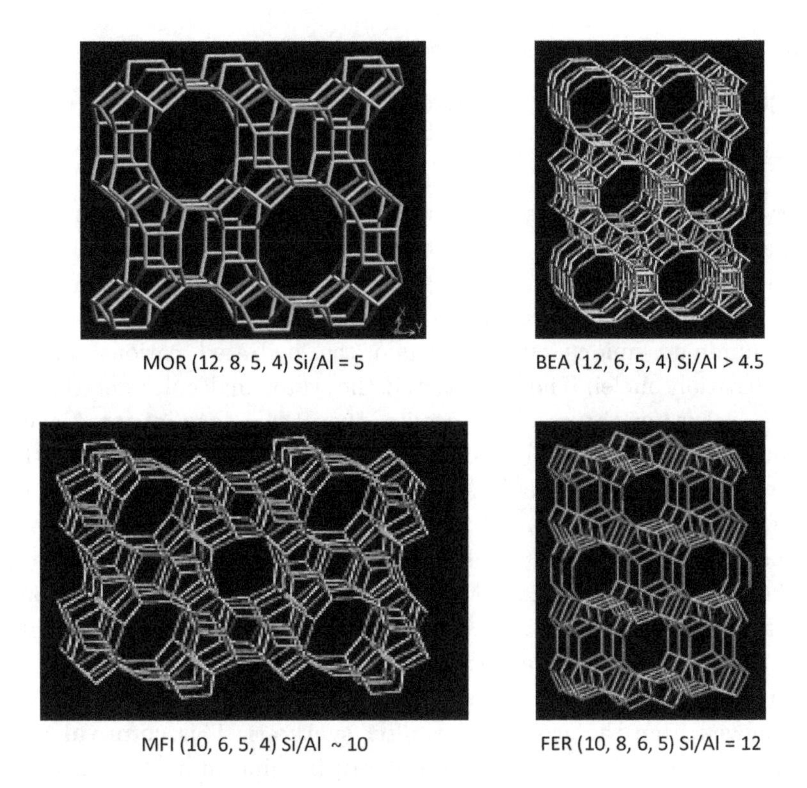

Fig. 7.4 Structures of mordenite (MOR), beta (BEA), ZSM-5 (MFI) and ferrierite (FER).

in Japan in 2011, bags filled with zeolite were dropped into the seawater near the power plant to adsorb the high levels of radioactive cesium.

Cation-containing zeolites are extensively used as desiccants, because of their high affinity for water, and for gas separation, where molecules are differentiated on the basis of their electrostatic interactions with the metal cations. Conversely, hydrophobic high-silica zeolites preferentially absorb organic solvents. Zeolites thus separate molecules according to their size, shape and polarity.

Zeolites are prepared by **hydrothermal synthesis**, which means at elevated temperature and pressure in aqueous solution. A solution of aluminium and silicon-containing salts and alkali or tetraalkylammonium ions is heated above 100 °C in an autoclave. For instance, sodium aluminate and sodium silicate in water or alumina, silica and sodium hydroxide in water are introduced into the autoclave.

$$NaAlO_2 + Na_2SiO_3 + NaOH + H_2O \rightarrow Na_a(AlO_2)_b(SiO_2)_c{\cdot}(NaOH)_d(H_2O)$$
$$\rightarrow Na_x(AlO_2)_x(SiO_2)_y(H_2O)_z.$$

After closing the autoclave and heating to 110–150 °C, the pressure reaches several atm, as required for synthesis. That is why natural zeolites are found in regions with extinct volcanoes and alkaline groundwater. Zeolite synthesis may take several hours to weeks. During this period, silicate and aluminate anions cluster around the hydrated cations and form crystallisation nuclei. The cations and the water molecules determine the pore structure of the resulting zeolite, they act as **structure-directing agents (SDA)**. The Si/Al ratio is an important parameter in the synthesis. Zeolite A is obtained when Si/Al ∼ 1 and zeolite Y at a higher Si/Al ratio.

Whereas FAU, MOR and LTA can be prepared in the presence of inexpensive inorganic cations, BEA, MFI and FER can only be produced in the presence of alkali as well as quaternary ammonium cations as structure-directing agents. In the resulting zeolite, the N atoms of the alkyl ammonium cations are located at the junctions of the zeolite pores and the alkyl chains fill the pores. The organic ammonium salts must be removed from the zeolite pores after synthesis. This **removal of the organic cations** is done by **calcination**, burning in air, which leads to NO_x and CO_2 (Eq. (7.1)) and, thus, to environmental pollution, which must be eliminated by cleaning the gas.

$$NR_4Z + O_2 \rightarrow HZ + H_2O + CO_2 + NO_x. \tag{7.1}$$

The organic SDA (OSDA) determines the resulting zeolite. For instance, ZSM-5 is obtained with tetrapropyl ammonium salts, and zeolite beta with tetraethyl ammonium salts. Zeolite seed crystals can also provide nuclei for crystal growth [7.1, 7.2]. They steer not only the growth of zeolite crystals in the aluminosilicate gel in the absence of organic templates, but may also reduce the duration of the crystallisation. Furthermore, calcination of the sample is unnecessary, because an OSDA is not used, thus avoiding environmental pollution and lowering production costs.

The main application of zeolites in catalysis is as hydrogen-exchanged acid zeolites (HZ), because the framework-bound protons are highly acidic. Zeolites produced in the presence of R_4N^+ cations react to protonic zeolites during the removal of the organic cations by calcination (Eq. (7.1)). Zeolites with a high Si/Al ratio made in the presence of inorganic cations (ZSM-5) can be brought into the proton form by M^+–H^+ ion exchange with an acid solution

$$NaZ + H^+ \rightarrow HZ + Na^+. \quad (7.2)$$

Zeolites with a low Si/Al ratio (FAU) are not stable in acid solution and have to be brought into the proton form in an indirect way. This is done by first applying a **Na^+–NH_4^+ exchange**

$$NaZ + NH_4^+ \rightarrow NH_4Z + Na^+, \quad (7.3)$$

followed by calcination to remove ammonia

$$NH_4Z \rightarrow HZ + NH_3. \quad (7.4)$$

Proton zeolites HZ behave like solid acids in hydrocarbon transformations, such as fluid catalytic cracking (zeolite Y), isomerisation of pentane and hexane (MOR), isomerisation of butene (FER), alkylation of benzene (BEA) and the synthesis of gasoline and alkenes from methanol (ZSM-5). In the cracking of large hydrocarbons in gas oil to smaller hydrocarbons at elevated temperature and H_2 pressure (**hydrocracking**), the Ni–MoS_2/USY catalyst consists of ultra-stable Y zeolite (USY) as the acid support with MoS_2 particles, promoted by Ni atoms, on its surface. H–USY is prepared by bringing NaY, the parent form of faujasite, into the ammonium form NH_4Y by Na^+–NH_4^+ exchange, followed by calcination of NH_4Y to HY. Then the HY zeolite is heated in steam to moderately "destroy" some of the zeolite lattice and, thus, remove some of the Al (and Si). This creates H–USY, which has mesopores, a higher Si/Al ratio than NaY and greater stability.

Zeolites can also serve as oxidation or reduction catalysts if metal ions are introduced into the framework. For instance, Ti-containing ZSM-5 zeolite (TS-1) catalyses the oxidation of phenol to hydroquinone and catechol by H_2O_2 [7.3] and the epoxidation of alkenes to epoxides by H_2O_2 or alkyl peroxide (ROOH) (Section 9.3.2). Sn-BEA is an example of a zeolite redox catalyst, which contains the metal in an extra-framework position and catalyses the Baeyer–Villiger reaction of a ketone to a lactone (R–CO–R + H_2O_2 $\rightarrow$ R–O–CO–R + H_2O) [7.4]. Another example is the use of copper zeolites in the selective catalytic reduction of NO_x by NH_3 to N_2 (SCR) [7.5] (Section 9.2.3).

Because impurities may form during the synthesis of a zeolite and because the Si/Al ratio of zeolites can vary, quality control of the zeolite synthesis product is important and normally a variety of techniques is applied. Elemental analysis gives the Si/Al/M ratios (M is the cation used in synthesis) and can be carried out by atomic absorption spectroscopy (AAS), X-ray fluorescence spectroscopy (XRF), inductively coupled plasma optical emission spectroscopy (ICP-OES) or by energy dispersive X-ray (EDX) analysis. XRD determines whether the desired zeolite has crystallised and whether crystalline impurities (e.g. other zeolites) and amorphous material have formed. Thermogravimetric analysis (TGA-DTA) determines the contents of water and SDA and the temperature at which they can be removed. After freeing the pores of water and SDA, N_2 adsorption determines the pore volume and microporous and mesoporous surface area.

The Si/Al ratio and the distribution of the Si and Al atoms in the zeolite framework can be determined by ^{29}Si NMR measurements. The ^{29}Si chemical shift depends on the number of the tetrahedral nearest-neighbour Al atoms. There are signals for Si(nAl) environments with $n = 1$–4 (Fig. 7.5) [7.6]. Loewenstein predicted that Al atoms would have tetrahedral Si atoms, but no Al atoms, as nearest neighbours; otherwise two negatively charged Al atoms would be neighbours. A Si atom, on the other hand, can have between zero and four Al tetrahedral nearest-neighbour atoms. Thus, the Si/Al ratio of the Si and Al atoms in the zeolite framework from the ^{29}Si NMR spectrum is:

$$(\mathrm{Si/Al})_{\mathrm{NMR}} = \frac{\sum_{n=0}^{4} \mathrm{I(Si, nAl)}}{\sum_{n=0}^{4} 0.25n \cdot \mathrm{I(Si, nAl)}},$$

where I(Si, nAl) is the intensity of the ^{29}Si NMR peak of the Si atom with n nearest-neighbour Al atoms. The ^{29}Si NMR spectrum of zeolite Y

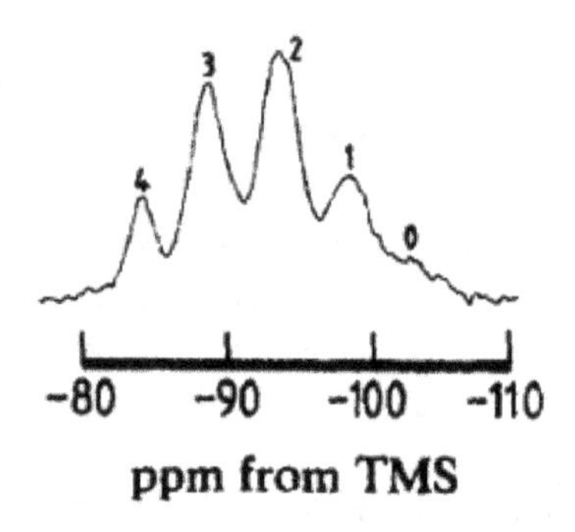

Fig. 7.5 ^{29}Si MAS NMR spectrum of zeolite Y with Si/Al = 1.87. Reprinted with permission from [7.6]. Copyright (1982) the Royal Society of Chemistry.

shows that with a Si/Al = 1.87 ratio all five possible peaks are visible (Fig. 7.5). Because all the Al atoms must have four Si atoms as next-nearest neighbours (Loewenstein's rule), when Si/Al = 1, all the Si atoms must have four Al atoms as next-nearest neighbours and only the peak at −83 ppm would show up. If the $(Si/Al)_{NMR}$ ratio differs from the Si/Al ratio determined by a bulk technique such as X-ray fluorescence or chemical analysis, then **extra-framework** Al or Si must be present.

The chemical properties of zeolites are determined by their microporous nature. The shape and size of their pore system have a steric influence on the reaction and controls the access of reactants and products. Therefore, zeolites act as **shape-selective** catalysts. The shape-selective properties of zeolites also explain their use in molecular adsorption. The preferential adsorption of some but not all molecules, is the basis of molecular sieving. In some cases the size and shape of the pores control access to the zeolite, while in other cases some molecules diffuse more quickly through the channels, while the diffusion of others is severely hindered. For instance, in the purification of *p*-xylene by silicalite (Al-free ZSM-5), *p*-xylene can easily diffuse through the channels of silicalite, but *o*-xylene and *m*-xylene are hindered in their diffusion. When toluene is methylated with methanol over a H–ZSM-5 catalyst, all three xylene isomers can be formed at channel crossings but, if one also narrows the pore-mouth openings, only *p*-xylene molecules can diffuse out of the ZSM-5 pores. A disadvantage of shape-selective catalysis is that one relies on diffusion hindering of the molecules. This means that the molecules have a long residence time inside the zeolite crystals and thus a high chance for subsequent reactions. Larger molecules may be formed which block the pores. To improve diffusion and ensure that the micropores behave as desired, mesoporosity has been introduced into zeolites. This can be done in several ways:

- create nanosized zeolite crystals, which have mesopores-sized intercrystalline space,
- combine an alkylammonium salt as structure-directing agent for the micropores with a detergent for directing the creation of mesopores (as in the synthesis of MCM-41),
- first synthesise zeolite crystals and then leach material away with base [7.7],
- create materials consisting of thin zeolite layers by delamination (exfoliation) of layered zeolites. ITQ-2 is produced by delamination of MCM-22 [7.8],

- directly synthesise a thin two-dimensional zeolite. Two-dimensional ZSM-5 was synthesised with a di-ammonium surfactant molecule ($C_{22}H_{45}$–$N^+(CH_3)_2$–C_6H_{12}–$N^+(CH_3)_2$–C_6H_{13}) [7.9, 7.10].

For every Al cation in a zeolite there is one proton bonded to an O anion, bridging an Al and a Si cation (Si–OH–Al). This OH group has the character of a Brønsted acid and IR spectroscopy is an important tool to determine these OH groups. Figure 7.6 presents the IR spectra of silica, silica–alumina and the zeolites ZSM-5 and mordenite in the OH region (3,800–3,000 cm^{-1}). The stronger the acid, the weaker the bond between the H atom and the O atom and the lower the O–H frequency. All the samples show a peak at about 3,750 cm^{-1} due to the very weakly acidic silanol (SiOH) groups on the surfaces of these materials. The zeolites have an additional OH peak at 3,619 (ZSM-5) and 3,616 cm^{-1} (MOR) due to Brønsted acid OH groups on the zeolite surface. The silica–alumina shows a silanol peak but no Brønsted acid OH peak. Nevertheless, silica–aluminas have substantial acidity.

When molecules adsorb on a zeolite surface, they interact with the OH groups and weaken the O–H bond. As a consequence, the IR bands shift. The original IR peaks of the OH groups around 3,600 cm^{-1} disappear and a new broad band appears at lower frequency. Figure 7.7 shows the differences between the diffuse reflectance IR spectra of four zeolites (H–MOR, H–Y, H–ZSM-5 and H–FER) measured in the presence of CO and in its absence [7.12]. A peak that disappears from the IR spectrum, because

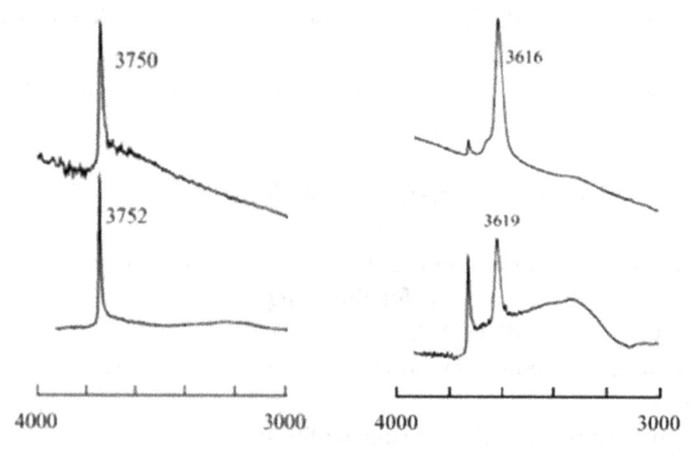

Fig. 7.6 IR spectra of silica and silica–alumina (left) and ZSM-5 and mordenite (MOR) (right). Reprinted with permission from [7.11]. Copyright (2003) Elsevier.

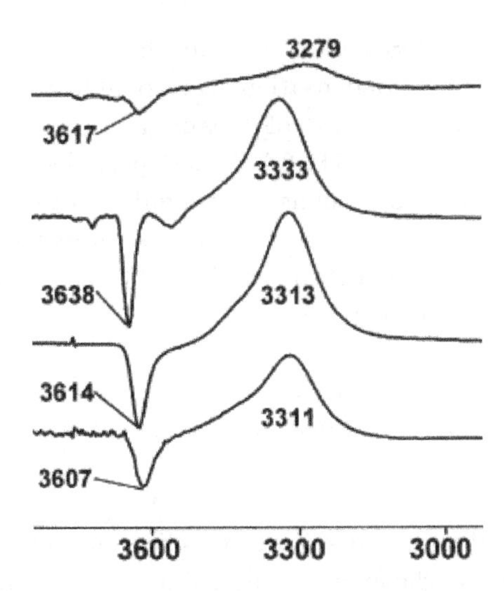

Fig. 7.7 Difference IR spectra of the zeolites H–MOR, H–Y, H–ZSM-5 and H–FER (from top to bottom) obtained by subtracting the spectra of the pure zeolites from the spectra of the zeolites with adsorbed CO. Reprinted with permission from [7.12]. Copyright (1998) Springer Science + Business.

of the adsorption of CO, shows up as a negative peak in the difference IR spectrum (around 3,600 cm^{-1} in Fig. 7.7). A new peak shows up as a positive peak in a difference spectrum (around 3,300 cm^{-1}). Weaker bases (CO, acetonitrile) induce smaller shifts in the OH IR peaks than stronger bases (pyridine). The shifts reveal the acid strength of the OH groups.

7.1.2 *Amorphous Silica–Alumina*

Before zeolites became industrially available, amorphous silica–aluminas (ASAs, $SiO_2 \cdot Al_2O_3$) were used as solid acid catalyst. ASA is a non-crystalline material that contains Si cations that are tetrahedrally coordinated by O anions and Al cations that are tetrahedrally as well as octahedrally coordinated by O anions. ASA has similar high surface area as alumina (Section 2.3.4) and silica (Section 2.4) and in addition possesses acid sites. The acidity of ASA is weaker than that of zeolites but its synthesis is cheaper, because no templates are needed, as in the synthesis of zeolites (Section 7.1.1). Furthermore, in contrast to zeolites (which have micropores), ASA has mesopores, which allow to react large molecules. Because ASA has a lower acidity than zeolites, it is used in applications that do not need strong acidity [7.13], e.g. in the refinery in the hydrocracking of hydrocarbons (Ni–MoS_2/ASA) and in the chemical industry in the production of pyridine by condensation of crotonaldehyde and formaldehyde with ammonia in the presence of steam and air ($CH_2{=}CH{-}CHO + 2\,CH_2O + NH_3 \rightarrow C_5H_5N + 3\,H_2O$). In the past decade interest in ASA has grown because ASA is thought to be similar to the amorphous Si and Al containing material that is created when zeolites are dealuminated by steam treatments, as in ultra-stabilised Y zeolite (Section 7.1.1).

ASA can be prepared by [7.13]

- deposition of alumina onto silica or of silica onto alumina,
- cogelation and coprecipitation of solutions of silicon and aluminium compounds.

The deposition of alumina on silica or of silica on alumina can be carried out with solutions of cheap inorganic salts [7.14] or of metal alkoxides such as aluminium iso-propoxide and tetraethyl orthosilicate, $Al(OC_3H_7)_3$ and $Si(OC_2H_5)_4$ [7.15]. In the latter case it is called grafting. Cogelation can be done with salts, e.g. with sodium silicate and sodium aluminate (aluminium trichloride can be used to lower the Na content and increase the acidity of the ASA), or with alkoxides of silica and aluminium [7.14]. After washing away salts that are formed during preparation, the ASA material is dried and calcined. During the calcination bonds are formed between the alumina and silica. The ASA structure depends on the synthesis method. The deposition methods lead to a core-shell structure of silica or alumina surrounded by a shell of alumina or silica respectively. The interface contains interconnected SiO_4 and AlO_4 tetrahedra, which are supposed to be responsible for the Brønsted acidity of ASA. The cogelation method has a better chance of producing homogeneous ASA, but even then ASA contains silica as well as alumina domains, which is due to the difference in the rate of hydrolysis of the metal alkoxides.

The silica domains in ASA do not contribute to acidity, the alumina domains contain Lewis acid sites (CUS Al^{3+} cations, Section 2.3.5.1) but where are the Brønsted acid sites located? They might be present as Si–OH–Al bridges (Fig. 7.8a), as in zeolites [7.16, 7.17]. Protonic zeolites show a peak of acidic OH groups in the 3,650–3,500 cm^{-1} IR region and peaks at 5–3.8 ppm in ^1H–NMR spectra but such peaks were not observed in the spectra of ASAs. On the other hand, conjugate Brønsted acid sites, Si–O^-–Al, BH^+ that can form by interaction of neighbouring ≡Si–OH and –Al≡ groups with a basic molecule B, were detected in the IR spectrum

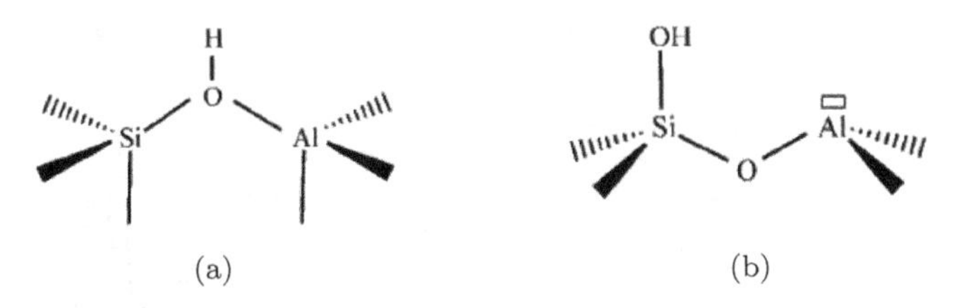

Fig. 7.8 Possible structure of acidic OH groups at the surface of ASA. (a) a silanol as in zeolites, (b) a silanol interacting with a neighbouring Al CUS site.

of ASA on which pyridine was adsorbed (Fig. 7.9) [7.18]. Pyridine forms a complex on a Lewis acid site and adsorbs as a pyridinium ion on a Brønsted site. Thus, the IR spectrum of pyridine on ASA shows peaks due to the adsorption of pyridine on Brønsted sites at 1,640, 1,547 and 1,492 cm^{-1} and on Lewis sites at 1,622, 1,578, 1,492 and 1,451 cm^{-1} [7.16]. Hydrogen-bonded pyridine is seen at 1,597, 1,492 and 1,451 cm^{-1}. After evacuation at 300 °C, only peaks due to adsorption on Lewis sites remain, indicating that pyridine binds weaker to the Brønsted sites than to the Lewis sites of ASA. Direct proof for the existence of few, but relatively strong acidic sites (as strong as zeolitic OH groups) has been provided by exchanging the acidic protons, but not the neutral silanol protons, by deuterons (H/D exchange). In this way, the overwhelming effect of the silanols on the OH region in the IR spectrum could be eliminated and OD bands of strong Brønsted acid sites in ASA could be detected. They had similar acidity as OH groups in zeolites but were present in much lower number [7.17].

Nevertheless, as Fig. 7.9 shows, ASA contains a considerable number of Brønsted sites, but they disappear by heating and, thus, are not as strong as on zeolites. These sites may consist of silanol groups next to a CUS Al^{3+} cation. The silanol groups do not form a chemical bond as in zeolites (Fig. 7.8a), but become acidic by the influence of neighbouring Al^{3+} cations (Fig. 7.8b) [7.19]. These can be three- or fourfold-coordinated Lewis-acid Al^{3+} cations [7.16, 7.20]. A combination of spectroscopic, programmed desorption and catalytic measurements demonstrated that the ASA surface contains a broad range of OH sites: a small number of strong Brønsted acid sites, many more weaker Brønsted acid sites (such as shown in Fig. 7.1b), non-acidic Al–OH and Si–OH sites, weak five-coordinated Al Lewis acid sites, at the interface between alumina domains and the ASA phase, and stronger CUS Al Lewis acid sites, which are grafted onto the silica surface [7.18]. Dynamic nuclear polarisation surface-enhanced spectroscopy (cf. Section 2.3.5.2) confirmed that Si and Al are preferentially connected

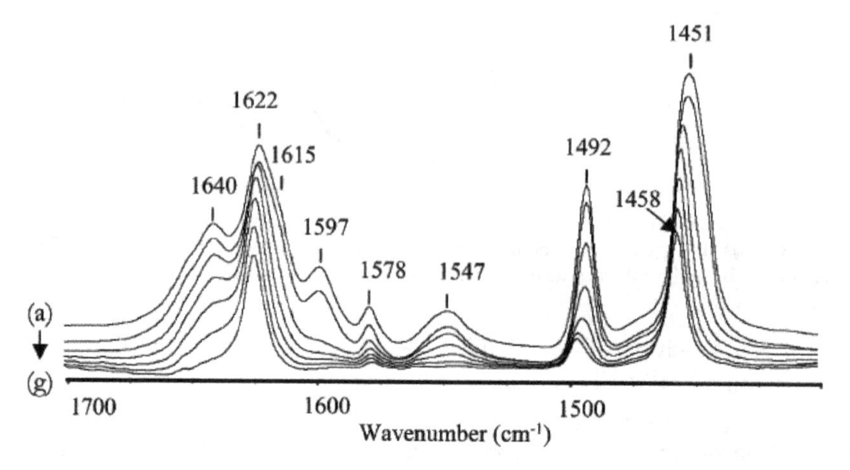

Fig. 7.9 IR spectra of pyridine adsorbed on ASA (40 wt% SiO_2–60 wt% Al_2O_3) activated at 300 °C. The pyridine was adsorbed at 50 °C and evacuated from 20 °C (a) to 300 °C (g). Reprinted with permission from [7.16]. Copyright (2006) the American Chemical Society.

by four-coordinated Si to four-coordinated Al in the mixed layer between silica and alumina in ASA, which enables the formation of bridging hydroxyl Brønsted acid sites [7.21].

7.2 Reactions of Hydrocarbons

7.2.1 *Reactions of Alkenes and Alkanes*

Acidic zeolites catalyse the same reactions as superacids. For many years, therefore, reactions of hydrocarbons catalysed by solid acids were assumed to occur by means of carbenium ion intermediates, in the same way as in strong liquid acids. Carbenium ions (carbocations) contain a carbon atom with three substituents and a positive charge (R_1–C^+–$(R_2)R_3$). Because of the high charge density, carbenium ions are very reactive and can be observed only under special conditions. Superacids have high dielectric constants and solvate carbenium ions. Zeolites and other solid acids (e.g. ASA), on the other hand, have a low dielectric constant and do not stabilise charged intermediates. Therefore, carbenium ions are not intermediates in reactions catalysed by solid acid catalysts but are transition-state structures between alkoxy intermediates. Only unsaturated molecules of high basicity (e.g. 1-methyl-indene and 2,4-dimethyl-cyclopentadiene) can be protonated to stable carbenium ions on solid acids [7.22]. Carbocations that are not

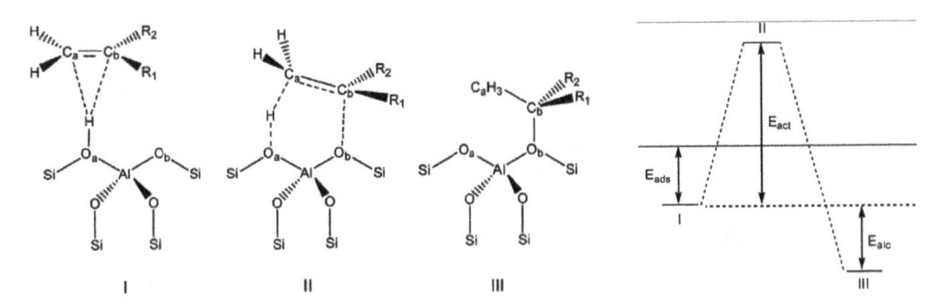

Fig. 7.10 Structure and energy of an alkene adsorbed on a Brønsted acid site (I), the transition state for the alkene protonation (II) and the resulting alkoxide (III). Reprinted with permission from [7.23]. Copyright (1994) Elsevier.

stabilised by conjugation are unstable with respect to deprotonation or formation of an alkoxy species with oxygen atoms of the solid acid. The alkoxy compounds are the actual intermediates of carbocations in solid acid media. Protonation of an alkene by an acid hydroxyl group starts with π adsorption of the alkene on the hydroxyl group (structure I in Fig. 7.10). The concerted reaction goes through a transition state that resembles a carbenium ion (II) and ends in a stable covalent alkoxide species (III).

The stability of alkoxides depends on their structure. The stability order is n-alkoxide > iso-alkoxide > tert-alkoxide, because branched alkoxides have more steric hindrance with the surface of the solid acid (zeolite wall) than linear alkoxides. The activation energy for protonation of alkenes and formation of alkoxides is determined by electrostatic effects because of the charge separation in the carbenium ion-like transition states. Thus, the order of the activation energies is $E_a(\text{primary}) > E_a(\text{secondary}) > E_a(\text{tertiary})$.

A DFT calculation of the adsorption of isobutene on the wall of the zeolite ferrierite confirmed that steric interactions between hydrocarbon and the zeolite wall are important [7.24]. The isobutoxide structure **5** in Fig. 7.11 has a lower energy ($-73\,\text{kJ}\cdot\text{mol}^{-1}$) than the t-butoxide structure **3** ($-48\,\text{kJ}\cdot\text{mol}^{-1}$), but isobutene that is π-bonded to the zeolite wall (structure **2**) has even lower energy ($-78\,\text{kJ}\cdot\text{mol}^{-1}$). The t-butyl carbenium ion **4** has the highest energy ($-21\,\text{kJ}\cdot\text{mol}^{-1}$).

Whereas the energy of alkoxy compounds depends weakly on the primary, secondary or tertiary structure of the carbon atom linked to the oxygen atom, the energy of carbenium ions depends strongly on the structure of the carbocation. The activation energy and, thus, the rates of the reactions catalysed by solid acids can be determined qualitatively

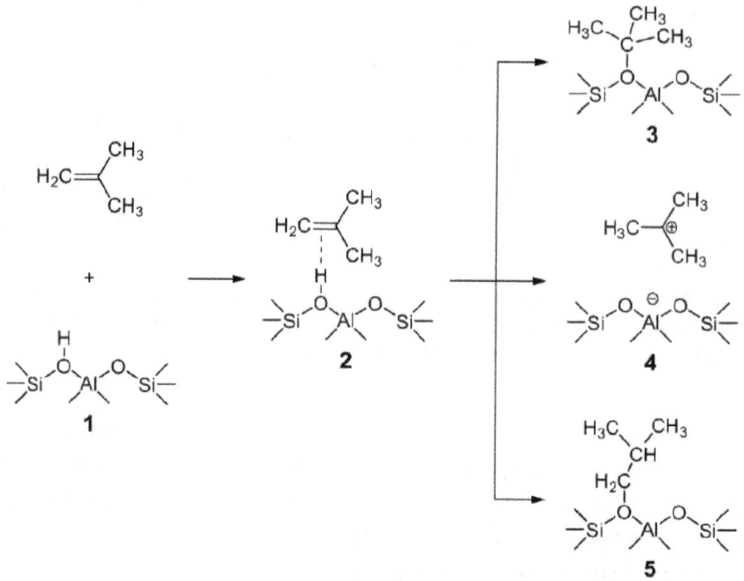

Fig. 7.11 Reaction of 2-methylpropene (isobutene) with solid acid **1** to a hydrogen-bonded π complex **2**, a *t*-butoxide **3**, a *t*-butyl carbenium ion **4** or an isobutoxide **5**. Reprinted with permission from [7.25]. Copyright (2006) the PCCP Owner Societies.

by assuming that the reaction takes place by carbenium ions. In-depth study must be based on theoretical models. These calculations are, however, complex, because the energy of covalent hydrocarbon structures as well as the energy of ionic carbenium ion structures must be calculated.

A reaction rate is determined by the difference in the free energies of the transition state and the reactants in the gas or liquid state plus the acid site, and does not depend on the stability of the adsorption complex [7.26]. Because carbenium ion-like structures function as transition-state structures and have a strong effect on energy, a comparison of carbenium ion structures gives qualitative information about acid-catalysed reactions. Carbenium ions can be prepared by protonation of alkenes, by exchange of an H atom between a hydrocarbon and a carbenium ion, by isomerisation of a carbenium ion and by β scission of a carbenium ion (Fig. 7.12) [7.27, 7.28]. In β scission, the second C–C bond counted from the charged C atom breaks under the formation of a new carbenium ion and an alkene. It plays an important role in the formation of isobutene by dimerisation of butene, isomerisation of the resulting octene and β scission of isooctene to isobutene (Section 7.2.2).

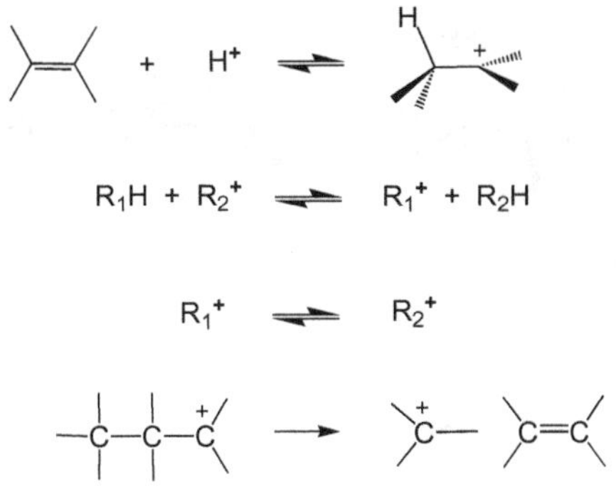

$$R_1H + R_2^+ \rightleftharpoons R_1^+ + R_2H$$

$$R_1^+ \rightleftharpoons R_2^+$$

Fig. 7.12 Four ways of preparing hydrocarbon carbocations.

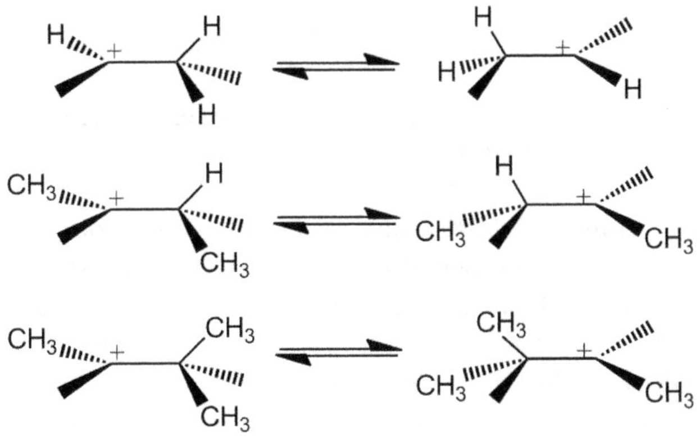

Fig. 7.13 1,2 shift of a hydrogen atom or methyl group in carbocations.

In strong liquid acids, the order of stability of carbenium ions is primary < secondary < tertiary ($CH_3^+ \ll RCH_2^+ \ll R_2CH^+ \ll R_3C^+$) with an energy difference of about $100\,kJ{\cdot}mol^{-1}$ between primary and secondary ions and $42\,kJ{\cdot}mol^{-1}$ between secondary and tertiary ions [7.27]. Carbenium ions undergo several intramolecular reactions. A very fast intramolecular reaction is the 1,2 shift of a hydrogen atom or methyl group (Fig. 7.13). 1,2-Shifts are fast, because the proton or methyl cation moves during the 1,2 shift through the π cloud of the double bond and stays bonded (Fig. 7.14).

When the energy of the initial and final states is the same, then the activation energy of the 1,2 shift is small (about $14\,kJ{\cdot}mol^{-1}$) and the

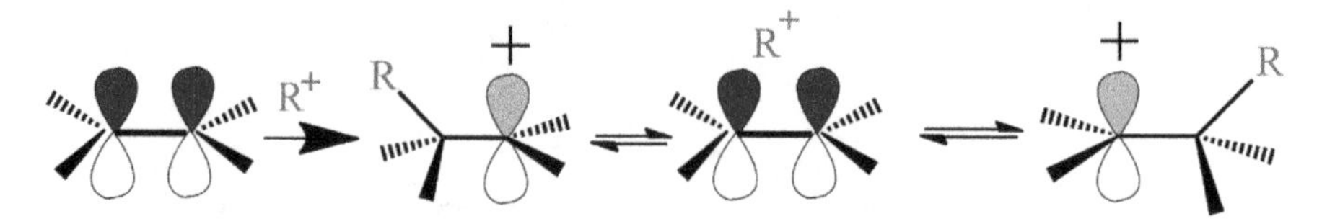

Fig. 7.14 Reaction of an alkene with a proton or carbocation R^+ by means of a π-bonded intermediate.

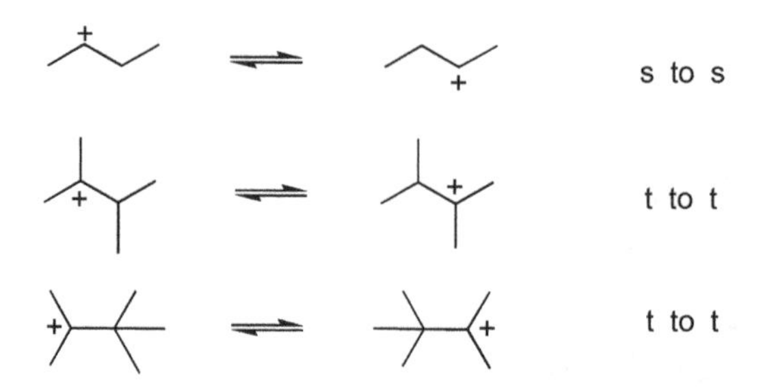

Fig. 7.15 1,2-Shifts of a hydrogen atom or methyl group in secondary to secondary and tertiary to tertiary carbenium ions.

corresponding shifts are fast. Thus, reactions of secondary to secondary and tertiary to tertiary carbenium ions are fast (Fig. 7.15).

Reactions, in which the initial and final states have a different energy, may be slower. For example, the reaction of the tertiary carbenium ion of 2-methylpentane to the tertiary carbenium ion of 3-methylpentane requires a positive charge on the C3 atom before the methyl group can shift. To create the positive charge on the C3 atom, a hydrogen atom must shift. The total reaction goes from a tertiary to a secondary carbenium ion, then to another secondary carbenium ion and finally to a tertiary carbenium ion. Because of the difference in the energy of the secondary and tertiary carbenium ions of about $42\,\mathrm{kJ{\cdot}mol^{-1}}$, the activation energy of the whole reaction is not 14 but $14 + 42 = 56\,\mathrm{kJ{\cdot}mol^{-1}}$ (Fig. 7.16).

In the above isomerisations, neither the degree of branching of the hydrocarbon, nor the structure of the carbocation (primary, secondary or tertiary) changed. Isomerisations, in which the degree of branching changes, have an activation energy that differs from the differences in the energy between primary, secondary and tertiary carbenium ions. If the isomerisation of the tertiary carbenium ion of 2-methylbutane to the secondary carbenium ion of pentane were to occur by a 1,2 shift of an H atom and a positive charge between the C1 and C2 atoms, then the

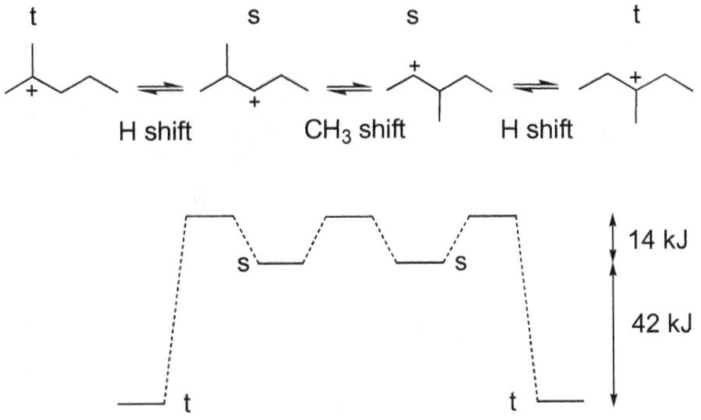

Fig. 7.16 Isomerisation of the tertiary carbenium ion of 2-methylpentane to the tertiary carbenium ion of 3-methylpentane through an intermediate secondary carbenium ion.

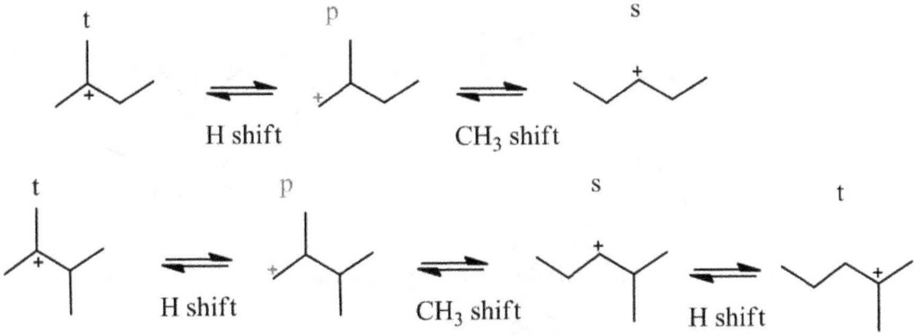

Fig. 7.17 Isomerisation of the tertiary carbenium ion of 2-methylpentane to the secondary carbenium ion of pentane, of the tertiary carbenium ion of 2,3-dimethylbutane to the tertiary carbenium ion of 2-methylpentane by 1,2 H and CH_3 shifts.

intermediate carbenium ion would be a primary cation (Fig. 7.17). Such a reaction would have a very high activation energy of about $42 + 100 + 14 = 156\,kJ{\cdot}mol^{-1}$. However, the experimental value is only $77\,kJ{\cdot}mol^{-1}$! Similarly, it would follow that the activation energy of the isomerisation of the tertiary carbenium ion of 2,3-dimethylbutane to the tertiary carbenium ion of 2-methylpentane (Fig. 7.17) is also about $156\,kJ{\cdot}mol^{-1}$, whereas actually it is only $80\,kJ{\cdot}mol^{-1}$. Another reaction in which the degree of branching changes is the isomerisation of the tertiary carbenium ion of methylcyclopentane to the secondary carbenium ion of cyclohexane; the activation energy is also relatively low ($71\,kJ{\cdot}mol^{-1}$).

Isomerisations during which the degree of branching of the hydrocarbon and the secondary or tertiary structure of the carbocation do not change,

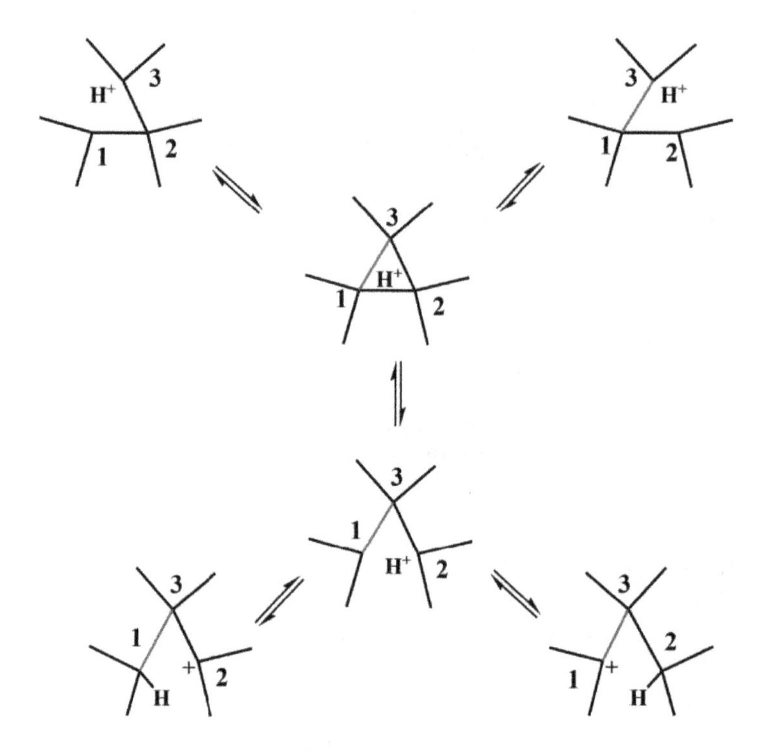

Fig. 7.18 Isomerisation of a carbenium ion by protonated cyclopropane intermediates.

are very fast ($E_a = 14\,\text{kJ·mol}^{-1}$) or fast ($E_{\text{act}} = 56\,\text{kJ·mol}^{-1}$). When the degree of branching changes, isomerisations are slower and their activation energies are lower than expected based on primary carbenium ions as the intermediate structures ($E_{\text{act}} \sim 75$ instead of $\sim 140\,\text{kJ·mol}^{-1}$).

The unexpectedly low activation energy of isomerisations, during which the degree of branching changes, is explained by protonated cyclopropane intermediates (Fig. 7.18). A carbocation lacks electrons and, thus, attracts electrons. The C2 carbocation (Fig. 7.18, left below) attains more electrons by reacting with carbon atom 1 (interaction between a Lewis acid, lacking electrons, and a Lewis base, with excess electrons). The proton is shared by carbon atoms 1 and 2 (Fig. 7.18, lower middle figure) and a cyclopropane ring forms with a weak C1–C2 bond (Fig. 7.18, upper middle figure). When the proton shifts to the C2–C3 bond or the C1–C3 bond, the other bonds in the cyclopropane ring are weakened. The actual protonated cyclopropane structure is an average of all structures indicated in Fig. 7.18, and the delocalisation of the proton leads to a decrease in the energy of

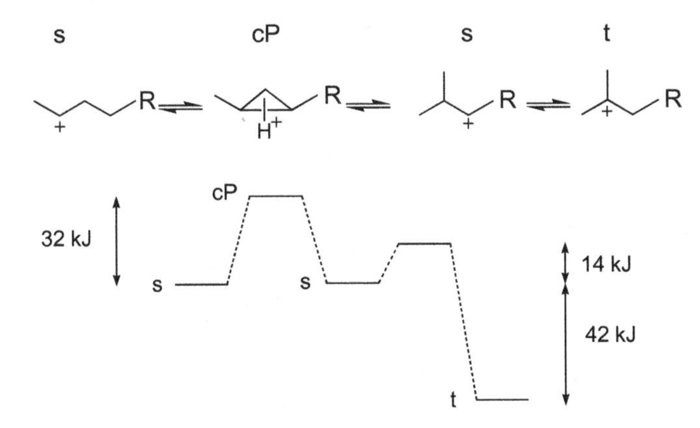

Fig. 7.19 Energy diagram of the isomerisation of the secondary carbenium ion of a linear alkane to the tertiary carbenium ion of a 2-methylalkane.

the cyclopropane structure. As a result, the energy difference between a protonated cyclopropane intermediate and a secondary carbenium ion is only $32\,\text{kJ}\cdot\text{mol}^{-1}$, which explains the activation energy of isomerisation, during which the degree of branching changes (about $75\,\text{kJ}\cdot\text{mol}^{-1}$). Thus, in the reaction of the secondary carbenium ion of pentane (Fig. 7.19, $R = CH_3$) to the tertiary carbenium ion of 2-methylbutane, the activation energy is $32\,\text{kJ}\cdot\text{mol}^{-1}$, whereas it is $74\,\text{kJ}\cdot\text{mol}^{-1}$ for the reverse reaction.

DFT calculations confirm the conclusions described in the foregoing on the basis of completely ionic transition states [7.26, 7.29, 7.30]. A 1,2-hydride shift of an alkoxide group requires the separation of charge and occurs via a transition state that resembles a carbenium ion plus the conjugate basic form of the oxide cluster. The activation energy for a branching reaction is calculated to be higher than that of a non-branching reaction. The DFT calculations also confirm that the activation energy for a branching reaction is lower via a protonated dialkylcyclopropane intermediate than via a 1,2-methyl or ethyl shift [7.29], although the inclusion of long-range dispersive forces decreases the difference [7.30].

Experiments with ^{13}C-labelled butene and pentene confirmed the protonated cyclopropane mechanism [7.27]. Scrambling of the label (distribution over all carbon atoms) occurred in the reaction of butene; however, isomerisation to isobutene did not occur. After formation of the protonated cyclopropane ring by closure of the C1–C3 bond between the C3 carbocation and the labelled C1 atom (Fig. 7.20), the ring reopens in the same way or opens in two other ways. Opening of the C1–C2 bond of the cyclopropane

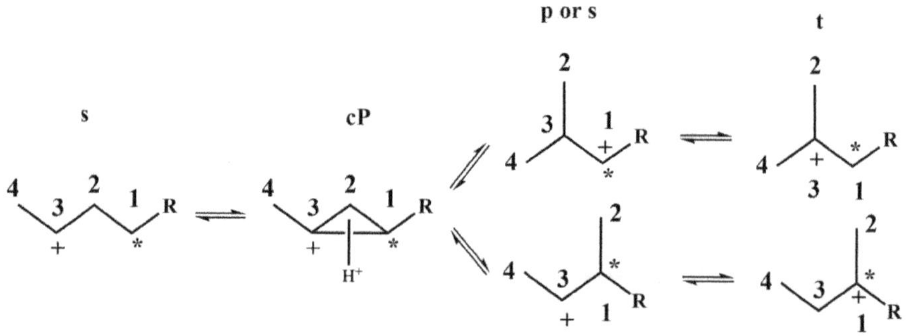

Fig. 7.20 Isomerisation of the secondary carbenium ion of a linear alkane to the tertiary carbenium ion of a 2-methylalkane by means of a cyclopropane intermediate.

ring (upper path) is impossible for R = H, because a primary carbenium ion would form on C1 or C2 requiring a very high activation energy. Opening of the C2–C3 bond (lower path) is possible, because a secondary C3 carbocation forms. Thus, for butane (R = H), the ^{13}C label can shift from a primary to a secondary position and the positive charge is again on a secondary carbon atom. Scrambling is fast, because the activation energy is low (32 kJ·mol^{-1}). For pentane, both scrambling and isomerisation to isopentane can occur, because opening of the C1–C2 (upper path) and of the C2–C3 (lower path) bonds give a secondary but not a primary carbenium ion when R = CH_3. Thus, a branched C_5 molecule forms.

NMR experiments with but-1-ene, ^{13}C-labelled in the C1 position, demonstrated that a 1,2-H shift is easier than the formation of a cyclopropane ring [7.31]. On non-acidic zeolite Na–FER, 1-^{13}C-but-1-ene did not react at room temperature (Fig. 7.21a), but on acidic H–FER it reacted within minutes to cis and trans 1-^{13}C-but-2-ene, in which the label remained on the outer carbon atom (Fig. 7.21b). It took one week for the label to shift from the outer C1 atom to the inner C2 atom (Fig. 7.21c). Increasing the temperature to 100 °C led to oligomerisation of butene but not to the formation of isobutene (Fig. 7.21d), demonstrating that opening the cyclopropane ring by the formation of a primary carbenium ion is very difficult. Kinetic measurements showed that the activation energy E_a was 41 kJ·mol^{-1} for the 1,2-H shift and 88 kJ·mol^{-1} for the scrambling of the ^{13}C label from position 1 to position 2.

Another reaction, in which the degree of branching increases, is the isomerisation of cyclohexane to methylcyclopentane (Fig. 7.22). This reaction plays an important role in the ring closure of alkanes to aromatic

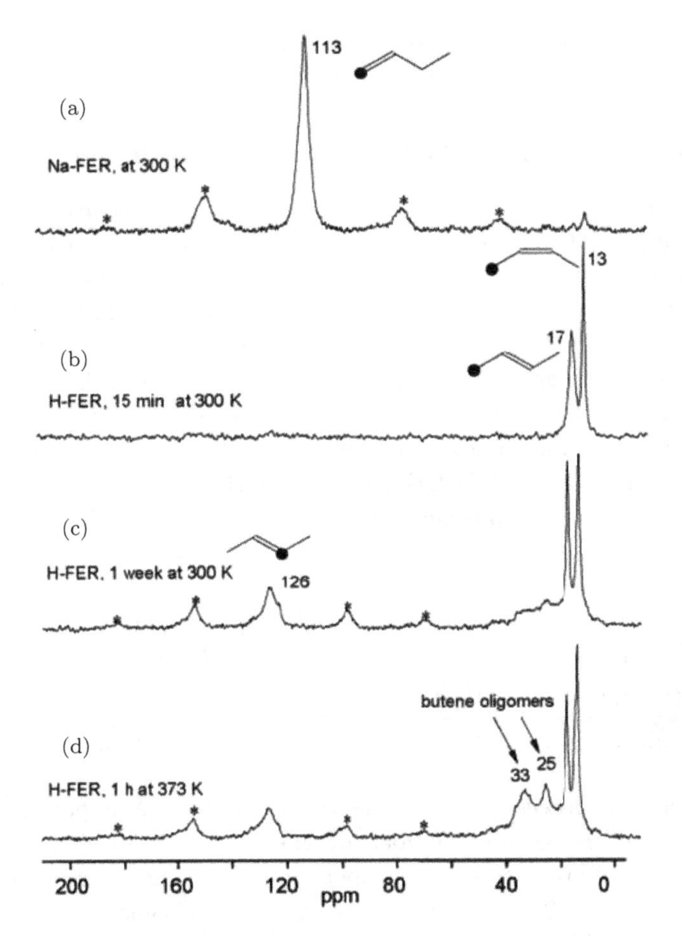

Fig. 7.21 ^{13}C CP/MAS NMR spectra of Na–FER and HFER loaded with [1-^{13}C]-*n*-but-1-ene. (a) Na–FER maintained at 300 K, (b) HFER at 300 K for 15 min, (c) HFER at 300 K for one week, (d) HFER at 373 K for 1 h, followed by measurement at 300 K. Asterisks indicate NMR spinning sidebands. Reprinted with permission from [7.31]. Copyright (2002) Elsevier.

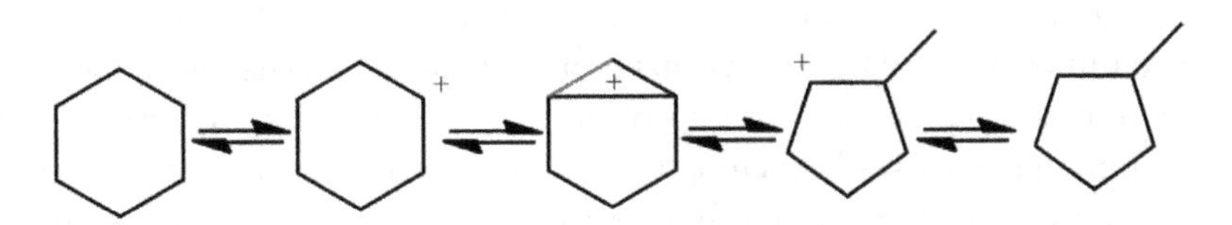

Fig. 7.22 Isomerisation of cyclohexane to methylcyclopentane by means of a cyclopropane intermediate.

molecules, as in catalytic reforming (Section 7.5.3), and also occurs by means of a protonated cyclopropane.

7.2.2 *Isomerisation of Pentane, Hexane and Butene*

Carbenium ions are transition states in chemical reactions. As discussed above, they form by protonation of alkenes and by transfer reactions between hydrocarbons and carbenium ions. In the reverse reaction, carbenium ions can react by dissociation to a proton and an alkene (CHR_2–$C^+R_2 \rightarrow CR_2{=}CR_2 + H^+$) or react by abstraction of a hydride atom from a hydrocarbon to an alkane ($R_1^+ + R_2H \rightarrow R_1H + R2^+$). Carbenium ions can also react by the breaking of a C–C bond in the beta position to the carbocation, forming an alkene and another carbenium ion ($CH_3CH(CH_3)$–$CH_2C^+HCH_3 \rightarrow CH_3C^+H(CH_3) + CH_2{=}CHCH_3$). As a result of these reactions, carbenium ions act as intermediates in the isomerisation of alkanes and alkenes:

$$\begin{array}{c} R_1^+ \rightarrow R_2^+ \\ R_1H + R_2^+ \rightarrow R_1^+ + R_2H \\ \hline \text{alkene} \rightarrow \text{isoalkene} \end{array} \qquad \begin{array}{c} \text{alkene} + H^+ \rightarrow R_1^+ \\ R_1^+ \rightarrow R_2^+ \\ R_2^+ \rightarrow \text{isoalkene} + H^+ \\ \hline \text{alkene} \rightarrow \text{isoalkene} \end{array}$$

In refinery processes, isomerisation reactions are important in the upgrading of the octane number of gasoline by reforming alkanes to isoalkanes and aromatics and isomerisation of pentane and hexane to branched isomers. They are also important in the isomerisation of butane to isobutane and butene to isobutene, the alkylation of butenes to isooctane, the hydroisomerisation of long alkanes to isoalkanes, and the hydrocracking of long to shorter hydrocarbons.

The **isomerisation of pentane** to isopentane and the impossibility of isomerising butene directly to isobutene were discussed above. The **isomerisation of hexane** to mono- and di-branched isomers is a process similar to the isomerisation of pentane (Fig. 7.23). *n*-Hexane becomes a carbenium cation by hydride transfer and isomerises to the methylpentane carbenium ion by means of a protonated cyclopropane structure. The carbenium ion increases branching to the 2,3-dimethylbutane carbenium ion by another protonated cyclopropane structure. Finally, 2,2-dimethylbutane forms after a 1,2-methyl shift. All carbenium ions become branched hexanes by proton transfer with hexanes. Branching increases the octane number

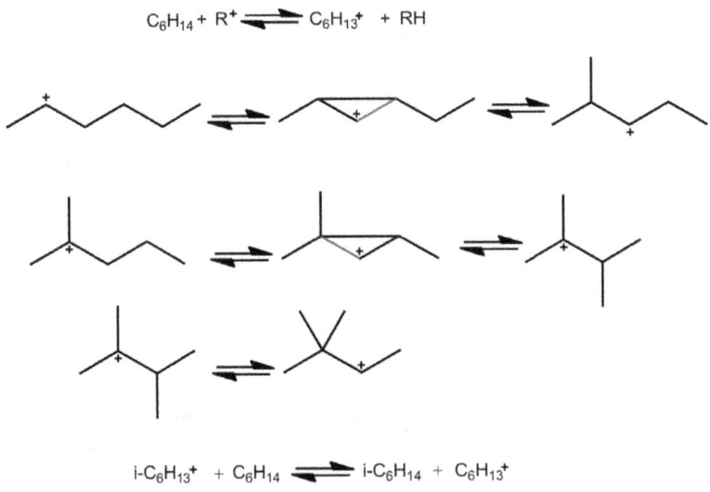

$$C_6H_{14} + R^+ \rightleftharpoons C_6H_{13}^+ + RH$$

$$i\text{-}C_6H_{13}^+ + C_6H_{14} \rightleftharpoons i\text{-}C_6H_{14} + C_6H_{13}^+$$

Fig. 7.23 Isomerisation of hexane to methylpentanes and dimethylbutanes.

from 31 for hexane to 74 for 2-methylpentane and 76 for 3-methylpentane, to 94 for 2,2-dimethylbutane and 105 for 2,3-dimethylbutane. Isomerisation increases the octane number of pentane from 62 to 94 for methylbutane. Isomerisation of pentane and hexane is an important reaction in the production of high-octane gasoline (Section 7.6).

The skeletal isomerisation of *n*-alkenes is important in upgrading refinery and petrochemical feeds. While for larger alkenes, shape-selective solid acid catalysts are applied successfully, the skeletal **isomerisation of *n*-butene** to isobutene was not achieved with conventional acid catalysts until the 1990s. However, ferrierite and other zeolites with medium-sized pores and silica–alumina phosphates (SAPOs) and metal aluminophosphates (MeAPOs) isomerised *n*-butene with yields close to the thermodynamic limit (50 mol% isobutene). Remarkably, the selectivity to isobutene improved continuously with time on stream. ^{13}C labelling of *n*-butene demonstrated that the initial reaction is bimolecular, while later it is monomolecular [7.32].

The bimolecular production of isobutene from *n*-butene occurs in three steps:

- dimerisation of butene to octene;
- isomerisation of octene to isooctene;
- β scission of isooctene to isobutene.

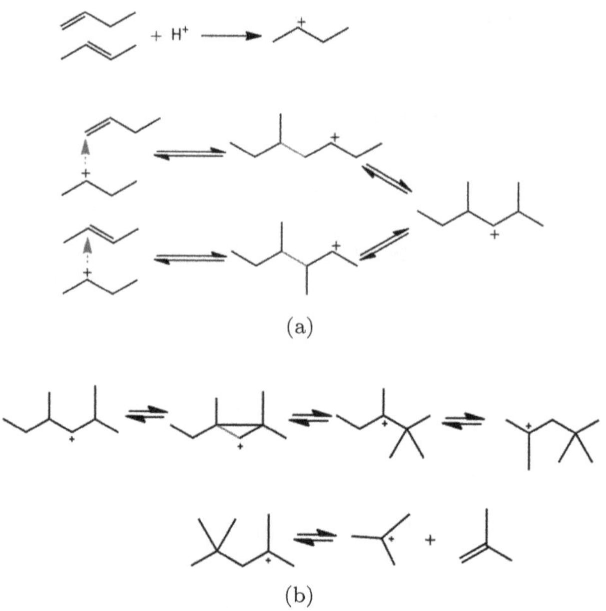

Fig. 7.24 Formation of isobutene by dimerisation to an octyl cation (a) and isomerisation to the isooctylcation followed by β scission (b).

The selective formation of isobutene is possible because of the fast isomerisation of octyl carbenium ions to tertiary octyl cations and the much faster cracking of the isooctyl cation than of other tertiary octyl cations. As a consequence, if there were no steric constraints, then isomerisation would continue until the triple-branched octyl cations formed (Fig. 7.24).

Of the five isomeric octyl cations (Fig. 7.25), only the 2,2,4-trimethyl octyl cation gives a tertiary carbenium ion upon β scission. This is the only cation that goes from a tertiary to a tertiary carbenium ion and does not need extra activation energy; it is the fastest cracking step. Beta scission of the 2,3,3-trimethylpentyl cation gives an ethyl cation while the 2,2,3-trimethylpentyl cation and 2,3,4-trimethylpentyl cation with C3 as carbocation give a methyl cation, all high-energy primary carbenium ions.

Only the 2,3,4-trimethylpentyl cation with C2 as carbocation gives a secondary propyl cation upon β scission. The 2,4-dimethylhexyl cation with C4 as carbocation also gives a secondary propyl cation upon β scission. In both cases, the other product molecule is isopentene. Similar small amounts of C_3 and C_5 by-products form in the isomerisation of butene, indicative of a bimolecular mechanism. Isobutene is the main product in liquid acids,

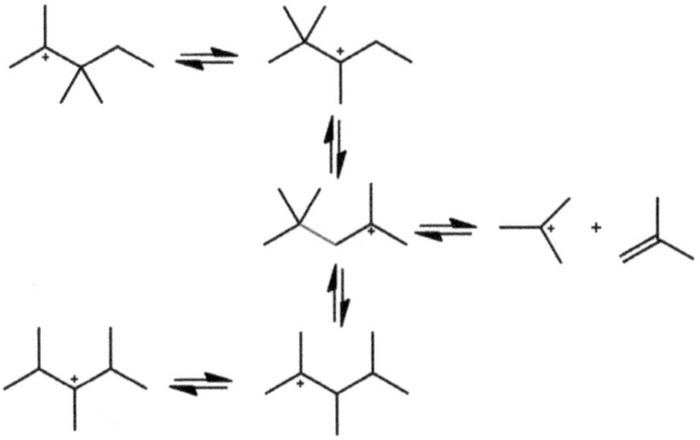

Fig. 7.25 Isomerisation of octyl cations to the isooctyl cation and β scission to isobutene and the isobutyl cation.

while in solid acids octyl cations crack to C_3 and C_5 molecules as well, because the formation of trimethylpentyl cations is sterically hindered in the zeolite pores.

How monomolecular production of isobutene occurs from n-butene [7.24] is still under discussion. An actual monomolecular mechanism occurring by a primary carbenium ion is unlikely. Thus, **pseudo-monomolecular mechanisms** have been proposed, such as the reaction of butene with cations of tertiary carbon atoms in coke precursors (e.g. cyclopentadienyl species) [7.33] or in methylpolyaromatics [7.34]. Isomerisation by a methyl and H shift followed by cracking gives isobutene. The initial formation of isobutene was bimolecular but, at a later stage, it was monomolecular, when a substantial amount of carbon had accumulated in and on the zeolite, because the concentration of the tertiary carbon atoms or methylpolyaromatics is constant. This was considered to be a strong indication of a pseudo-monomolecular mechanism (Fig. 7.26).

However, ferrierite with a high Si/Al ratio has a high isobutene selectivity and low propene-pentene selectivity, independent of time on stream, and only small amounts of carbon are formed. Therefore, another pseudo-monomolecular reaction was proposed, involving an autocatalytic mechanism with one molecule of n-butene and one molecule of isobutene [7.35]. This reaction of n-butene with isobutene, adsorbed on a protonic site (Fig. 7.27), is favoured by the long residence time of isobutene in the 10-ring pores due to constraints in the desorption of the branched molecule from the narrow channels.

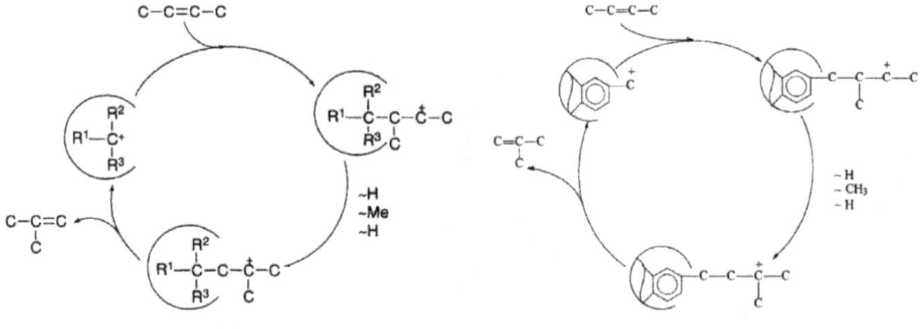

(a) (b)

Fig. 7.26 Pseudo-monomolecular mechanisms of butene isomerisation to isobutene over coked FER samples. Left figure reprinted with permission from [7.33]. Copyright (1995) the Royal Society Chemistry. Right figure reprinted with permission from [7.34]. Copyright (1998) Elsevier.

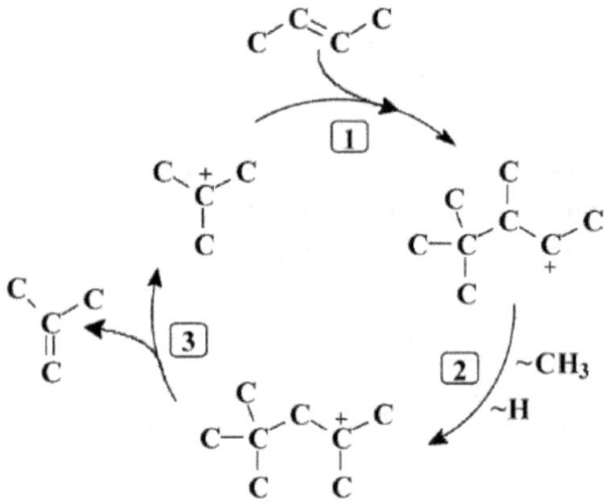

Fig. 7.27 Autocatalytic mechanism of n-butene isomerisation through t-butyl carbenium ions. Reprinted with permission from [7.35]. Copyright (2005) Elsevier.

Recently, the pure monomolecular reaction has been reconsidered again as a possible explanation of the isomerisation of n-butene. Calculations were carried out by the quantum chemical MP2 method for the active $AlSi_4O_4H$ site and a molecular mechanics model for the surrounding 37 zeolite T atoms to account for the van der Waals interaction, which plays a significant role in the interaction of the hydrocarbon with the zeolite [7.36]. The active centre ($AlSi_4O_4H$) and the probe molecule were allowed to relax, while the surrounding 37 T atoms remained fixed within the crystallographic structure. The calculations showed that the zeolite framework can cause a significant decrease of the activation energy, because in the carbenium ion-type transition state, which is more ionic than its alkoxy intermediate precursor, there are stronger stabilising interactions with the zeolite framework. As a result, the calculated activation energy of the monomolecular reaction was only 71 kJ·mol^{-1}, not much higher than the experimental value of 59 kJ·mol^{-1}. On the other hand, a calculation with a similar method indicated that the activation energy for the autocatalytic mechanism with one molecule of n-butene and one molecule of isobutene (Fig. 7.26) is lower than that for the pure monomolecular reaction [7.37].

β Scission of C_7 and C_6 carbocations occurs more slowly than scission of C_8 carbocations. Figure 7.28 shows possible β scissions of C_8, C_7 and C_6 carbocations. As discussed above, several triple-branched C_8 cations exist, but only the isooctyl cation begins as a stable tertiary cation and gives a tertiary cation upon scission (Fig. 7.25). This is impossible for C_7 and C_6 cations, for which the reaction starts or ends with a less stable secondary cation, while tertiary hexyl cations lead to primary cations. Only β scission

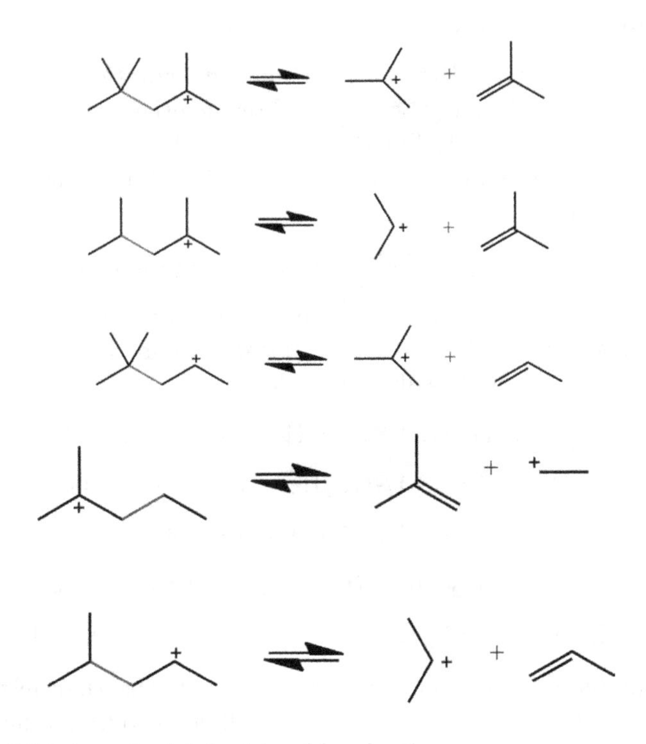

Fig. 7.28 β Scissions of octyl, heptyl and hexyl cations.

of the secondary 2-methylpentyl cation gives a secondary rather than a primary carbocation, but the equilibrium concentration of the secondary 2-methylpentyl cation is low, because it is in fast equilibrium with the much more stable tertiary 2-methylpentyl cation.

This explains why acid-catalysed cracking of hydrocarbons with eight or more carbon atoms is fast, whereas cracking of smaller hydrocarbons is slow. Therefore, industrial isomerisation of pentane and hexane is performed with acidic catalysts, while complete isomerisation of higher hydrocarbons is not possible; the losses would be too high due to the cracking to small hydrocarbons. Acid catalysts are suitable only for the partial conversion of higher hydrocarbons, as in the dewaxing of lubrication oils. The feed for lubrication oils, which must have a high boiling point, contains a high percentage of linear alkanes. These crystallise to wax at low temperature and are detrimental to the lubrication of car engines in the winter. Isomerisation of some of the alkanes to isoalkanes lowers the freezing point substantially.

7.3 Alcohols from Alkenes

Alcohols can be produced by direct or indirect hydration of alkenes with the aid of an acid catalyst [7.38]. In the direct synthesis of ethanol, ethene and steam are fed over a H_3PO_4/SiO_2 catalyst at 250–300 °C and 7 MPa. The reaction is exothermic, so conversion is only 5% and a large recycle of ethene is required.

$$C_2H_4 + H_2O \rightarrow C_2H_5OH \quad \Delta H = -46\,\text{kJ·mol}^{-1}.$$

In the indirect synthesis, ethene is in contact with concentrated sulfuric acid at 70–80 °C and 1–1.5 MPa in an absorption column, where esterification produces mono- and diethylsulfate:

$$CH_2{=}CH_2 + HO\text{–}SO_2\text{–}OH \rightarrow C_2H_5O\text{–}SO_2\text{–}OH$$

$$CH_2{=}CH_2 + HO\text{–}SO_2\text{–}OC_2H_5 \rightarrow C_2H_5O\text{–}SO_2\text{–}OC_2H_5.$$

Subsequent hydrolysis gives ethanol and dilute sulfuric acid:

$$C_2H_5O\text{–}SO_2\text{–}OH + H_2O \rightarrow C_2H_5OH + H_2SO_4$$

$$C_2H_5O\text{–}SO_2\text{–}OC_2H_5 + 2\,H_2O \rightarrow 2\,C_2H_5OH + H_2SO_4.$$

Because the temperature is lower, conversion is higher than in the direct synthesis, but the recovery of concentrated sulfuric acid from dilute sulfuric acid requires substantial amounts of energy.

2-Propanol (**isopropanol**) is produced from propene by acid hydration, in the same way as ethanol from ethene. **1-Propanol**, on the other hand, is produced by **hydroformylation** of ethene:

$$CH_2{=}CH_2 + H_2 + CO \rightarrow CH_3CH_2CHO.$$

The hydroformylation reaction occurs by homogeneous catalysis with a Rh hydrido carbonyl phosphine complex $RhH(P\emptyset_3)_3CO$ ($\emptyset$ is an aryl group) [7.39]. During reaction, the $RhH(P\emptyset_3)_3CO$ complex forms the $RhH(P\emptyset_3)_2CO$ complex. The elementary steps in the hydroformylation reaction are the addition of ligands (molecules or groups) to and the elimination of ligands from the coordination sphere of the Rh complex (Fig. 7.29). After adding a ligand, the number of ligands surrounding the Rh atom increases from four to five, while after elimination the number is again four. The first reaction is the addition of an alkene to the four-coordinated $RhH(P\emptyset_3)_2CO$ complex, forming a five-coordinated Rh complex with π-bonded alkene:

$$RhH(P\emptyset_3)_2CO + CH_2{=}CHR \rightarrow RhH(P\emptyset_3)_2CO(\pi{-}CH_2{=}CHR).$$

Addition of CO to the metal alkene complex would give a six-coordinated Rh complex, which would be sterically crowded. Therefore, the addition of CO induces a transformation of the π-bonded alkene to a σ-bonded

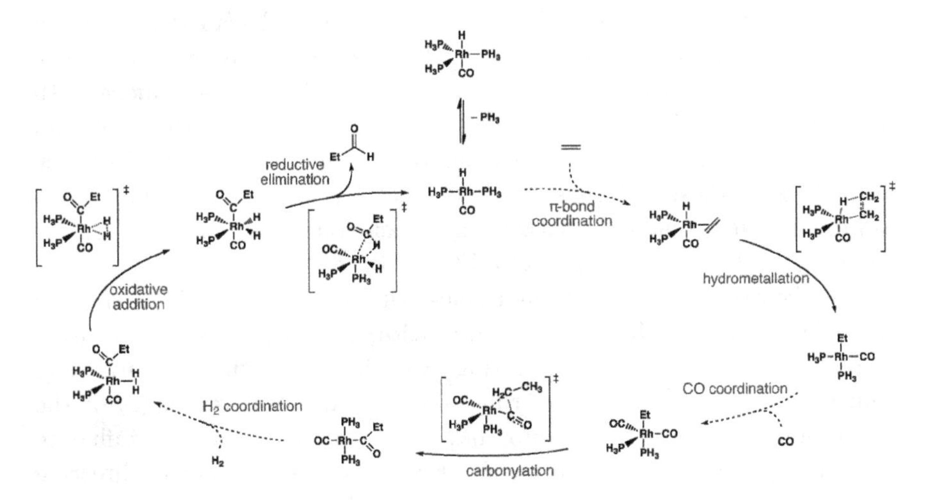

Fig. 7.29 Mechanism of the hydroformylation of ethene to propionaldehyde. Figure courtesy of Dr. Nick Greeves, licensed under CC BY-SA 3.0 www.chwmtube3d.com/oxo.Reaction.html.

alkyl group by an intramolecular reaction with the H atom to form a five-coordinated Rh complex:

$$RhH(P\varnothing_3)_2CO(\pi{-}CH_2{=}CHR) + CO \rightarrow Rh(P\varnothing_3)_2(CO)_2(CH_2CH_2R).$$

In the next step, the alkyl group migrates to one of the CO ligands, forming a four-coordinated Rh complex with an acyl group:

$$Rh(P\varnothing_3)_2(CO)_2(CH_2CH_2R) \rightarrow (RCH_2CH_2CO)Rh(P\varnothing_3)_2CO.$$

Finally, addition of hydrogen to the metal acyl complex liberates an aldehyde molecule and reforms the four-coordinated Rh hydrido carbonyl phosphine complex:

$$(RCH_2CH_2CO)Rh(P\varnothing_3)_2CO + H_2 \rightarrow RhH(P\varnothing_3)_2CO + RCH_2CH_2CHO.$$

In a separate reactor, hydrogenation of propionaldehyde results in 1-propanol:

$$CH_3CH_2CHO + H_2 \rightarrow CH_3CH_2CH_2OH.$$

7.4 Alkylation of Aromatics

7.4.1 *Ethylation and Propylation of Benzene*

Small aromatic molecules (benzene, toluene and xylene (BTX)) and alkenes ($C_2^=$, $C_3^=$ and $C_4^=$) are the main raw materials in the petrochemical industry. Liquid or solid acids are used as catalysts in several transformations of these molecules, for example in the alkylation of benzene with ethene or ethanol and propene or propanol, in the alkylation of toluene with methanol and in the isomerisation of *m*-xylene. Both the ethylation and propylation of benzene are of great importance in industry. Ethylbenzene is used to synthesise styrene, the monomer for making polystyrene and isopropylbenzene (cumene) is used to synthesise phenol and acetone ($C_6H_5{-}CH(CH_3)_2 + O_2 \rightarrow C_6H_5OH + CH_3COCH_3$).

The reaction of aromatic molecules with alkenes and alcohols is an example of the **Friedel–Crafts alkylation** reaction, in which an electrophile reacts with an aromatic ring. Substituents of the aromatic ring, which donate electrons to the ring, stabilise the positive charge of the electrophile. Thus, methyl groups have an ortho-para-directing influence and nitro groups, which withdraw electrons from the ring, a meta-directing influence. Strong acids function as catalysts by converting the reactant to an electrophile. In organic laboratories, $AlCl_3$ ($RCl + AlCl_3 \rightarrow R^+ + AlCl_4^-$) and H_2SO_4 ($ROH + H_2SO_4 \rightarrow R^+ + HSO_4^- + H_2O$) are still used as the catalyst and alkyl halides and alcohols as reagents, but in industry zeolites are the preferred catalysts and alkenes the preferred reagents. Thus, ethylbenzene and cumene (2-propylbenzene) are produced by alkylation of benzene with ethene and propene, respectively. Since the 1980s, zeolite catalysts (H–ZSM-5, beta, Y, MCM-22) are used in gas-phase and liquid-phase processes. In the original gas-phase Mobil–Badger process a ZSM-5 catalyst was used at about 400 °C, a pressure of a few MPa and a high benzene to alkene ratio of 5–20, to minimise polyalkylation of the benzene and tedious separation steps [7.40]. Nevertheless, poly-alkylbenzenes are always produced, because alkyl groups increase the basicity of the aromatic molecule and further alkylation is, thus, always faster than the first alkylation. The poly-alkylbenzenes are recycled and react with benzene in the alkylation reactor or in a separate transalkylation reactor if the catalyst in the alkylation reactor does not catalyse transalkylation efficiently. Zeolite Y, beta or omega is used in transalkylation (Section 7.4.3). Figure 7.30 gives a flow diagram of an ethylbenzene process.

In the 1990s, Mobil–Badger (now ExxonMobil–Raytheon) developed the liquid-phase EBMax process with an MCM-22 catalyst in the alkylation reactor and an undisclosed catalyst in the transalkylation unit [7.40, 7.41]. MCM-22 is much more selective than USY and beta and produces significantly less diethylbenzene, which allows the plant to run for several years even at a relatively low $B/C_2^=$ ratio of about 4. By 2017, EBMax technology had been installed with a total output of 20 MT/a. Another process is based on catalytic distillation (CDTech). A zeolite Y catalyst is in the upper section of a reactor column and the lower section contains

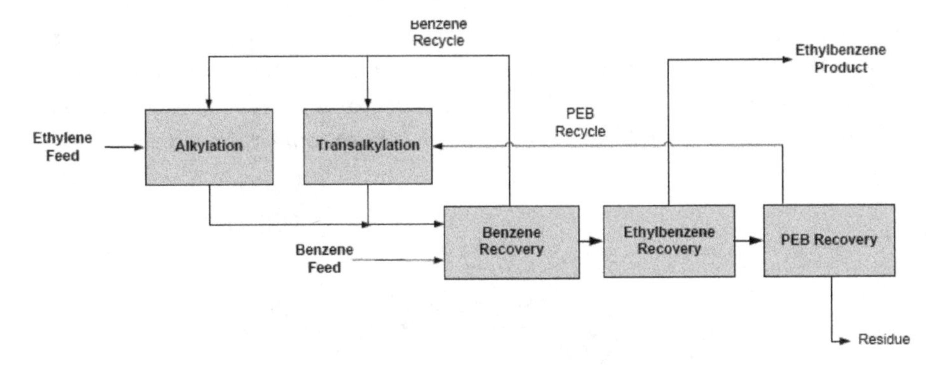

Fig. 7.30 An ethylbenzene process diagram with an alkylation reactor, transalkylation reactor and three distillation units (PEB = poly-ethylbenzene).

distillation trays. Ethene is fed into the middle section of the reactor column, below the catalyst bed, and benzene is introduced at the top of the reactor column. The counter-current flow reduces coke formation and limits the production of unwanted by-products.

Whereas textbooks on organic chemistry indicate that Friedel–Crafts reactions occur by the attack of an electrophile on an aromatic ring, reactants and intermediates on zeolites tend to be covalent structures rather than ionic structures (Section 7.2). This raises the question as to how alkylation of benzene in the pores of a zeolite occurs. Experimental and theoretical studies indicate that ethene and benzene may react on zeolites by a direct one-step and an indirect two-step mechanism. In the one-step mechanism, the alkene interacts with the Brønsted acid site, while benzene adsorbs edge-on next to the adsorbed alkene (Fig. 7.31). In the two-step mechanism, first an ethoxide intermediate is formed by reaction of ethene with an OH group of the zeolite, which then reacts in a second step with benzene. In industry, a high benzene to ethene ratio ($\gg$1) is used and the probability of finding ethene at an unoccupied intersection is low, because of the much stronger adsorption of benzene than of ethene in the channel intersection. Therefore, under industrial conditions, the two-step mechanism is less relevant than the one-step scheme because protonated ethene would react with benzene directly, and a surface ethoxide would not form.

The interactions of alkylated aromatics with zeolite pore channels are dominated by van der Waals interactions and DFT calculations do not correctly account for such interactions, because they underestimate adsorption energies and energy barriers. Also the long-range crystal potential and steric effects caused by confinement within the pores of these materials cause problems. More accurate DFT calculations, in which periodic boundary conditions were applied, were combined with MP2 energy calculations on a series of cluster models of increasing size to account for contributions from higher-order correlation effects [7.42]. These sophisticated calculations

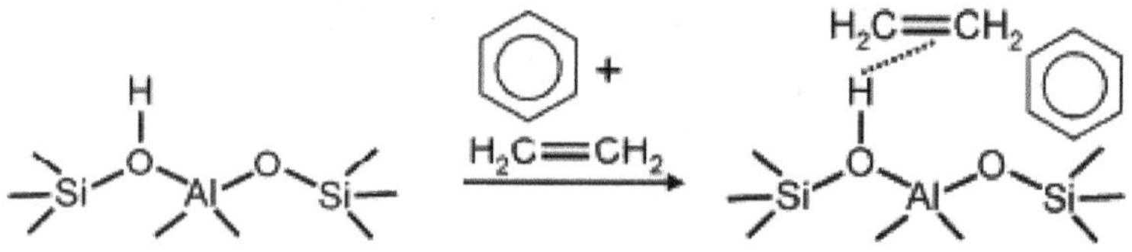

$$\text{Si–O(H)–Al–O–Si} \xrightarrow[\ H_2C{=}CH_2\ ]{C_6H_6\ +} \text{Si–O(H}\cdots H_2C{=}CH_2\text{)–Al–O–Si} + C_6H_6$$

Fig. 7.31 One-step ethylation of benzene by ethene.

reached better agreement with experimental results. For instance, without these corrections, DFT wrongly predicted that the adsorption strength decreases in the sequence ethene > benzene > ethylbenzene, but the full calculation predicts the correct sequence. The calculations demonstrated that a reliable description of hydrocarbon reactions in zeolites is only possible with a method that includes dispersion, limits the self-interaction correction error and takes long-range electrostatic effects into account.

At the beginning of the reaction between ethene and benzene, the acid proton is attached to an oxygen atom bridging an Al and a Si atom. The proton interacts with ethene, which is adsorbed as a π complex. After attack of the acid proton on ethene, a positively charged carbenium ion forms, which performs an electrophilic aromatic substitution on benzene. In the transition structure, the proton is already attached to one carbon atom of ethene, the bond between the carbon atoms of ethene has lost its double bond character, whereas the other carbon atom has not yet formed a bond with benzene. After the transition structure, the bond between the ethyl fragment and benzene forms and a proton of benzene returns to the zeolite to restore electrical neutrality. Ethylbenzene then adsorbs on an acid site of the zeolite cluster.

The mechanism for the formation of cumene is similar to that for the formation of ethylbenzene, but oligomerisation is a problem in cumene synthesis, because propene reacts much faster than ethene over Brønsted acid catalysts. Most cumene is still produced from benzene and propene with a solid phosphoric acid (SPA) catalyst at about 200 °C, 3–4 MPa and a C_6H_6/C_3H_6 ratio of about six. More than 40 SPA plants are licensed worldwide. The SPA catalyst consists of polyphosphoric acid supported on kieselguhr (also called diatomaceous earth, a naturally occurring siliceous sedimentary rock composed of the fossilised remains of hard-shell algae). To maintain the desired level of activity, small amounts of water are fed continuously into the reactor. The water liberates H_3PO_4, which causes downstream corrosion. Approximately 4–5 wt.% of the product consists of poly-isopropylbenzenes, which cannot be recycled to the alkylation reactor, because H_3PO_4 does not catalyse transalkylation. The formation of poly-alkylated compounds and propene oligomers is minimised by operating at a high C_6H_6/C_3H_6 ratio in a reactor with several small catalyst beds and spreading the propene feed over the beds. In the past decades zeolite-based cumene processes have been introduced. Their operating and maintenance costs are lower, there is no corrosion and no need for the disposal of H_3PO_4 supported on kieselguhr. Zeolites are regenerated and disposed of by digestion or on landfill sites. Zeolite-based cumene processes are similar to ethylbenzene processes and use a separate transalkylation reactor. The catalysts are zeolite Y (CDTech), MCM-22 (Mobil–Raytheon), beta (UOP) and dealuminated (wide-pore) mordenite (Dow) [7.41].

7.4.2 *Methylation of Toluene*

In contrast to the ethylation and propylation of benzene, the methylation of benzene to toluene is not an industrial process; the refinery already produces more toluene from oil than is required in the chemical industry. Because the demand for xylene is much higher than for toluene, a substantial amount of toluene is transformed to xylene by disproportionation (2 T $\rightarrow$ B + X), transalkylation with trimethylbenzene (T + TMB $\rightarrow$ 2 X) and methylation with methanol (Fig. 7.32).

The most important xylene isomer is *p*-xylene, which is used mainly for the production of terephthalic acid, an intermediate of polyethylene terephthalate (PET). However, a thermodynamic mixture of xylenes is often obtained and this has an ortho : meta : para ratio of about 22 : 53 : 25. Production of *p*-xylene can be increased by isomerisation of the xylenes on an acid catalyst, followed by separation and reintroduction of the *o*- and *m*-xylene to the isomerisation step. Mechanisms of the methylation of toluene will be discussed in this section and mechanisms for the disproportionation of toluene in the next section.

As in the ethylation of benzene, two mechanisms have also been proposed for the methylation of toluene, an indirect two-step mechanism,

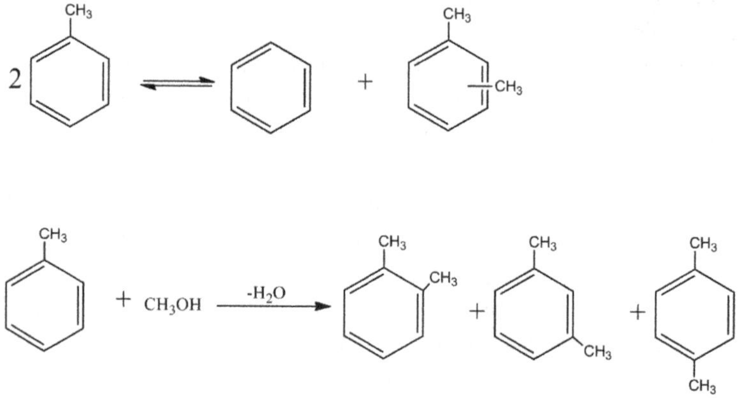

Fig. 7.32 Production of xylenes by disproportionation of toluene and methylation of toluene.

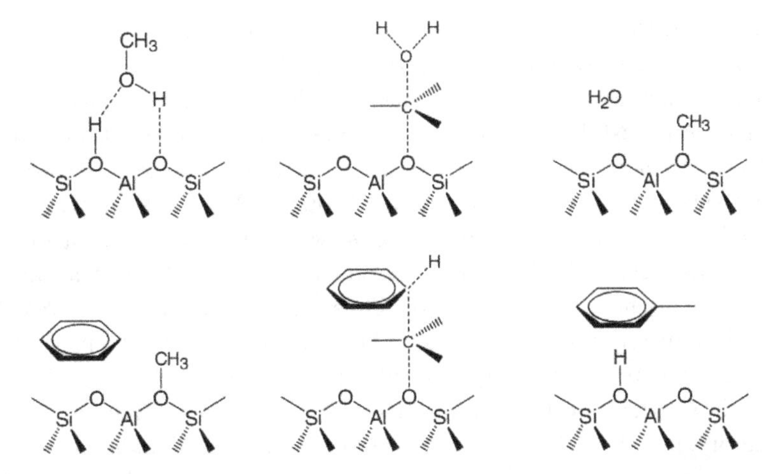

Fig. 7.33 Indirect methylation of benzene: A methoxide group forms, which then methylates benzene. Reprinted with permission from [7.43]. Copyright (2011) Springer Science + Business Media.

involving a surface-bound methoxide intermediate, and a direct one-step mechanism, in which physisorbed methanol directly interacts with the aromatic molecule. FTIR studies provided evidence for the indirect, stepwise mechanism (Fig. 7.33) [7.43]. Exposure of H–ZSM-5 to CD_3OH at 150 °C created a high concentration of surface-bound deuterated methoxide groups. Upon subsequent exposure to benzene at 220 °C, the deuterated methoxide groups reacted to $C_6H_5CD_3$, indicating that benzene is methylated by methoxide groups. Similarly, ^{13}C NMR measurements indicated that toluene is methylated by methoxy groups, because methoxy signals decreased and xylene signals appeared. Unfortunately, these studies were not performed *in situ*, but in a stepwise fashion and Svelle *et al.* concluded that a definitive conclusion with regard to the prevailing pathway has not yet been reached [7.43].

Rate constants, calculated for the methylation of benzene, were larger for H–ZSM-5 than for H-beta at 350 °C, in agreement with experimental rate constants. Because the pores of H–ZSM-5 are narrower than those of H-beta, both the adsorption energy and the entropy of benzene and methanol on the active site and the zeolite walls are more negative for H–ZSM-5 than for H-beta, but the stronger interactions with H–ZSM-5 than with H-beta are more important than the greater loss of entropy [7.44].

To obtain *p*-xylene with a purity of $> 99\%$ (as required in the production of PET), *m*-xylene, *o*-xylene and ethylbenzene must be separated from *p*-xylene, and the resulting *o*- and *m*-xylene isomers must be subsequently isomerised to the equilibrium *o*, *m*, *p* mixture. Since the equilibrium amount of para isomer in the xylenes is only about 25%, separation and isomerisation must be repeated many times. The discovery that a higher selectivity to *p*-xylene is obtained when medium pore zeolites are used as catalysts was an important breakthrough. Reactions in the narrow pores of zeolites have steric restrictions, which may result in a different product than that obtained in solution or on the surface of a catalyst, where there are no constraints [7.45]. Thus, with microcrystalline ZSM-5 an equilibrium mixture of xylenes (25% para) was obtained in the methylation of toluene, whereas with large crystals of ZSM-5 about 47% *p*-xylene was obtained. An increase to 90% selectivity was achieved when the catalyst was impregnated with phosphorus and boron compounds. The larger *o*- and *m*-xylene isomers diffuse more slowly than *p*-xylene, the diffusivity of which is three orders of magnitude greater than that of *o*- and *m*-xylenes. The xylene isomers that form in the zeolite pores must escape from the zeolite crystal; this is more difficult when the pore openings are narrowed by the deposition of phosphorus and boron compounds near the pore mouths. Thus, while *p*-xylene can easily diffuse from the zeolite crystal, *o*- and *m*-xylene molecules hardly diffuse away and continue to be isomerised as pore openings are narrow (Fig. 7.34). This results in selective production

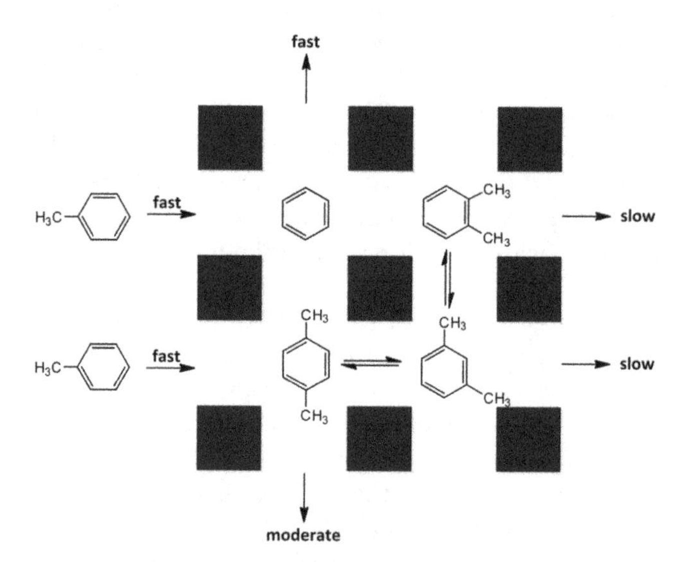

Fig. 7.34 Shape-selective enhancement of the para selectivity in the methylation of toluene. Reprinted with permission from [7.46]. Copyright (1982) Elsevier.

of *p*-xylene. However, isomerisation of *p*-xylene to *o*- and *m*-xylene can still occur on acid sites on the external surface. Zeolites with small crystals have a relatively large external surface area and, therefore, a para selectivity close to equilibrium, while zeolites with large crystals have a relatively small external surface area, which results in a significantly higher *p*-xylene concentration. Treatment with alkylchlorosilane or organic bases, deposition of tetraethyl orthosilicate (TEOS) and coking deactivate the external zeolite surface, which improves the *p*-xylene selectivity.

7.4.3 *Isomerisation, Disproportionation, Transalkylation*

When *p*-xylene is produced by the methylation of toluene, *o*- and *m*-xylene are co-produced, even with modified H-ZSM-5 catalysts. This requires **isomerisation** of the isomers to an equilibrium mixture of *o*-, *m*- and *p*-xylene, followed by separation and recycling of the *o*- and *m*-xylene. Another method of producing *p*-xylene is the **disproportionation** of two toluene molecules to one benzene and one xylene molecule (2 T → B + X) or the **transalkylation** of toluene with trimethylbenzene (T + TMB → 2 X). In principle, **disproportionation** and transalkylation are similar bimolecular reactions, while the isomerisation of aromatics occurs by an intramolecular mechanism as well as by an intermolecular mechanism.

The isomerisation of xylene over acidic catalysts is mainly intramolecular, involving a carbenium ion intermediate, which undergoes a 1,2-methyl shift, similar to that of alkyl carbenium ions (Section 7.2). Hence, *o*-xylene converts to *m*-xylene, *m*-xylene to *o*- and *p*-xylene, and *p*-xylene to *m*-xylene. The direct reaction of *o*- to *p*-xylene is not possible (Fig. 7.35).

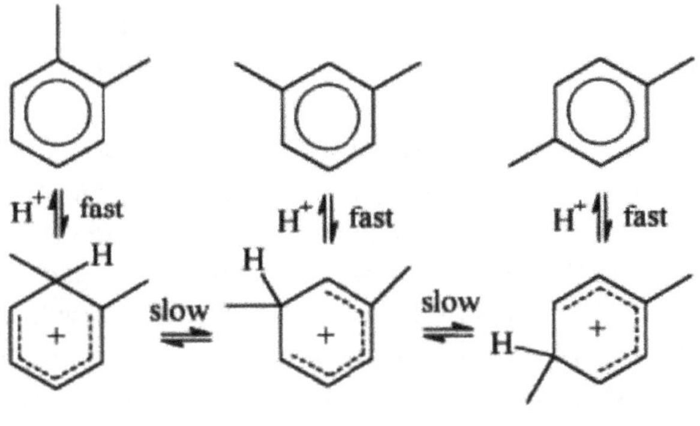

Fig. 7.35 Mechanism of the isomerisation of xylenes.

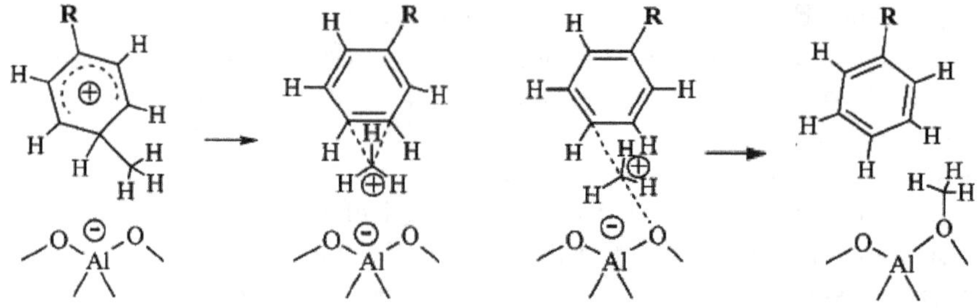

Fig. 7.36 Left: Protonated toluene (Wheland structure) and transition state in the 1,2-methyl shift mechanism. Right: Transition state and methoxy intermediate in indirect isomerisation. Reprinted with permission from [7.47]. Copyright (2001) the American Chemical Society.

Two mechanisms of the 1,2 shift have been proposed: a direct shift of the methyl group from a carbon atom to the neighbouring atom (Fig. 7.36, left) and the indirect shift of the methyl group, first to the zeolite, forming a methoxide (Fig. 7.36, right), followed by rotation of the aromatic ring and transfer of the methyl group to the next carbon atom. A DFT study indicates that both mechanism are possible [7.47].

Toluene is protonated before it isomerises. The protonated toluene (Wheland structure) does not correspond to a transition state but to an inflection point on the reaction pathway toward isomerisation. A phenoxy intermediate (in which the protonated toluene is covalently bonded to an oxygen atom) has a somewhat higher energy than the Wheland structure [7.47].

At 200 °C, a ZSM-5 catalyst with small crystals exhibits xylene isomerisation kinetics consistent with a stepwise 1,2-methyl shift mechanism. However, with increasing crystallite size and increasing temperature, the kinetics can be fitted only when there is a contribution by a direct *o*- to *p*-xylene path [7.46]. This is due to the influence of diffusion on the intrinsic kinetics. Molecules stay in the zeolite pores of the larger crystallites long enough for several isomerisation steps to occur, because higher temperature and/or longer pores favour the chemical isomerisation reaction over diffusion. The products outside the zeolite crystals are, thus, in part the result of multistep reactions. Since *p*-xylene diffuses much faster than *o*- and *m*-xylene, enhancement of *p*-xylene will occur. In the presence of diffusional effects, the **apparent kinetics** may change from a single-series reaction to a more complex reaction, which incorporates an apparent direct 1,4-methyl shift from *o*- to *p*-xylene. This effect is referred to as **disguised kinetics**. We already encountered disguised kinetics in the pseudo-monomolecular isomerisation of butene to isobutene (Section 7.2.2).

Xylene isomerises by a monomolecular reaction as well as by a bimolecular disproportionation reaction of two xylene molecules to toluene and trimethylbenzene (TMB) followed by transalkylation between TMB and another xylene molecule [7.48–7.50]. Because the product xylene molecule can be a different isomer than the reactant xylene, isomerisation of xylene takes place, for example,

$$2\,o\text{-}C_6H_4(CH_3)_2 \rightarrow C_6H_5CH_3 + 1,2,4\text{-}C_6H_3(CH_3)_3$$

$$o\text{-}C_6H_4(CH_3)_2 + 1,2,4\text{-}C_6H_3(CH_3)_3$$
$$\rightarrow 1,2,4\text{-}C_6H_3(CH_3)_3 + m\text{-}C_6H_4(CH_3)_2.$$

The main trimethylbenzene isomer formed in these reactions is 1,2,4-$C_6H_3(CH_3)_3$ and not 1,2,3-$C_6H_3(CH_3)_3$, because its formation is less sterically hindered. The reverse reaction of the first step, $C_6H_5CH_3 + C_6H_3(CH_3)_3 \rightarrow 2\,C_6H_4(CH_3)_2$, may also give an isomerised xylene molecule, but since the concentration and basicity of toluene are lower than for xylene, the contribution of the reversed first step is much smaller than that of the second step.

Disproportionation is the transfer of an alkyl group from a molecule to an identical molecule, as in

$$2\,C_6H_5R \rightarrow C_6H_6 + C_6H_4R_2,$$

whereas transalkylation is the transfer of an alkyl group from one molecule to a different molecule, as in

$$C_6H_5R' + C_6H_3R_3 \rightarrow RC_6H_4R' + C_6H_4R_2.$$

Both reactions are bimolecular alkyl transfer reactions.

7.5 Gasoline Production

7.5.1 *Fluid Catalytic Cracking and Hydrocracking*

In a refinery, crude oil is distilled at atmospheric pressure in fractions with different boiling point ranges: gas (<25 °C), naphtha (40–180 °C), kerosene (180–250 °C), diesel (250–350 °C), gasoil (350–550 °C) and residue. The residue is further distilled in a vacuum distillation unit into light and heavy gasoil and vacuum residue. Crude oil does not contain enough lower boiling fractions to satisfy demand for fuels such as gasoline (produced from naphtha, Section 7.5.3), kerosene and diesel. Therefore, in a refinery higher boiling fractions such as gasoil or vacuum gas oil are transformed into lower boiling point fractions, to produce more of the valuable fuel fractions.

In cars with Otto engines, gasoline is compressed and then ignited by a spark. Self-ignition (detonation), without a spark, must be avoided and, therefore, the gasoline must have a high **octane number**. The more the fuel can be compressed before self-ignition, the higher the octane number. Fuels with a higher octane number are used in cars with a higher-compression engine and yield higher power. In contrast, fuels with high cetane number (and low octane number) are used in diesel engines. Diesel engines do not compress the fuel, but compress only air and then inject fuel into the air, which was heated by compression. Ignition takes place by self-ignition. Branched alkanes and aromatic molecules have a high octane number and low cetane number, whereas the reverse is true for linear alkanes. This often means that the naphtha fraction of crude oil (with boiling point between 40 and 180 °C) does not have a sufficiently high octane number and that linear alkanes must be changed into isomers or naphthenes (cyclic hydrocarbons). This will be described in Section 7.5.3. Thus, naphtha becomes gasoline.

Fuels can not only be produced from oil but also from coal in the Bergius process (Section 6.3) and from synthesis gas in the Fischer–Tropsch (FT) process (Section 6.4.1.3). The synthesis gas for the FT process can come from coal, oil or natural gas. Another possibility to make fuels from synthesis gas is to transform synthesis gas in methanol and the methanol in hydrocarbons. The last step is called the methanol-to-hydrocarbons (MTH) process and will be described in Section 7.5.2.

The transformation of higher-boiling crude oil fractions into lower boiling point fractions occurs by cracking hydrocarbons with a higher molecular weight (200–600) into smaller molecules by bringing the feed into contact with a hot, powdered solid acid catalyst at elevated temperature. The cracking takes place by fluid catalytic cracking (FCC) or by hydrocracking. In the USA and China about one-third of the crude oil that is handled in refineries is processed in FCC units to produce high-octane gasoline and other fuels. FCC units are less common in Europe, because there the demand for fuels is satisfied by hydrocracking and catalytic reforming. FCC is a continuous process that takes place in a riser reactor, while the catalyst is regenerated in a regenerator (Fig. 7.37). Hot catalyst powder, which comes from the regenerator (see below), leaves the regenerator at about 715 °C and flows to the bottom of the riser reactor, where it meets and vapourises the gasoil feed. The hydrocarbon vapour "fluidises" the

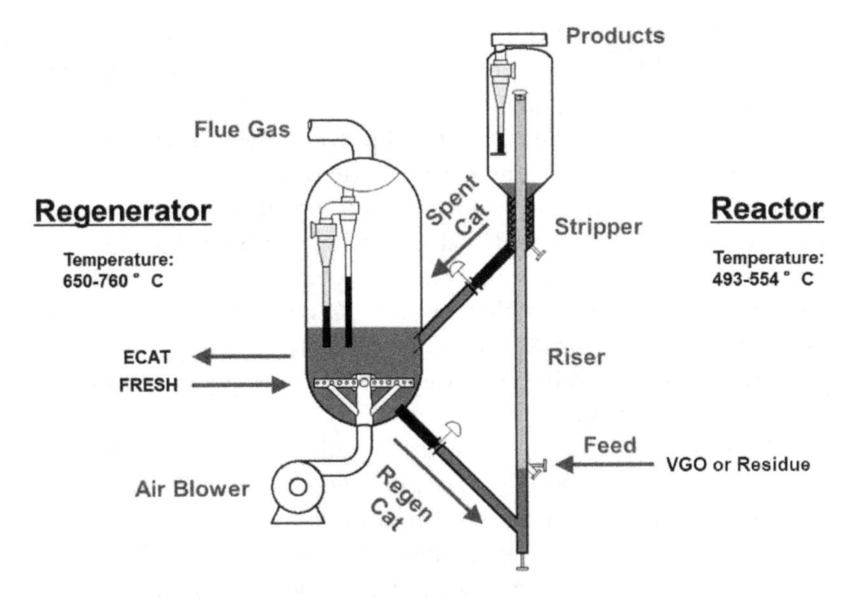

Fig. 7.37 Schematic depiction of FCC process with reactor and regenerator. Reprinted with permission from [7.51]. Copyright (2015) Royal Society of Chemistry.

powdered catalyst (entering the reactor at about 535 °C and 0.17 MPa) and, within a few seconds, the gasoil molecules are cracked into smaller gaseous molecules. Because cracking increases the number of molecules and, thus, the pressure, the gas and solid flow upwards in the riser reactor.

The selectivity to gasoline in the FCC process is in the order of 50%. The catalyst cycles between 500 and 760 °C and moves at great speed, so that the catalyst is exposed to harsh reaction conditions and, as a result, deactivates. FCC catalyst particles have an average lifetime in the order of about one month. A small portion of the complete catalyst inventory of the regenerator is removed daily and replaced by fresh catalyst in order to maintain a steady level of activity. The catalyst in this steady state is called equilibrium catalyst. Depending on the size of the FCC unit and the operational parameters, catalyst withdrawal rates can be between 1 and 30 tons per day.

Cyclones separate the cracked product gas from the spent catalyst, which flows downward to the catalyst regenerator. The catalyst must be regenerated because cracking reactions produce alkenes, which can further crack or polymerise and form carbonaceous material (coke) that deposits on the catalyst and very quickly reduces the catalyst reactivity. The catalyst is regenerated by burning off the deposited coke with air at about 715 °C and 0.24 MPa. A fluid-bed reactor is better suited to handle the large amount of heat that is produced than a fixed-bed reactor. The generated heat is absorbed by the regenerated catalyst and provides the heat required for the gasification of the gasoil feed and the endothermic cracking reactions that take place in the riser reactor. The hot catalyst that leaves the regenerator at about 715 °C flows to the feedstock injection point below the riser reactor. About 5 kg catalyst per kg of feedstock circulate between the regenerator and the reactor. This means that in an FCC unit that processes 100,000 barrels per day about 75,000 T catalyst circulates per day. About 10% of this catalyst mass is used in the process, so the total worldwide production of FCC catalyst is about 840,000 T per year.

FCC catalysts are fine powders with a bulk density of 0.80–0.96 g/cm^3 and an average particle size of 60–100 μm. These properties enable them to become fluidised in the hot gas stream of the riser reactor. FCC catalysts should have good stability to high temperature and to steam, high activity, large pore sizes to accommodate large feed molecules, good resistance to attrition and low coke production. Modern, highly active FCC catalysts produce less coke. This creates a problem because then not enough heat is provided in the regenerator required in the riser reactor for the cracking. To improve full combustion to CO_2, and thus create more heat, modern FCC catalysts contain some Pt to enhance combustion of CO. FCC catalysts are composed of four major components: a crystalline zeolite, a matrix, a binder and a filler. A modified form of zeolite Y is used as the active component (15–50 wt.% of the catalyst) in FCC catalysts, e.g. rare earth exchanged Y (REY) and ultra-stable Y (USY). Rare earth elements stabilise the zeolite framework and increase its thermal and hydrothermal stabilities. USY is zeolite Y in which some of the framework aluminium has been selectively removed to enhance thermal and hydrothermal stability. The alumina matrix component also contributes to catalytic activity. The binder (usually silica sol) and filler (usually a clay, e.g. kaolin) provide the physical strength and integrity of the catalyst.

The principle reactions that occur on the acid sites in the FCC riser reactor are the isomerisation of alkenes A to mono-branched alkenes (m-br-A), then to di-branched alkenes (di-br-A) and finally to tri-branched alkenes (tri-br-A) (Section 7.2.1) and the fast cracking of the tri-branched alkenes by β-scission (cf. Fig. 7.23).

$$\text{n-A} \rightarrow \text{m-br-A} \rightarrow \text{di-br-A} \rightarrow \text{tri-br-A} \rightarrow \text{br-A}_1 + \text{br-A}_2.$$

The small number of alkenes present in the feed is enough to start the reaction, because the product alkenes undergo hydride exchange with alkanes in the feed, giving new reactant alkenes. Sometimes the FCC process is used to coproduce propene (for polymerisation to polypropene). In that case, ZSM-5 is added to the FCC catalyst. With its stronger acidity than zeolite Y, ZSM-5 enhances selectivity to propene and also increases the octane number of the produced gasoline, because of better acid-catalysed isomerisation.

Hydrocracking is another process that cracks long alkanes to smaller alkanes. While FCC uses an acidic catalyst, hydrocracking uses a bifunctional catalyst, containing metal and acid sites. The metal sites are responsible for dehydrogenation of the alkanes to alkenes, which are isomerised and cracked on the acid sites (as in the FCC process), while the resulting alkenes are hydrogenated to alkanes on the metal sites. Condensation reactions between alkenes can also occur, eventually leading to coke formation. By keeping the alkene concentration low, the metal prevents deactivation by coking. The ratio of the metal to acid sites determines the activity and selectivity of the hydrocracking catalyst. An increase in the metal loading increases the catalyst activity until a plateau is reached and thermodynamic equilibrium is reached between the alkane reactant and its corresponding alkene [7.52]. The selectivity to cracking also depends on the metal loading. At low metal loading, the number of metal sites is not sufficient to fast enough hydrogenate the branched alkenes that are formed on the acid sites, and they undergo cracking. Pt has a high (de)hydrogenation activity, and Pt on zeolite catalysts are good hydro-isomerisation catalysts. Transition metal sulfides have a lower (de)hydrogenation activity than Pt, which modifies their bifunctional balance and catalyst performance. For NiMo sulfide catalysts with relatively high zeolite loading, the insufficient (de)hydrogenation activity of the metal sulfide phase leads to a high cracking selectivity.

FCC and hydrocracking differ in the treatment of coke formation and, as a consequence, in the process layout. In FCC, the monofunctional acid catalyst not only cracks the large alkanes to smaller molecules but also forms alkenes that lead to coke formation. That coke is removed in a regenerator (Fig. 7.36). In hydrocracking, coke formation is prevented by hydrogenation of the alkenes on the metal sites of the bifunctional catalyst. The FCC process runs at relatively low pressure in the absence of hydrogen in a fluid-bed riser reactor to quickly transport the catalyst to the regenerator to remove the coke and re-establish good catalyst activity. The hydrocracking process, on the other hand, keeps catalyst activity by operating in a fixed-bed reactor at high hydrogen pressure (10–20 MPa). The use of sulfided NiMo as the (de)hydrogenation component in hydrocracking allows it to process dirty feedstocks, because this catalyst can remove impurities, as we shall see in Chapter 8.

7.5.2 *Methanol to Hydrocarbons*

In the past decades, methanol has attracted attention as an alternative feedstock to crude oil, because it can be produced from natural gas, coal and biomass. Methanol can be used directly as fuel, or indirectly by transformation in a hydrocarbon fuel. It is also used as feed in the preparation of small alkenes. Light alkenes are traditionally produced by refining crude oil and cracking of the resulting naphtha and gasoil fractions, but the low prices of coal and natural gas relative to oil led to the development of new technologies. The zeolite-catalysed methanol-to-hydrocarbons (MTH) reaction, discovered by Mobil (now Exxon–Mobil) in 1976, has led to a methanol-to-gasoline (MTG) technology, which makes use of inexpensive coal and natural gas as a feedstock for methanol.

The MTG process is one possibility for replacing oil refinery processes for the production of liquid fuels (Fig. 7.38). The other two are the Fischer–Tropsch process (Section 6.4.1.3) and the Bergius coal-liquefaction process (Section 6.3). The data in Table 7.1 demonstrate that the product distribution of the MTG process differs from the distribution obtained

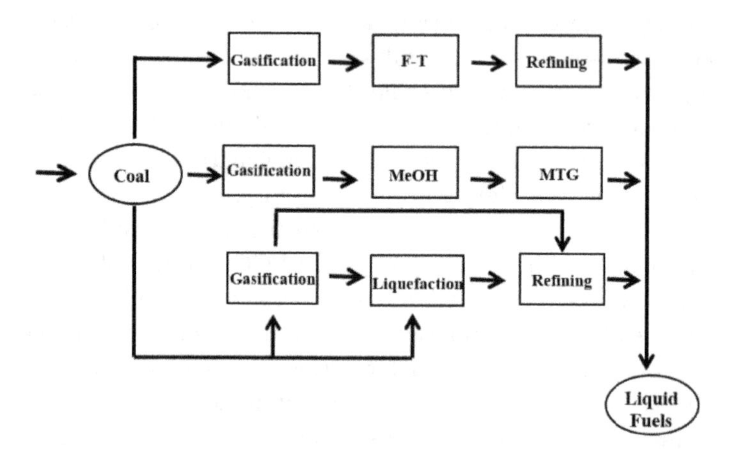

Fig. 7.38 Comparison of three processes for the production of liquid fuels from coal: Fischer–Tropsch (F–T), methanol-to-gasoline (MTG) and coal liquefaction (Bergius).

Table 7.1 Products (in %) of the low- and high-temperature FT processes reported by Sasol, of coal liquefaction (H–Coal) reported by HRI and MTG.

Product	Low T FT	High T FT	H–Coal	MTG
C_1	5	8	0	1
C_2–C_4	7	30	0	17
C_5–C_{16}	19	36	36	82
Distillates	22	16	43	0
Wax	46	5	20	0
Oxygenates	1	5	0	0

by low-temperature and high-temperature Fischer–Tropsch (FT) processes and by the coal liquefaction H–Coal process.

The product of the H–Coal (Bergius) process is similar to an aromatic-rich crude oil and requires hydrocracking/hydrotreating and other reforming processes before the liquid products are suitable transportation fuels. The FT process produces a broad spectrum of linear alkanes (Section 6.4.1.3), which require upgrading to gasoline, kerosene and diesel. The FT product mixture depends on the temperature and catalyst. The low-temperature FT process with a Co catalyst produces more heavy hydrocarbons, whereas the high-temperature FT process with a Fe catalyst produces a lighter mixture. The MTG process produces a mixture of hydrocarbons with boiling point similar to that of gasoline (C_5–C_{12}), which can go directly to the gasoline pool in the refinery. Because both the FT and MTG process start with clean gaseous reactants (synthesis gas and methanol, respectively), they produce a sulfur-free product mixture.

Methanol can be transformed in a mixture of hydrocarbons as well as in a mixture of small alkenes. UOP (Honeywell) together with Norsk Hydro [7.53] and The Dalian Institute of Chemical Physics (DICP) of the Chinese Academy of Sciences [7.54] developed a methanol-to-olefins (MTO) process with ethene and propene as the main products. The MTG process begins with methanol, made from synthesis gas (cf. Section 6.4.2), and converts it to a pre-equilibrium mixture containing dimethyl ether (DME) and water (Fig. 7.39) over a dehydration catalyst (e.g. Al_2O_3). This mixture is processed to alkenes (MTO) or further to high-octane gasoline (MTG). By means of this technology, methanol is an excellent feedstock for the entire petrochemical industry and is an excellent substitute for crude oil.

ZSM-5 is the catalyst in the MTG process (43% C_4^+, 45% $C_3^=$, 7% $C_2^=$), while the molecular sieve SAPO-34 is the catalyst in the MTO process

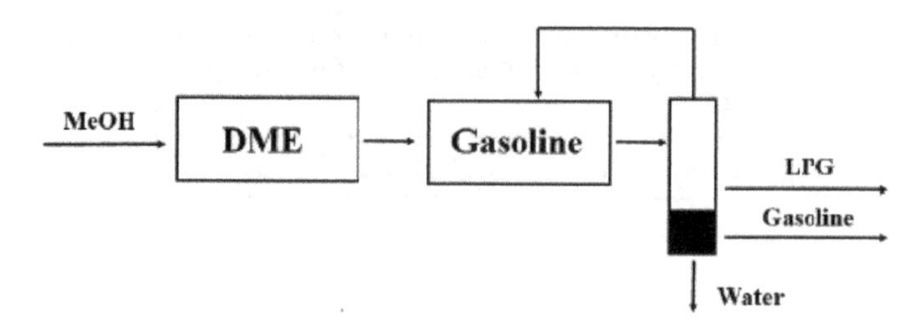

Fig. 7.39 Production of gasoline from dimethylether. Reprinted with permission from [7.55]. Copyright (2012) John Wiley & Sons, Inc.

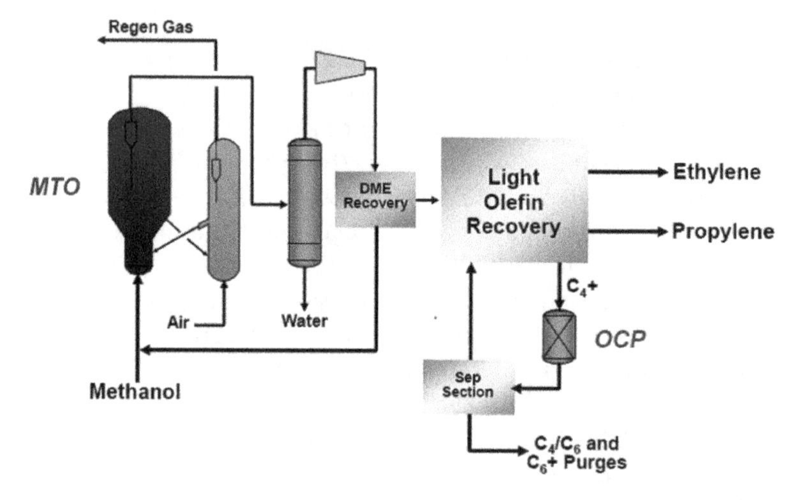

Fig. 7.40 UOP/Norsk Hydro MTO process combined with the total process for cracking higher alkenes. UOP Honeywell Middle East Chemical Week Conference, 16–19 October 2011.

(11% C_4^+, 34% $C_3^=$, 51% $C_2^=$) due to its smaller pore openings (0.38 nm versus 0.55 nm for ZSM-5) [7.55]. In the MTO processes of UOP/Norsk Hydro and DICP, methanol is converted in a fluidised-bed reactor operating by turbulent fluidisation with a selectivity of ethene and propene of about 80 wt.%. The spent catalyst is transported to a bubbling fluidised-bed regenerator to restore activity, and the regenerated catalyst is then fed back to the reactor (Fig. 7.40). As in the FCC process (Section 7.5.1), in the MTO process fluid bed technology is used to take care of the large amount of heat produced in the reactor and regenerator.

Furthermore, a methanol-to-propene (MTP) fixed-bed process has been developed, mainly for the production of propene. The Lurgi MTP process contains several parallel fixed-bed reactors (one for regeneration and the others for operation) with a zeolite ZSM-5 catalyst [7.56]. In 2020, 22 MTO and MTP plants with a total capacity of around 15 MT/year were on stream in China (at a total production of 31 MT ethylene and 44 MT propene in 2020). China has large coal reserves and MTO has the potential to become China's most important process for the production of light alkenes [7.54] and, thus, polymers.

The chemistry of the MTG and MTO reactions has much in common; different products are determined by the catalyst and process conditions. In the first stage of the MTG and MTO reactions, equilibration is reached among methanol, dimethyl ether (DME) and water (Fig. 7.38), and this mixture reacts to hydrocarbons. When H-ZSM-5 is the catalyst in the MTH process, C_1 to C_{10} hydrocarbons are obtained, the lower hydrocarbons being alkanes and the higher hydrocarbons mainly methylated benzenes (up to tetramethylbenzene). Small alkenes are formed in the MTO process over H-SAPO-34. Originally, the MTH and MTO products were explained by classic acid catalysis: Oligomerisation of ethene gives alkenes, which isomerise to isoalkenes, which are cracked to smaller alkenes. Cyclisation of heptadiene gives methylcyclohexene, which is dehydrogenated by hydrogen transfer with alkenes to give aromatics and alkanes. However, several aspects are not explained by simple acid catalysis, for example the formation of ethene (which would require formation of a primary carbenium ion), the occurrence of an induction period before the rate of formation of hydrocarbons becomes appreciable and the formation of fully labelled polymethylbenzenes in the co-reaction of $^{13}CH_3OH$ and benzene [7.58]. Therefore, the MTH and MTO products are explained by the hydrocarbon-pool mechanism. The mechanism is not only based on acid catalysis but also on the assumption that methanol reacts inside the zeolite to higher hydrocarbons, which do not escape from the zeolite and are, therefore, rarely found at the reactor exit (Fig. 7.41) [7.57]. The MTH and MTO reactions are, thus, examples of **disguised kinetics** [7.57]. Larger alkenes form by sequential methylation of alkenes, and aromatics form by ring closure of hexenes and heptenes, followed by hydrogen transfer between cyclic molecules and alkenes, explaining the concurrent production of aromatics and alkanes. The aromatics are methylated by methanol to a size that enables polymethylbenzene to diffuse out of the zeolite pores. Just one alkene molecule, formed during the induction phase, would be sufficient to catalyse methanol conversion. All alkenes produced thereafter are the result of repeated methylation, oligomerisation and cracking.

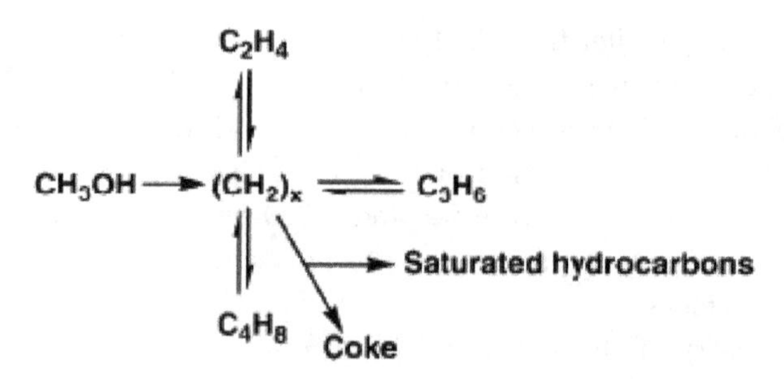

Fig. 7.41 Hydrocarbon-pool mechanism. Reprinted with permission from [7.58]. Copyright (1994) Elsevier.

The higher hydrocarbons inside the zeolite form the so-called hydrocarbon pool; they act as skeletons (scaffolds) that stabilise intermediates and transition states. Methanol conversion proceeds by repeated methylation and dealkylation and does not take place on inorganic sites on the walls of the zeolite; it takes place on organic sites, on hydrocarbon species that work in concert with acid sites and are regenerated in catalytic cycles. Hence, an active MTH–MTO catalyst can be considered to be an organic-inorganic hybrid material consisting of the inorganic zeolite and the organic hydrocarbon pool [7.58]. Small polymethylated aromatic cations are assumed to be the main constituents of the hydrocarbon pool. It takes time for this pool to build up in the initial stage of the MTH reaction, which explains the induction period. Once the hydrocarbon pool has formed, it catalyses the formation of alkenes. The hydrocarbon-pool mechanism is similar to the pseudo-monomolecular mechanism in the isomerisation of butene to isobutene (Section 7.2.2); the catalyst is made up of a combination of organic and inorganic material.

A paring (peeling) mechanism and a side-chain mechanism (through exocyclic methylene) are responsible for the MTO products (Fig. 7.42) [7.59]. After methylation of hexamethylbenzene to the heptamethylbenzenium cation, the 1-isopropyl-1,2,3,4,5-pentamethylcyclopentadienyl cation forms by paring (ring contraction) (Fig. 7.42, left cycle). Dealkylation gives propene and the pentamethylcyclopentadienyl cation, the ring of which expands to tetramethylbenzene and reacts with methanol back to hexamethylbenzene. In the side-chain mechanism (Fig. 7.42, right cycle), the heptamethylbenzenium cation deprotonates to a molecule with an exocyclic double bond (outside the molecule). Methylation of the double bond leads to an ethyl group. The ethyl group can be dealkylated to ethene

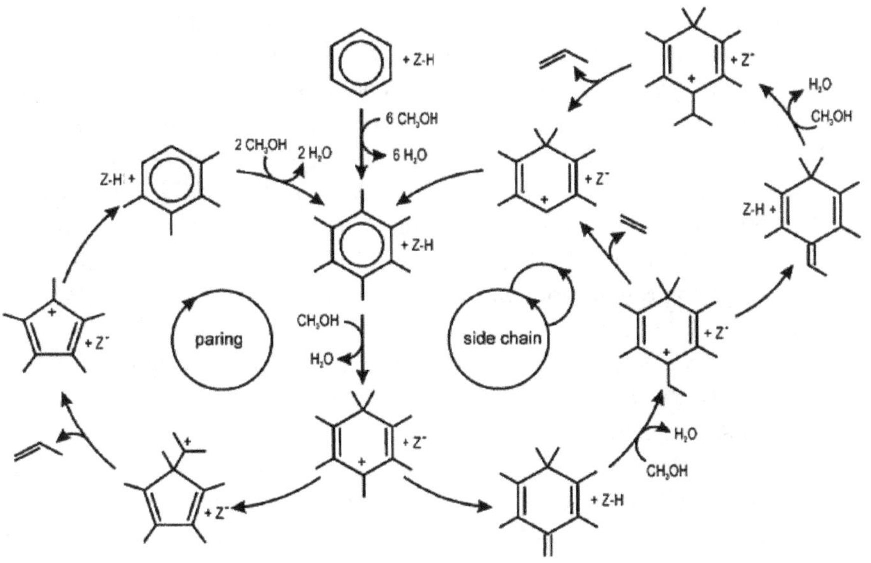

Fig. 7.42 Paring and side-chain mechanisms in MTO conversion based on aromatic hydrocarbon pool species. Reprinted with permission from [7.59]. Copyright (2013) John Wiley & Sons, Inc.

or methylated to an isopropyl group, which can be dealkylated as propene. A methyl shift reforms the original hexamethylbenzene.

7.5.3 *Reforming of Hydrocarbons by Bifunctional Catalysis*

To improve the octane number of naphtha and obtain high quality gasoline, hydrocarbons are isomerised to branched hydrocarbons. In the past, they were also cyclised to cyclic hydrocarbons and further dehydrogenated to aromatics because benzene and toluene have high octane numbers. Unfortunately, benzene is cancerogenic and its contents in gasoline must be minimal.

The isomerisation of alkanes is catalysed by a combination of metal and acid catalytic sites [7.60]. Neither a metal catalyst nor an acid catalyst alone can catalyse this reaction. Thus, the conversion of hexane to methylpentanes and dimethylbutane at 373 °C and 0.1 MPa was only 0.9% over Pt/SiO_2 (metal on a weakly acidic support) and 0.3% over ASA (an acidic support), but 6.8% over Pt/ASA. Also other hydrocarbon reactions need a bifunctional catalyst. For instance, dehydrogenation needs a metal catalyst, while an acid is required to achieve a change in ring size, as confirmed by the reaction of cyclohexane. As a consequence, 85% benzene and 2% methylcyclopentane formed over Pt/SiO_2, while over Pt/ASA, with Pt on an acidic support, 40% benzene and 40% methylcyclopentane formed. The reactions of *m*-xylene to ethylbenzene, heptane to isoheptane and hexane to benzene all need a metal as well as an acid catalyst to give reasonable conversion. As shown in Table 2.1, Ni-MoS_2/ASA catalysts are used in the refinery in the hydrocracking of large hydrocarbons to smaller hydrocarbons. The Ni–MoS_2 (Section 8.2.1) is responsible for preparing alkenes by dehydrogenation and the acidic ASA for making carbenium cations and cracking them by β scission (Section 7.2.2).

The reaction of an alkane to an aromatic molecule depends on the interplay between a dehydrogenation catalyst (metal or metal sulfide) and a cyclisation and isomerisation catalyst (acid). The sequence metal-catalysed dehydrogenation, protonation, acid-catalysed cyclisation and isomerisation and metal-catalysed dehydrogenation explains how hexane reacts to benzene over a bifunctional catalyst (Fig. 7.43). Hexane first dehydrogenates on a metal site to hexatriene, which is protonated and reacts to the hexadienyl carbenium ion ($CH_2{=}CHCH{=}CHCH^+CH_3$). This ion undergoes ring closure to the 1-methylcyclopentenyl-3 carbenium ion (methylcyclopentene protonated at the 3 position). This isomerises to the 1-methylcyclopentenyl-2 carbenium ion, which then enlarges the ring, passing through a protonated cyclopropane transition state, to the

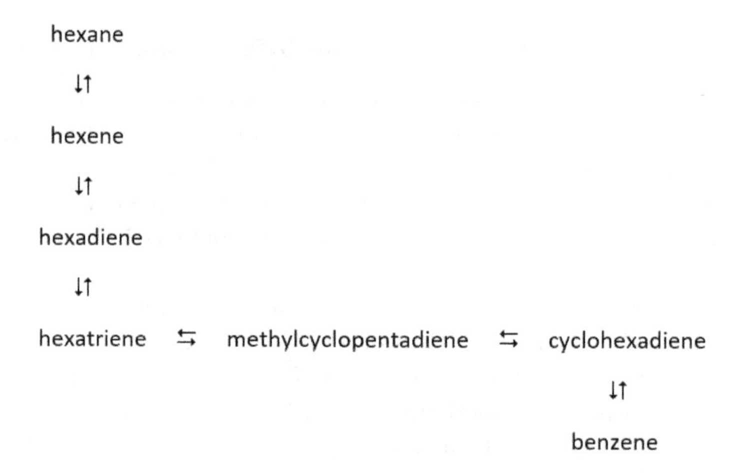

Fig. 7.43 Transformation of hexane to benzene over a bifunctional catalyst. Metal-catalysed reactions are indicated by vertical arrows (↑↓) and acid-catalysed reactions by horizontal arrows (⇆).

cyclohexadienyl carbenium ion. The ring enlargement is similar to the reverse reaction of cyclohexane to methylpentane shown in Fig. 7.21. Benzene results after deprotonation on an acid site and dehydrogenation on a metal site.

In Section 4.3 it was shown that the two catalyst components of a bifunctional catalyst do not have to be positioned close together to achieve bifunctional catalysis. Experiments on the isomerisation of heptane to isoheptane confirmed this [7.60]. The conversion of heptane was very low over ASA (an acid catalyst) and Pt/C (a metal on a neutral support) and low over Pt/SiO_2 (a metal on a very weak acidic support) (Fig. 7.44), as expected for a reaction that needs both a metal and an acid catalyst. A mixture of Pt/SiO_2 and ASA, both with 1,000 μm particles, showed higher conversion and when the particle size of both catalysts was decreased to 70 and 5 μm the conversion increased dramatically and became about equal to that of a Pt/ASA catalyst with Pt deposited on ASA (Fig. 7.44). This demonstrates that, for the isomerisation of heptane, it is not necessary to mix both catalyst components on a molecular scale; a distance of a few micrometers is sufficient, as predicted by Eq. (4.25).

In the bifunctional reaction, a heptane molecule diffuses into a Pt/SiO_2 particle, where it is converted to heptene, diffuses out of the Pt/SiO_2

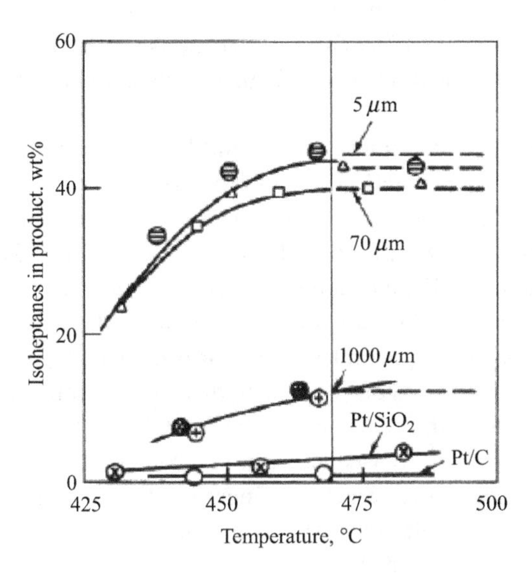

Fig. 7.44 Conversion of n-heptane in the isomerisation over Pt/C, Pt/SiO_2 and mechanical mixtures of Pt/SiO_2 and ASA with particle sizes of 1,000, 70 and 5 μm. Reprinted with permission from [7.60]. Copyright (1962) Elsevier.

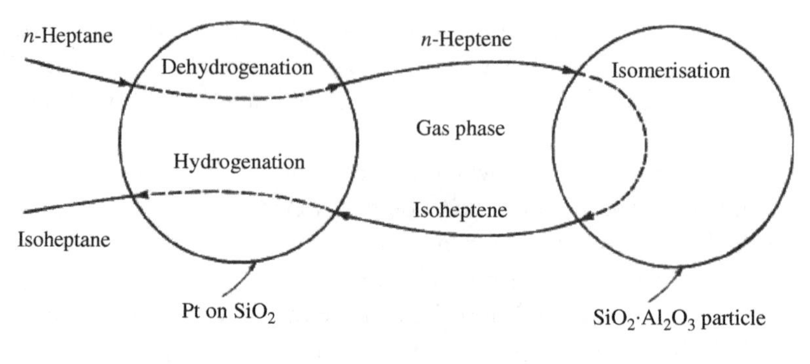

Fig. 7.45 Isomerisation of heptane to isoheptane over a bifunctional catalyst. Pt/SiO_2 is responsible for dehydrogenation and hydrogenation and ASA ($SiO_2 \cdot Al_2O_3$) for isomerisation [7.61].

particle and into an ASA particle, where the heptene is isomerised to isoheptene (Fig. 7.45). Further diffusion of isoheptene to a Pt/SiO_2 particle (which may be another Pt/SiO_2 particle) leads to hydrogenation to isoheptane.

Bifunctional catalysis is of great importance in several refinery processes. The isomerisation of pentane and hexane is carried out over a $Pt/Cl\text{-}Al_2O_3$ or a Pt/zeolite catalyst to increase the low octane number of n-pentane and n-hexane to the high octane number of their branched isomers (Fig. 7.46).

Because the alkane isomerisation equilibrium is favoured by low temperature, at which the reaction rate is low, conversion is limited in a once-through operation. Therefore, in isomerisation units, the branched and unreacted alkanes are separated and the alkanes recycled. Zeolite A is used in the separation and Pt/mordenite in isomerisation (Fig. 7.47).

An even larger application of bifunctional catalysis is the catalytic reforming process, an important process for preparing high-octane gasoline. In this process a mixture of C_5–C_{10} hydrocarbons, obtained by primary distillation of crude oil, reacts over a $Pt\text{-}Re/Cl\text{-}Al_2O_3$ or $Pt\text{-}Sn/Cl\text{-}Al_2O_3$ catalyst. Alkanes react to isoalkanes and aromatics, while cyclic alkanes react to aromatics. Both reactions increase the octane number of the product. Figure 7.48 left gives the reactions of the C_6 molecules in the feed, over a metal catalyst M such as Pt. In principle, the isomerisation of hexane to methylpentane and the dehydrocyclisation to benzene can be performed with a metal catalyst but a substantial part of the hexane

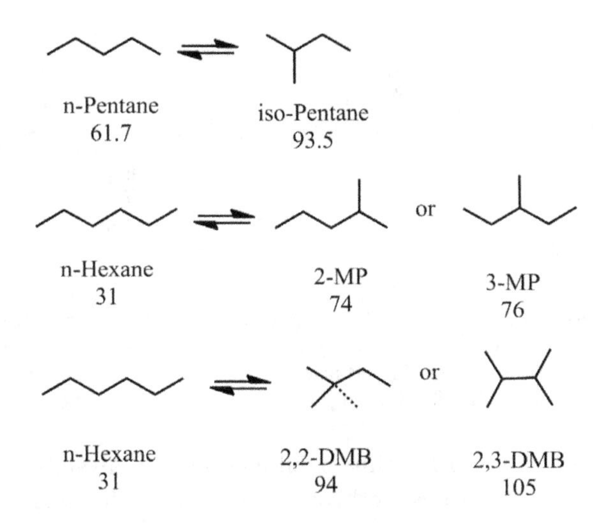

Fig. 7.46 Isomerisation of pentane and hexane to isomers with higher octane number.

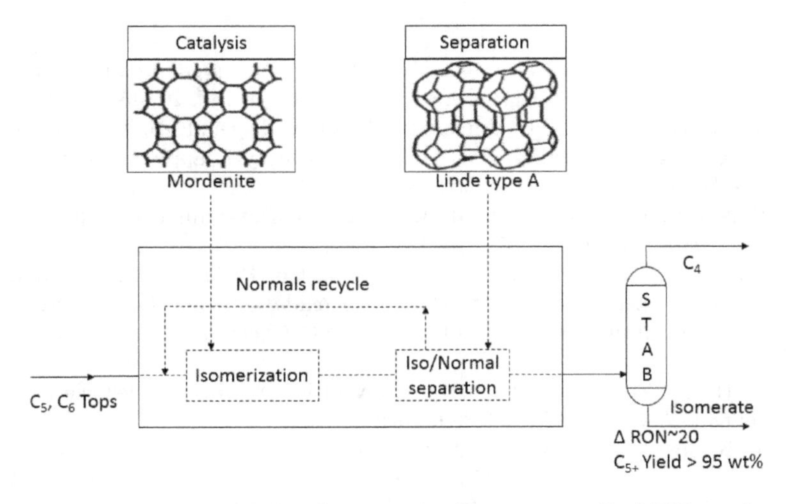

Fig. 7.47 Isomerisation of light alkanes to isoalkanes over a Pt/MOR catalyst and separation of product and reactant molecules over zeolite A (UOP LLC, www.uop.com).

would crack to smaller hydrocarbons and the yield of the liquid product would be low. To avoid metal-catalysed cracking, a second metal (Re, Sn, Ge) is added to the Pt catalyst. The second metal dilutes the Pt and decreases the number of neighbouring Pt surface atoms, which suppresses C–C bond hydrogenolysis and, thus, the formation of small hydrocarbons. Unfortunately, the isomerisation and dehydrocyclisation reactions are also

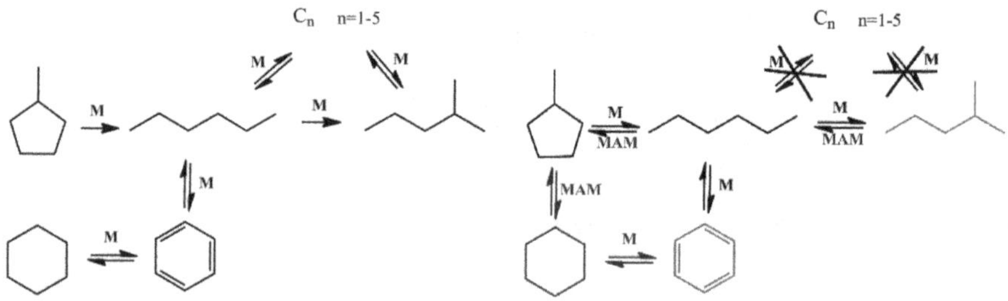

Fig. 7.48 Reaction scheme of the catalytic reforming of hexane. The left scheme shows the reactions that occur over a metal catalyst. The right scheme shows the reactions that occur over a bifunctional (metal M and acid A) catalyst. The bifunctional catalyst also contains a second metal (Re, Sn, Ge) to prevent C–C bond breaking.

suppressed. To make up for this, a bifunctional catalyst is employed, in which Pt is responsible for (de)hydrogenation and the acid for changing the structure of the hydrocarbon (Fig. 7.48, right).

References

7.1 B. Xie, J. Song, L. Ren, Y. Ji, J. Li, F.-S. Xiao, Organotemplate-free and fast route for synthesizing Beta zeolite, *Chem. Mater.* 20, 4533–4535, 2008.

7.2 T. De Baerdemaeker, B. Yilmaz, U. Müller, M. Feyen, F.-S. Xiao, W. Zhang, T. Tatsumi, H. Gies, X. Bao, D. De Vos, Catalytic applications of OSDA-free Beta zeolite, *J. Catal.* 308, 73–81, 2013.

7.3 B. Notari, Synthesis and catalytic properties of titanium containing zeolites, *Stud. Surf. Sci. Catal.* 37, 413–425, 1987.

7.4 M. Boronat, A. Corma, M. Renz, G. Sastre, P. M. Viruela, A multisite molecular mechanism for Baeyer-Villger oxidations on solid catalysts using environmentally friendly H_2O_2 as oxidant, *Chem. Eur. J.* 11, 6905–6915, 2011.

7.5 C. H. F. Peden, Cu/Chabazite catalysts for "lean-burn" vehicle emission control, *J. Catal.* 373, 384–389, 2019.

7.6 J. Klinowski, S. Ramdas, J. M. Thomas, C. A. Fyfe, J. S. Hartman, A re-examination of Si, Al ordering in Zeolites NaX and NaY, *J. Chem. Soc. Farad. Trans. II* 78, 1025–1050, 1982.

7.7 D. Verboekend, T. C. Keller, M. Milina, R. Hauert, J. Pérez Ramírez, Hierarchy brings function: Mesoporous clinoptilolite and L zeolite catalysts synthesised by tandem acid–base treatments, *Chem. Mater.* 25, 1947–1959, 2013.

7.8 A. Corma, V. Fornes, S. B. Pergher, Th. L. M. Maesen, J. G. Buglass, Delaminated zeolite precursors as selective acidic catalysts, *Nature* 396, 353–356, 1998.

7.9 M. Choi, K. Na, J. Kim, Y. Sakamoto, O. Terasaki, R. Ryoo, Stable single-unit-cell nanosheets of zeolite MFI as active and long-lived catalysts, *Nature* 461, 246–249, 2009.

7.10 K. Na, M. Choi, W. Park, Y. Sakamoto, O. Terasaki, R. Ryoo, Pillared MFI zeolite nanosheets of a single-unit-cell thickness, *J. Am. Chem. Soc.* 132, 4169–4177, 2010.

7.11 J. N. Kondo, K. Domen, IR observation of adsorption and reactions of olefins on H-form zeolites, *J. Mol. Catal. A* 199, 27–38, 2003.

7.12 F. Thibault-Starzyk, A. Travert, J. Saussey, J-C. Lavalley, Correlation between activity and acidity on zeolites: A high temperature infrared study of adsorbed acetonitrile, *Top. Catal.* 6, 111–118, 1998.

7.13 G. Busca, Silica-alumina catalytic materials: A critical review, *Catal. Today* 357, 621–629, 2020.

7.14 E. J. M. Hensen, D. G. Poduval, P. C. M. M. Magusin, A. E. Coumans, J. A. R. van Veen, Formation of acid sites in amorphous silica-alumina, *J. Catal.* 269, 201–218, 2010.

7.15 M. Caillot, A. Chaumonnot, M. Digne, C. Poleunis, D. P. Debecker, J. A. van Bokhoven, Synthesis of amorphous aluminosilicates by grafting: Tuning the building and final structure of the deposit by selecting the appropriate synthesis conditions, *Miropor. Mesopor. Mater.* 185, 179–189, 2014.

7.16 G. Crépeau, V. Montouillout, A. Vimont, L. Mariey, T. Cseri, F. Maugé, Nature, structure and strength of the acidic sites of amorphous silica alumina: An IR and NMR study, *J. Phys. Chem. B* 110, 15172–15185, 2006.

7.17 E. J. M. Hensen, D. G. Poduval, D. A. J. M. Ligthart, J. A. R. van Veen, M. S. Rigutto, Quantification of strong Brønsted acid sites in aluminosilicates, *J. Phys. Chem. C* 114, 8363–8374, 2010.

7.18 E. J. M. Hensen, D. G. Poduval, V. Degirmenci, D. A. J. M. Ligthart, W. Chen, F. Maugeì, M. S. Rigutto, J. A. R. van Veen, Acidity characterisation of amorphous silica-alumina, *J. Phys. Chem. C* 116, 21416–21429, 2012.

7.19 M. Trombetta, G. Busca, S. Rossini, V. Piccoli, U. Cornaro, A. Guercio, R. Catani, R. J. Willey, FT-IR studies on light olefin skeletal isomerisation catalysis III. Surface acidity and activity of amorphous and crystalline catalysts belonging to the SiO2–Al2O3 system, *J. Catal.* 179, 581–596, 1998.

7.20 C. Chizallet, P. Raybaud, Pseudo-bridging silanols as versatile Brønsted acid sites of amorphous aluminosilicate surfaces, *Angew. Chem. Int. Ed.* 48, 2891–2893, 2009.

7.21 M. Valla, A. J. Rossini, M. Caillot, C. Chizallet, P. Raybaud, M. Digne, A. Chaumonnot, A. Lesage, L. Emsley, J. A. van Bokhoven, C. Copeìret, Atomic description of the interface between silica and alumina in aluminosilicates through dynamic nuclear polarisation surface-enhanced NMR spectroscopy and first-principles calculations, *J. Am. Chem. Soc.* 137, 10710–10719, 2015.

7.22 J. B. Nicholas, J. F. Haw, The prediction of persistent carbenium ions in zeolites, *J. Am. Chem. Soc.* 120, 11804–11805, 1998.

7.23 M. Boronat, P. M. Viruela, A. Corma, Reaction intermediates in acid catalysis by zeolites: Prediction of the relative tendency to form alkoxides or carbocations as a function of hydrocarbon nature and active site structure, *J. Am. Chem. Soc.* 126, 3300–3309, 2004.

7.24 C. Tuna, T. Kerber, J. Sauer, The tert-butyl cation in H-zeolites: Deprotonation to isobutene and conversion into surface alkoxides, *Angew. Chem. Int. Ed.* 49, 4678–4680, 2010.
7.25 C. Tuna, J. Sauer, Treating dispersion effects in extended systems by hybrid MP2: DFT calculations-protonation of isobutene in zeolite ferrierite, *Phys. Chem. Chem. Phys.* 8, 3955–3965, 2006.
7.26 M. A. Natal-Santiago, R. Alcalá, J. A. Dumesic, DFT study of the isomerisation of hexyl species involved in the acid-catalysed conversion of 2-methyl-pentene-2, *J. Catal.* 181, 124–144, 1999.
7.27 D. M. Brouwer, "Reactions of alkylcarbenium ions in relation to isomerisation and cracking of hydrocarbons", in *Chemistry and Chemical Engineering of Catalytic Processes*, Eds. R. Prins, G. C. A. Schuit. Sijthoff & Noordhoff, Alphen aan den Rijn, pp. 137–160, 1980.
7.28 B. C. Gates, *Catalytic Chemistry*, John Wiley, New York, 1992.
7.29 T. Demuth, X. Rozanska, L. Benco, J. Hafner, R. A. van Santen, H. Toulhoat, Catalytic isomerisation of 2-pentene in H-ZSM-22 — A DFT investigation, *J. Catal.* 214, 68–77, 2003.
7.30 Y.-L. Wang, X.-X. Wang, Y.-A. Zhu, K.-K. Zhu, D. Chen, X.-G. Zhou, Shape selectivity in acidic zeolite catalysed 2-pentene skeletal isomerisation from first principles, *Catal. Today* 347, 115–123, 2020.
7.31 A. G. Stepanov, M. V. Luzgin, S. S. Arzumanov, H. Ernst, D. Freude, n-Butene conversion on H-Ferrierite studied by ^{13}C MAS NMR, *J. Catal.* 211, 165–172, 2002.
7.32 P. Meriaudeau, R. Bacaud, L. Ngoc Hung, Anh. T. Vu, Isomerisation of butene in isobutene on ferrierite catalyst: A mono- or bimolecular process?, *J. Mol. Catal. A* 110, L177–179, 1996.
7.33 M. Guisnet, P. Andy, N. S. Gnep, C. Travers, E. Benazzi, Origin of the positive effect of coke deposits on the skeletal isomerisation of n-butenes over a H-FER zeolite, *J. Chem. Soc. Chem. Commun.* 1685–1686, 1995.
7.34 P. Andy, N. S. Gnep, M. Guisnet, E. Benazzi, C. Travers, Skeletal isomerisation of n-butenes II. Composition, mode of formation, and influence of coke deposits on the reaction mechanism, *J. Catal.* 173, 322–332, 1998.
7.35 B. de Ménorval, P. Ayrault, N. S. Gnep, M. Guisnet, Mechanism of n-butene skeletal isomerisation over HFER zeolites: A new proposal, *J. Catal.* 230, 38–51, 2005.
7.36 C. Wattanakit, S. Nokbin, B. Boekfa, P. Pantu, J. Limtrakul, Skeletal isomerisation of 1-butene over ferrierite zeolite: A quantum chemical analysis of structures and reaction mechanisms, *J. Phys. Chem. C* 116, 5654–5663, 2012.
7.37 D. Gleeson, The skeletal isomerisation in ferrierite: A theoretical assessment of the bi-molecular conversion of cis-butene to iso-butene, *J. Mol. Catal. A* 368, 107–111, 2013.
7.38 P. J. Chenier, *Survey of Industrial Chemistry*, VCH, New York, 1992.
7.39 P. W. N. M. van Leeuwen, *Homogeneous Catalysis*, Kluwer, Dordrecht, 2004.

7.40 T. F. Degnan, C. M. Smith, C. R. Venkat, Alkylation of aromatics with ethylene and propylene: Recent developments in commercial processes, *Appl. Catal. A* 221, 283–294, 2001.
7.41 C. Perego, P. Ingallina, Recent advances in the industrial alkylation of aromatics: New catalysts and new processes, *Catal. Today* 73, 3–22, 2002.
7.42 N. Hansen, T. Kerber, J. Sauer, A. T. Bell, F. J. Keil, Quantum chemical modeling of benzene ethylation over H-ZSM-5 approaching chemical accuracy: A hybrid MP2: DFT study, *J. Am. Chem. Soc.* 132, 11525–11538, 2010.
7.43 S. Svelle, M. Visur, U. Olsbye, Saepurahman, M. Bjørgen, Mechanistic aspects of the zeolite catalysed methylation of alkenes and aromatics with methanol: A review, *Top. Catal.* 54, 897–906, 2011.
7.44 J. Van der Mynsbrugge, M. Visur, U. Olsbye, P. Beato, M. Bjørgen, V. Van Speybroeck, S. Svelle, Methylation of benzene by methanol: Single-site kinetics over H–ZSM-5 and H-beta zeolite catalysts, *J. Catal.* 292, 201–212, 2012.
7.45 W. W. Kaeding, C. Chu, L. B. Young, B. Weinstein, S. A. Butter, Selective alkylation of toluene with methanol to produce para-xylene, *J. Catal.* 67, 159–174, 1981.
7.46 L. B. Young, S. A. Butter, W. W. Kaeding, Shape selective reactions with zeolite catalysts III. Selectivity in xylene isomerisation, toluene-methanol alkylation, and toluene disproportionation over ZSM-5 zeolite catalysts, *J. Catal.* 76, 418–432, 1982.
7.47 X. Rozanska, R. A. van Santen, F. Hutschka, J. Hafner, A periodic DFT study of intramolecular isomerisation reactions of toluene and xylenes catalysed by acidic mordenite, *J. Am. Chem. Soc.* 123, 7655–7667, 2001.
7.48 M. Guisnet, N. S. Gnep, S. Morin, Mechanisms of xylene isomerisation over acidic solid catalysts, *Microp. Mesop. Mater.* 35, 47–59, 2000.
7.49 X. Rozanska, R. A. van Santen, F. Hutschka, A periodic density functional theory study of intermolecular isomerisation of toluene and benzene catalysed by acidic mordenite zeolite: Effect of the zeolite steric constraints, *J. Phys. Chem. B* 106, 4652–4657, 2002.
7.50 T. Demuth, P. Raybaud, S. Lacombe, H. Toulhoat, Effects of zeolite pore sizes on the mechanism and selectivity of xylene disproportionation — a DFT study, *J. Catal.* 222, 323–337, 2004.
7.51 E. T. C. Vogt, B. M. Weckhuysen, Fluid catalytic cracking: Recent developments on the grand old lady of zeolite catalysis, *Chem. Soc. Rev.* 44, 7342–7370, 2015.
7.52 N. Batalha, L. Pinard, C. Bouchy, E. Guillon, M. Guisnet, *J. Catal.* 307, 122–131, 2013.
7.53 J. Q. Chen, A. Bozzano, B. Glover, T. Fuglerud, S. Kvisle, Recent advancements in ethylene and propylene production using the UOP/Hydro MTO process, *Catal. Today* 106, 103–107, 2005.
7.54 P. Tian, Y. Wei, M. Ye, Z. Liu, Methanol to olefins (MTO): From fundamentals to commercialisation, *ACS Catal.* 5, 1922–1938, 2015.

7.55 U. Olsbye, S. Svelle, M. Bjørgen, P. Beato, T. V. W. Janssens, F. Joensen, S. Bordiga, K. P. Lillerud, Conversion of methanol to hydrocarbons: How zeolite cavity and pore size control product selectivity, *Angew. Chem. Int. Ed.* 51, 5810–5831, 2012.
7.56 H. Koempel, W. Liebner, Lurgi's methanol to propylene (MTP®) report on a successful commercialisation, *Stud. Surf. Sci. Catal.* 167, 261–267, 2007.
7.57 R. M. Dessau, R. B. LaPierre, On the mechanism of methanol conversion to hydrocarbons over HZSM-5, *J. Catal.* 78, 136–141, 1982.
7.58 I. M. Dahl, S. Kolboe, On the reaction mechanism for hydrocarbon formation from methanol over SAPO-34. 1. Isotopic labelling studies of the co-reaction of ethene and methanol, *J. Catal.* 149, 458–464, 1994.
7.59 K. Hemelsoet, J. Van der Mynsbrugge, K. De Wispelaere, M. Waroquier, V. Van Speybroeck, Unraveling the reaction mechanisms governing methanol-to-olefins catalysis by theory and experiment, *ChemPhysChem* 14, 1526–1545, 2013.
7.60 P. B. Weisz, Polyfunctional heterogeneous catalysis, *Adv. Catal.* 13, 137–190, 1962.
7.61 B. C. Gates, J. R. Katzer, G. C. A. Schuit, *Chemistry of Catalytic Processes*, McGraw-Hill Inc., New York, pp. 280–283, 1979.

Questions

7.1 Describe all steps in the preparation of zeolite HY, starting from a silicate solution and an aluminate solution.

7.2 Scrambling of the carbon atoms of pentane and isomerisation of pentane to isopentane on an acid catalyst both have the same rate. Are these reactions very fast, fast or slow?

7.3 What is the activation energy of the isomerisation of the tertiary carbenium ion of 2,3-dimethylbutane to the secondary carbenium ion of 2,2-dimethylbutane, in which the degree of branching remains two?

7.4 The activation energy of the 1,2-H shift from 1-^{13}C-but-1-ene to cis and trans 1-^{13}C-but-2-ene is 41 kJ·mol^{-1} and the activation energy of the shift of the ^{13}C label from the carbon atom C1 to the carbon atom C2 is 88 kJ·mol^{-1}. Explain the difference in activation energy.

7.5 ^{13}C labelling of n-butene demonstrates that the isomerisation of butene to isobutene over zeolite H–FER is initially bimolecular, while later the reaction is monomolecular. How was this determined?

7.6 Figure 7.19 shows that isomerisation of the 2-^{13}C-pentyl C4 carbocation leads to the 2-methyl-2-^{13}C-butyl and 2-methyl-3-^{13}C-butyl tertiary carbenium ions, both with the label in an inner position. How can the label shift to the outer C1 or C4 position of the methylbutyl carbenium ion starting from the same 2-^{13}C-pentyl C4 carbocation?

7.7 Why are pentane and hexane, but not heptane, isomerised in industry?

7.8 Why is 1-propanol made through hydroformylation, why is it not possible to prepare 1-propanol from propene in the same way as ethanol is made from ethene?

7.9 Ethylbenzene is a low-value by-product in the production of xylenes. Would it be possible to isomerise ethylbenzene directly to valuable xylenes? How could this be achieved by hydrogenation?

7.10 Acid-catalysed isomerisation of xylene can be determined from the change in the position of the two methyl groups, but when the methyl group of toluene shifts it leads to the same product as the reactant. Is it possible to use isotopic labelling to determine whether a methyl shift in toluene occurs?

7.11 Why does an alkylcarbenium ion break by β scission and not by α scission?

7.12 The reverse β scission reaction could, in principle, be used to produce a larger alkylcarbenium ion from a carbenium ion and an alkene. By repeating this reaction, a large alkylcarbenium ion would be formed and a long α-alkene by splitting off a proton. Why does this not work for ethene?

7.13 Why does ZSM-5 produce more propene than zeolite Y?

Chapter 8

Cleaning of Fuels by Hydrotreating

8.1 Hydrotreating

Hydrotreating is a refinery process for the cleaning of fuels. In this process, impurity atoms (S, N, O and metals such as V and Ni) are removed from molecules, which contain heteroatoms and are present in oil. The removal occurs with the aid of hydrogen and a catalyst and is called hydrotreating. The removal of these atoms is necessary not only because S and N atoms would lead to SO_2 and NO_x in combustion of the fuel in engines, but also because fuels are upgraded in the refinery (to improve the octane number of gasoline and the cetane number of diesel) over catalysts that do not tolerate these atoms. Most oil streams in a refinery must therefore be hydrotreated and, as a consequence, hydrotreating is the largest application of industrial catalysis based on the amount of material (oil fractions) processed per year. On the basis of the amount of catalyst used per year, hydrotreating catalysts constitute the third largest catalyst business after exhaust gas catalysts (Section 9.1.2) and fluid cracking catalysts (Section 7.5.1). S, N and aromatics contents of the feed influence the process design of hydrotreating units and therefore several different hydrotreaters are present in a refinery. Depending on the element that is removed, hydrotreating processes are distinguished in **hydrodesulfurisation** (HDS), **hydrodenitrogenation** (HDN), **hydrodeoxygenation** (HDO) and **hydrodemetalisation** (HDM) processes.

8.2 Hydrotreating Catalysts

8.2.1 *Metal Sulfides*

8.2.1.1 *Structure of sulfided Co–Mo/Al_2O_3 and Ni–Mo/Al_2O_3*

Industrial hydrotreating catalysts contain Mo and Co or Ni in sulfided state and supported on γ-Al_2O_3. When Mo sulfide is supported alone on γ-Al_2O_3, it has a much higher activity for the removal of S, N and O atoms than Co and Ni sulfide supported on γ-Al_2O_3, while sulfided Co–Mo/Al_2O_3 and Ni–Mo/Al_2O_3 catalysts have a substantially higher catalytic activity than Mo/Al_2O_3. Therefore, MoS_2 is considered to be the catalyst and Co and Ni the promoters of the Mo activity. Co–Mo catalysts have a lower hydrogenation activity than Ni–Mo catalysts and are, therefore, better suited for HDS and especially the HDS of FCC naphtha. FCC naphtha contains alkenes, which should not be hydrogenated to alkanes (Section 8.4.1). Ni–Mo is used when a high hydrogenation activity is required, as in HDN and in the hydrogenation (HYD) pathway of the HDS of 4,6-dialkyldibenzothiophene (Section 8.3.1). Tungsten has chemical properties similar to Mo, but it is more expensive. Sulfided Ni–W/Al_2O_3 is an even stronger hydrogenation catalyst than sulfided Ni–Mo/Al_2O_3. In addition to Mo or W and Co or Ni, hydrotreating catalysts often contain modifier elements such as P, B, F or Cl, which may influence the catalytic as well as the mechanical properties of the catalyst. A new application for metal sulfides may be the treatment of biomass, when this develops into an alternative energy source. Biomass is rich in oxygen-containing compounds and the oxygen atoms can be removed by similar processes (HDO) and catalysts as S- and N-containing molecules (Section 8.3.3).

Hydrotreating catalysts are prepared by pore volume impregnation (Section 2.5) of γ-Al_2O_3 with an aqueous solution of $(NH_4)_6Mo_7O_{24}$ followed by drying and calcination (heating in air). In a second step, the resulting material is impregnated with an aqueous solution of $Co(NO_3)_2$ or $Ni(NO_3)_2$ and then dried and calcined. In a one-step impregnation (as preferentially used in the industry), all inorganic materials are co-impregnated and the resulting catalyst precursor is dried and calcined [8.1]. Heptamolybdate anions are preferentially adsorbed from $(NH_4)_6Mo_7O_{24}$ solutions at high Mo concentrations and at pH $<$ PZC (the PZC of Al_2O_3 is about at pH $= 8$ and of SiO_2 it is at pH $= 2$, cf. Section 2.5). Thus, when applying a pore volume impregnation with an aqueous $(NH_4)_6Mo_7O_{24}$ solution (pH ≈ 5.5) at a molybdenum concentration high enough to reach a monolayer of molybdate on the Al_2O_3 surface (ca. 6 Mo atoms per nm^2 Al_2O_3 BET surface area), mainly $Mo_7O_{24}{}^{6-}$ is adsorbed on the surface. After drying, hydrotreating catalysts are calcined in air to about 500 °C. After subsequent exposure to air they rapidly become rehydrated and contain hydroxyl groups as well as adsorbed water.

When both Mo and Co or Ni are put on the support surface, polymolybdate complexes are present on the support surface and Co^{2+} cations are present in tetrahedral and octahedral positions in the surface and subsurface of γ-Al_2O_3 (Section 2.3.4) at low loading and as Co_3O_4 at high Co loading [8.2]. The Co^{2+} cations form cobalt-heteropolymolybdate and stay at the surface, close to the Mo ions, and are well prepared to form the active Co–Mo–S structure during sulfidation (see below). The promoter ions will then also interact less strongly with the support and thus, after sulfidation, can be used more efficiently.

γ-Al_2O_3 is the preferred support for industrial hydrotreating catalysts because it has good textural and mechanical properties. Hydrotreating catalysts have to be regenerated several times during their existence and this puts high demands on the catalyst support. Regeneration involves burning-off of the coke from the catalyst with air and resulfiding of the oxidised catalyst. Catalyst regeneration can be carried out on site in the HDS reactor, or off-site by specialised firms.

The oxidic catalyst precursors, which are formed during the impregnation, drying and calcination (and also regeneration) steps, are transformed into the actual hydrotreating catalysts by sulfidation in a mixture of H_2 and a sulfur-containing compound such as H_2S, dimethyl disulfide (DMDS, CH_3–S–S–CH_3) or an oil fraction. Optimum calcination and sulfidation temperatures for Al_2O_3 as a support are 400–500 °C. As a temperature-programmed sulfidation study showed, H_2S is taken up and H_2O is given off already at room temperature, indicating that a sulfur–oxygen exchange reaction from MoO_3 to MoS_3 takes place [8.3]. At higher temperature reduction to MoS_2 occurs with concomitant H_2 consumption and H_2S evolution. At still higher temperature further sulfidation takes place.

During sulfidation and during hydrotreating, the conditions are highly reducing and H_2S is always present. Thermodynamics then predicts that Mo should be in the MoS_2 form. Indeed, Extended X-ray Absorption Fine Structure (EXAFS) studies [8.4] demonstrated that after sulfidation the average Mo ion has the same environment as Mo atoms in MoS_2 [8.5]. MoS_2 has a layer lattice with only weak van-der-Waals sulfur–sulfur

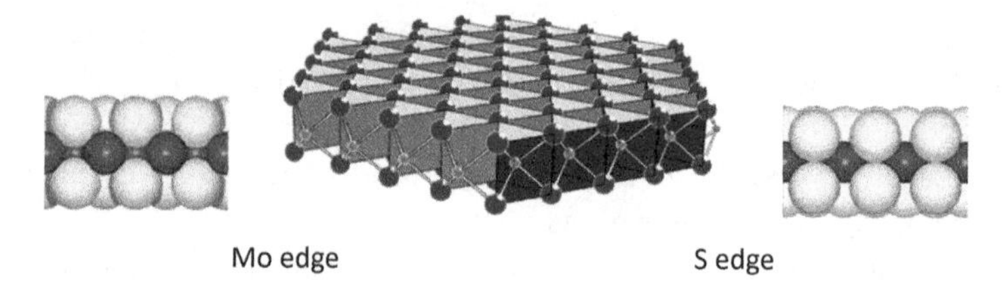

Fig. 8.1 The middle part of the figure shows a single MoS_2 slab with a layer of (grey) Mo atoms sandwiched by two layers of (red) S atoms. The side views show more clearly that the Mo edge is terminated by (blue) Mo atoms on top of four (yellow) S atoms and the S edge by Mo atoms coordinated by six S atoms.

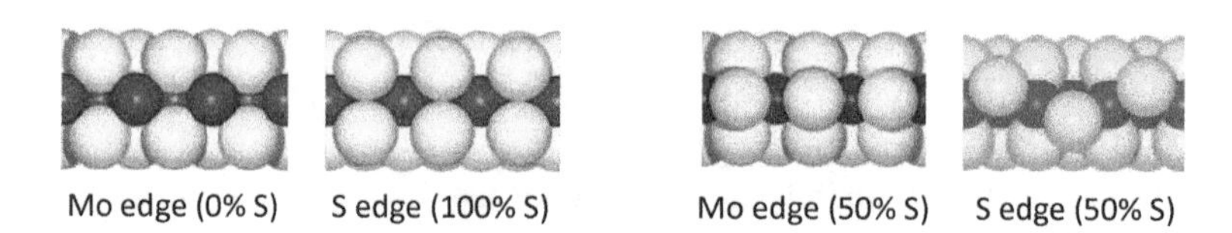

Fig. 8.2 Side views of the Mo and S edges before (left) and after (right) redistribution of the S atoms along the two edges and subsequent reconstruction of the edges. Reprinted with permission from [8.6]. Copyright (2004) Elsevier.

interactions between successive MoS_2 slabs, which explains why MoS_2 forms two-dimensional crystallites (Fig. 8.1). Crystals grow in the form of platelets with relatively large dimensions parallel to the basal (0001) sulfur planes and a small dimension perpendicular to the basal plane. Figure 8.1 shows a model of such a MoS_2 crystallite with S:Mo = 2 stoichiometry.

The side planes of the MoS_2 platelets, perpendicular to the basal plane, are normally the $(10\bar{1}0)$ and $(\bar{1}010)$ surfaces. They have the lowest energy because they have the lowest number of broken Mo–S bonds when a MoS_2 crystal is cleaved through the S atoms of the $(10\bar{1}0)$ planes. If all S atoms in the cleavage plane go to one of the two resulting crystal fragments, MoS_2 edges are created as shown in Fig. 8.2. In one type of edge the Mo atoms are sitting on top of four S atoms in a square planar fashion (Fig. 8.2, left, Mo edge, 0% S), the so-called **Mo edge**. In the other edge the Mo atoms are coordinated by six S atoms in trigonal pyramidal fashion (Fig. 8.2, left, S edge, 100% S); that edge is called the **S edge**. However, the Mo edges would be positively charged and the S edges would be negatively charged, giving an unstable situation. A more stable situation can be reached by distributing the S atoms equally over the edges. When S atoms are shifted to the plane of the Mo atoms, they can assume bridge positions between

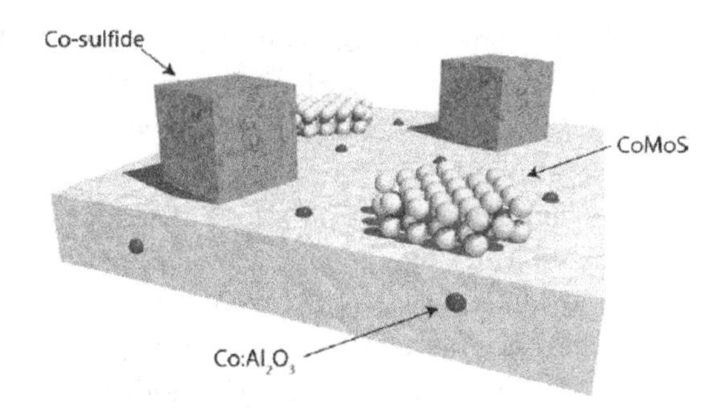

Fig. 8.3 Three forms of cobalt present in a sulfided $CoMo/\gamma\text{-}Al_2O_3$ catalyst: Co atoms on the edges of MoS_2 crystallites in the Co–Mo–S structure, Co atoms in Co_9S_8 crystallites and Co cations in the lattice of $\gamma\text{-}Al_2O_3$ ($Co{:}Al_2O_3$). Reprinted with permission from [8.7]. Copyright (2007) Elsevier.

Mo atoms. This increases the number of Mo–S bonds and, thus, lowers the energy (Fig. 8.2, right).

Cobalt may be present in three forms after sulfidation, as Co_9S_8 crystallites on the support, as Co atoms adsorbed on the edges of MoS_2 crystallites (the so-called Co–Mo–S phase, see below), and as Co^{2+} cations in tetrahedral sites in the $\gamma\text{-}Al_2O_3$ lattice (Fig. 8.3). In the following we will address Mo, Co and Ni as atoms when they are present in sulfided form because metal sulfides are more covalent than ionic compounds. Co and Ni in the $\gamma\text{-}Al_2O_3$ lattice, on the other hand, will be denoted as cations. Analogous to cobalt, nickel can be present as segregated Ni_3S_2, as Ni–Mo–S and as Ni^{2+} cations in the support. Depending on the relative concentrations of Co and Mo and on the pretreatment, a sulfided catalyst contains either a relatively large amount of Co_9S_8 or a large amount of the Co–Mo–S phase.

The structure of the catalyst in the sulfided state is predetermined by the structure of the oxidic precursor. As mentioned in the foregoing, cobalt can be present as Co_3O_4 on the surface of the $\gamma\text{-}Al_2O_3$ support and as Co^{2+} cations in octahedral and tetrahedral positions in the surface or subsurface of the support. Co_3O_4 transforms into Co_9S_8, Co^{2+} cations in octahedral support sites transform into the Co–Mo–S phase, and Co^{2+} cations in tetrahedral support sites remain in the support. A combination of Mössbauer spectroscopy and catalytic activity studies showed that the promoter effect of cobalt is related to the Co atoms in the Co–Mo–S phase and not to separate Co_9S_8 [8.2]. Small amounts of Co give preferentially Co–Mo–S and strongly increase the HDS activity. Co_9S_8 is formed only at higher Co loadings where the activity already levels off. An infrared study of the adsorption of NO molecules on sulfided Co–Mo/Al_2O_3 catalysts confirmed that the Co atoms are adsorbed on the MoS_2 edges [8.8]. At increasing cobalt loading the spectrum of NO adsorbed on Co sites increased in intensity, while that of NO adsorbed on Mo sites decreased in intensity because Co atoms at the edge decoration sites cover Mo atoms and block adsorption of NO on these Mo atoms. At low Co loadings and mild sulfiding temperatures all Co atoms can be positioned around the MoS_2 edges. Initially, the catalytic activity increases with increasing Co/Mo (or Ni/Mo) ratio. When all edge positions are occupied, the additional Co atoms will have either to go on top of Co atoms that are already present or have to form separate Co_9S_8 crystallites. High sulfiding temperatures also lead to large Co_9S_8 particles because of growth of the MoS_2 crystallites and, as a consequence, decrease in the MoS_2 edge area. Since the Co_9S_8 particles have a low catalytic activity and cover the MoS_2 particles, the HDS activity of Co–Mo catalysts decreases at high Co/Mo ratios. The maximum activity is reached at a Co/Mo ratio of 0.3–0.5. Commercial catalysts usually contain Co/Mo or Ni/Mo ratios slightly above 0.5, at Mo loadings of 10–15 wt.%.

8.2.1.2 *Active sites*

How are the metal atoms situated at the catalyst surface and what is the structure of the catalytic sites? As shown in Fig. 8.2 (right), energetically stable $(10\bar{1}0)$ and $(\bar{1}010)$ edges can be obtained by distributing the S atoms in the cleavage plane of a stoichiometric MoS_2 crystallite equally over the edges. However, the coordination number of the Mo atoms on the Mo edge is low; the Mo atoms are only surrounded by four S atoms. Addition of S atoms to the Mo edge is an exothermic process for S coverages up to 50% [8.6, 8.9]. **Scanning tunnelling microscopy** (STM) [8.10] of MoS_2 combined with DFT calculations showed that the structure of single-layer MoS_2 nanoclusters on a gold (111) surface depends on the synthesis conditions. Triangles are formed under high-sulfur conditions, $p(H_2S):p(H_2) = 500$, and truncated hexagons under low-sulfur conditions, $p(H_2S):p(H_2) = 0.07$, as in HDS [8.6]. The triangular MoS_2 nanoclusters are terminated by Mo edges saturated with S dimers. The hexagonal MoS_2 clusters are terminated by Mo edges covered with S monomers, with the S atoms in bridging positions between the Mo edge atoms (Fig. 8.2, right, Mo edge 50% S), while the Mo atoms at the S edges of MoS_2 are surrounded by six S atoms in trigonal prismatic fashion, as in the interior of the MoS_2 clusters (Fig. 8.2, left, S edge 100% S).

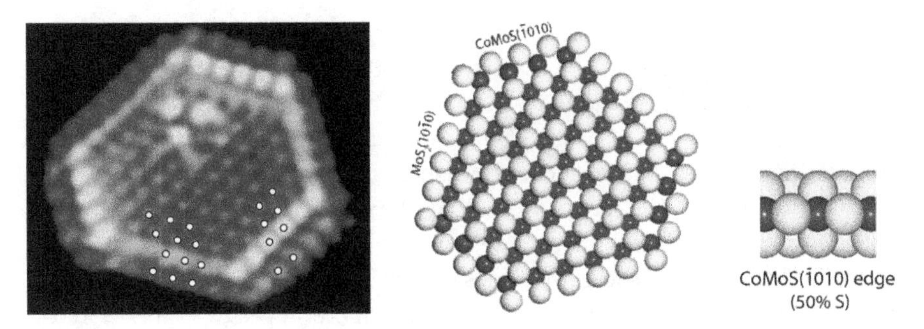

Fig. 8.4 STM image and ball model of a Co–Mo–S nanoparticle and side views of the Co–Mo–S ($\bar{1}010$) edge. S atoms are indicated in yellow, Mo atoms in dark blue and Co atoms in red. Reprinted with permission from [8.7]. Copyright (2007) Elsevier.

STM of Co–MoS_2 and Ni–MoS_2 nanoclusters on a gold (111) substrate proved that the Co and Ni atoms in Co- and Ni-promoted MoS_2 particles are located at the edges of the MoS_2 particles, substituting Mo atoms and forming the so-called **Co–Mo–S and Ni–Mo–S structures** [8.7]. The Co atoms in the Co–Mo–S catalyst were exclusively situated at the S edges, tetrahedrally coordinated by S atoms with 50% sulfur coverage (Fig. 8.4, right).

The location and coordination of the Ni atoms in the Ni–Mo–S structure is less clear as that of the Co atoms. Two types of Ni–Mo–S nanoparticles were observed, depending on synthesis conditions. At a synthesis temperature of 500 °C, Ni–Mo–S crystallites with a truncated trigonal shape similar to that of the Co–Mo–S crystallites were formed. The location and coordination of the Ni atoms was the same as of the Co atoms in Co–Mo–S, exclusively on the S edge with tetrahedral coordination. After synthesis at 400 °C, in addition to truncated trigonal Ni–Mo–S crystallites, also almost round Ni–Mo–S crystallites were formed. These particles not only contained Ni–Mo–S ($10\bar{1}0$)- and ($\bar{1}010$)-type edges but also ($11\bar{2}0$)-edges. Ni atoms were present all around these round Ni–Mo–S particles, partial as well as complete substitution of Mo atoms by Ni atoms was observed and Ni atoms coordinated by four, five and six S atoms were believed to be present [8.7]. In agreement with the experimental STM results, DFT calculations predicted that the Mo atoms at the S edges of MoS_2 and the Co atoms at the S edges of Co–Mo–S are tetrahedrally coordinated by four S atoms, two S atoms being in bridge position (Fig. 8.4, right). On the other hand, Ni was predicted to be present on the S as well as on the Mo edge. Under normal sulfiding conditions the Ni–Mo–S phase predominantly exposes the Ni edge with 0% S coverage, with the Ni atoms in square planar coordination, and a small fraction of Ni on the S edge with 50% S (similar to that of Co on the S edge of Co–Mo–S) [8.9].

The STM images of MoS_2, Co–Mo–S and Ni–Mo–S particles show a bright ridge, called **brim**, located adjacent to the outermost edge atoms. The Mo atoms in the brim are coloured yellow in Fig. 8.4 (left). DFT calculations have shown that these brim atoms have a high conductance, all around the edge of the two-dimensional slabs. MoS_2 is an semiconductor, but the brim structure is associated with metallic edge states and the edge states can be considered as one-dimensional conducting wires [8.11]. The one-dimensional metallic edge states may play an important role in catalysis. The fully sulfided Mo edges (i.e. Mo edges containing no CUS sites) of the triangular MoS_2 nanoclusters are quite reactive. The activity was associated with the electronic brim which is able to adsorb thiophene (C_4H_4S) and facilitate a subsequent hydrogenation and C–S cleavage.

As we will see in Section 8.3, hydrotreating reactions proceed via several reactions. Which metal atom, Mo or Co (Ni), is responsible for these reactions? Are surface Mo atoms the catalytically active sites and, if so, how are they promoted by Co and Ni, or are the Co and Ni atoms new sites which are supported on and influenced by MoS_2? In the past, it was generally assumed that the catalytically active sites in a hydrotreating catalyst are the Mo atoms at the edges of the MoS_2 crystallites, with at least one S vacancy at a site to allow the reacting molecule to bind chemically to the Mo atom [8.1, 8.2]. Sulfur atoms in the basal planes of MoS_2 are much more difficult to remove than S atoms at edges and corners and, therefore, exposed Mo atoms will be mainly present at edges and corners. Catalysis therefore occurs at MoS_2 edges and corners rather than on the basal plane, as verified by a surface-science study in which a MoS_2 single crystal, with a high basal plane to edge surface area ratio, was found to have a low HDS activity. Its activity increased after the S atoms were sputtered away from the basal plane and exposed to the Mo atoms [8.12]. DFT calculations, which included thermal and entropic effects (translations, vibrations and rotations) indicated that, under industrially relevant conditions (300–450 °C, 1–10 MPa H_2, $p(H_2S)/p(H_2) = 0.01$–0.05), the enthalpy change is positive but the free energy change is negative for H_2 chemisorption as well as for the creation of S vacancies on the Mo- as well as S-edge [8.13]. The vacancies at the MoS_2 edge are supposed to be the catalytically active sites where the S-containing molecule can adsorb in σ fashion. The H atoms on the S atoms at the edges then deliver the H atoms for the hydrogenation.

The HDS and HDN activities of a MoS_2/Al_2O_3 catalyst increase substantially on addition of Co or Ni. Several explanations for the promoter function of Co and Ni have been proposed [8.1, 8.2]. Infrared [8.8] and EXAFS [8.5] investigations demonstrated that the Mo atoms are covered by the S and promoter atoms. Thus, if only the Mo atoms were active, the activity should decrease upon addition of Co or Ni atoms. The most widely accepted model, the Co–Mo–S model (or Ni–Mo–S model for Ni–Mo catalysts), attributes the promotion effect to the Co atoms present in the Co–Mo–S phase at the MoS_2 edges [8.2]. In an HDS reaction, a S-containing molecule is supposed to adsorb on a site with a S vacancy (indicated by $*$) and react to give a hydrocarbon molecule and an adsorbed S atom (R-S $+ * \rightarrow$ R-S$*$ $\rightarrow$ R + S$*$). This S atom occupies the vacancy and must be removed by hydrogenation to H_2S before the catalytic cycle can start again ($S* + H_2 \rightarrow H_2S + *$). A sulfur atom between a Co (or Ni) and a Mo atom is less strongly bonded than a sulfur atom between two Mo atoms; therefore, it can be more easily removed. This would explain the promoter action of Co and Ni on Mo in HDS. The edge structures presented in Fig. 8.2 (right) may represent the actual sites during catalysis, but it is likely that restructuring occurs before or during the catalytic reaction. For instance, a bridging edge S atom may shift to non-bridging on-top position, thus creating a vacancy at a Co, Ni or Mo atoms at the edge. A reacting molecule can then adsorb with its S atom on this vacancy.

8.2.2 *Metal Phosphides*

Catalysts with metallic properties can bind H, C, N and O atoms and break and make H–H, C–C, C–N and C–O bonds. As we will see in Section 8.3, hydrogenation and hydrogenolysis (bond breaking) reactions are required in HDS, HDN and HDO and that makes materials which have metallic properties, such as metals, metal sulfides, metal carbides, metal nitrides and metal phosphides, potential catalysts for these reactions [8.14]. Of these materials, metals have the highest initial activity but form a deactivating sulfur layer on the metal surface. The noble metals Pt and Pd, either alloyed or in monometallic form, have a high HDS activity and are less susceptible than other metals to deactivation. Pt is highly active and selective to **direct desulfurisation (DDS)** in which C–S bonds are broken (Section 8.3.1), whereas Pd has a lower activity but possesses a high hydrogenation ability, especially for the refractory 4,6-dimethyldibenzothiophene. Metal carbides and nitrides suffer from the same deactivation problem as metals and even transform into bulk sulfides. Metal sulfides have lower initial HDS and HDN activities than metals but remain stable during reaction. That is why metal sulfides have become the materials of choice in industrial hydrotreating.

Many metal phosphides have metallic properties and, thus, are potential HDS, HDN and HDO catalysts. Unlike metals, metal carbides and metal nitrides, metal phosphides do not become poisoned by sulfur and do not deactivate in the presence of sulfur-containing molecules. Several transition-metal phosphides are active catalysts in HDS and HDN reactions [8.15] and of the monometal phosphides Fe_2P, CoP, Ni_2P, MoP and WP (also called binary phosphides), Ni_2P/SiO_2 has the highest HDS and HDN activity. Bimetallic phosphides (also called ternary phosphides) have activities in between those of the monometallic phosphides (e.g. MoP $<$ NiMoP $<$ Ni_2P). This is in contrast to metal sulfides, where Co and Ni are strong promoters for the activity of sulfided Mo and W catalysts (e.g. Co_9S_8 $<$ MoS_2 $<$ Ni-MoS_2), as described in Section 8.2.1.1.

More than one hundred monometallic transition-metal phosphides are known, with compositions from metal-rich M_4P to phosphorus-rich MP_{15} (for instance, nickel forms eight phosphides: Ni_3P, Ni_5P_2, $Ni_{12}P_5$, Ni_2P, Ni_5P_4, NiP, NiP_2 and NiP_3), and about one hundred bimetallic $(M)_x(M')_yP_z$ phosphides are known. The differences in composition lead to a broad range of structural, electronic, magnetic, optical and catalytic properties. Many of the metal-rich monometallic and bimetallic phosphides have metallic properties and, thus, are potential catalysts for HDS, HDN and HDO. Several metal phosphides proved to be active and stable under hydrotreating reaction conditions and, therefore, interest in these materials as HDS and HDN catalysts has increased substantially. In recent years, their HDO properties have also attracted attention [8.14] because of the importance of HDO processing for the upgrading of biomass feedstocks to renewable transportation fuels.

Metal phosphides have different structures than metal sulfides and many structures are known. The structures are three-dimensional, different from the two-dimensional slab structure of MoS_2. The structures of Ni_2P and MoP, which are good hydrotreating catalysts, are presented in Fig. 8.5. The P atoms in Ni_2P are surrounded by nine Ni atoms, six in a trigonal prism and three bonded to the P atom through the three rectangular faces of the trigonal prism. There are two types of Ni and P atoms in Ni_2P, which

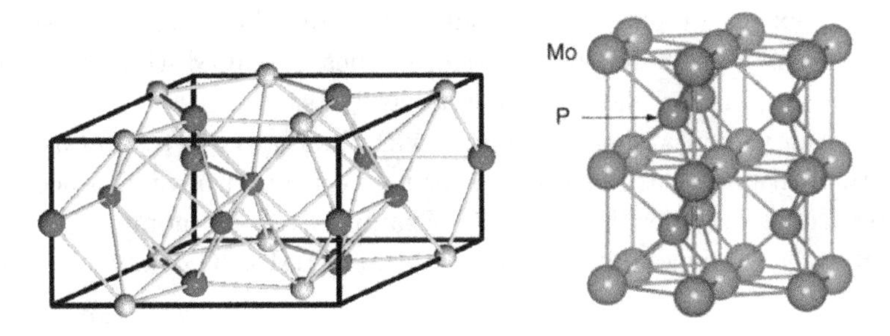

Fig. 8.5 Left: structure of Ni_2P, with grey Ni atoms surrounded by four red P atoms and blue Ni atoms surrounded by five red P atoms. Right: structure of MoP, with orange P atoms surrounded by six cyan Mo atoms in a trigonal prismatic way.

form two different trigonal prisms. The Ni atoms are either surrounded by four P atoms or by five P atoms. The Mo atoms in MoP are surrounded by six P atoms in a trigonal prism and the P atoms are surrounded by six Mo atoms in a trigonal prism.

Supported metal phosphides are often prepared by temperature programmed reduction of a metal salt and ammonium phosphate, impregnated on a catalyst support, between 400 and 1,000 °C in flowing hydrogen [8.16]. Reduction of the phosphate P–O bond requires high temperature, even with hydrogen atoms. Most probably, the first reaction is the reduction of the metal salt to metal particles. They dissociate hydrogen molecules to hydrogen atoms, which spill over to the phosphate and reduce it to phosphorus or phosphine (PH_3). P atoms or PH_3 react with the metal particles to metal phosphide particles. This explains why successively Ni_3P, $Ni_{12}P_5$ and Ni_2P were observed when a Ni salt was reduced in the presence of ammonium phosphate [8.17]. The relative ease of preparing supported phosphate materials has made the phosphate reduction method very popular for the preparation of catalysts. However, the high temperature needed to break the strong P–O bond in phosphate leads to large metal phosphide particles and consequently relatively low catalytic activity. Another disadvantage of the phosphate method is that it is difficult to prepare metal phosphides on alumina, the support which is preferred in industry. Phosphate strongly interacts with γ-Al_2O_3 to form $AlPO_4$ upon heating of alumina-supported phosphate. As a consequence, a large proportion of the phosphate is unavailable for reduction to phosphide and excess phosphate has to be added to form metal phosphide particles on γ-Al_2O_3 [8.15, 8.18]. Thus, only Ni_2P/Al_2O_3 catalysts with low dispersion and large excess of phosphorus could be made. Silica does not strongly interact with phosphate and, therefore, silica or MCM-41 is more often used as a support for phosphide-based catalysts.

Phosphorus-containing compounds with the P atom in a lower oxidation state than in phosphate (V) are easier to reduce and have also been used to prepare metal phosphides [8.17]. Examples are ammonium phosphite ($NH_4H_2PO_3$), ammonium hypophosphite ($NH_4H_2PO_2$) and phosphine (PH_3).

8.3 Reaction Mechanisms

8.3.1 *Hydrodesulfurisation*

Crude oils contain organosulfur compounds. The low-boiling fraction that is produced in the primary distillation of crude oil contains mainly mercaptans, sulfides and disulfides, which are very reactive and can easily be removed. The higher boiling fractions produced by primary distillation (heavy naphtha and diesel) and light naphtha, produced in the FCC process (Section 7.5.1), contain relatively more sulfur compounds and with a higher molecular weight, such as thiophene, benzothiophene and their alkylated derivatives. These compounds are more difficult to convert via hydrotreating than mercaptans and sulfides. The heaviest fractions that are mixed into the gasoline and diesel pools (heavy naphtha and diesel from the FCC and coking plants) contain mainly alkylbenzothiophenes, dibenzothiophene and alkyldibenzothiophenes [8.1].

During hydrotreating, the sulfur compounds react in the order: thiophene > alkylthiophene > benzothiophene > alkylbenzothiophene > dibenzothiophene, alkyldibenzothiophenes without substituents at the 4 and 6 positions > 4-alkyldibenzothiophenes > 4,6-dialkyl-dibenzothiophenes [8.19]. Deep desulfurisation of fuels implies that more and more of the least reactive sulfur compounds must be converted. Aliphatic and aromatic thiols are intermediates in ring-opening reactions of cyclic sulfur-containing compounds. They have a high reactivity, and the sulfur atom can be easily removed, which explains why they do not occur in oil. Aliphatic thiols may react through elimination and hydrogenation

$$\mathrm{R{-}CH_2{-}CH_2{-}SH} \xrightarrow{-H_2S} \mathrm{R{-}CH{=}CH_2} \xrightarrow{+H_2} \mathrm{R{-}CH_2{-}CH_3}$$

and through hydrogenolysis

$$R{-}CH_2{-}CH_2{-}SH \xrightarrow[+H_2]{-H_2S} R{-}CH_2{-}CH_3$$

Elimination on the surface of metal sulfides may take place in the same way as on the surface of metal oxides, by an acid-base catalysed β-hydrogen elimination reaction [8.1]. The surfaces of MoS_2 and Co- and Ni-promoted MoS_2 are metallic, however, and their acidity and basicity are much weaker than those of metal oxides. The elimination might occur by a slightly different mechanism, by adsorption of an alkylthiol RCH_2CH_2SH with the S atom in a vacancy on the metal sulfide surface. The resulting alkylthiolate RCH_2CH_2S* can eliminate HS by leaving the S atom on the surface and abstracting an H atom from the β-carbon atom. This H atom can be abstracted by a metal atom or a S atom, but will move to the S atom anyway. Hydrogenation and hydrogenolysis take place at the metal sulfide surface, probably in an analogous way as hydrogenolysis on metals, via C–S and H–H bond scissions and C–H and S–H bond formations [8.20]. β-H elimination from aliphatic thiols normally occurs at a faster rate than hydrogenolysis. Hydrodesulfurisation of **thiophenol** mainly gives benzene [8.21] and may occur by hydrogenolysis.

H_2

$- H_2S$

SH

Thiophene, benzothiophene, dibenzothiophene and their alkylated derivatives are the major sulfur-containing molecules in oil. The major reaction path of the HDS of thiophene under industrial conditions, at elevated H_2 pressure, is by hydrogenation of thiophene to tetrahydrothiophene [8.24]. Tetrahydrothiophene can react to butadiene through two successive β-H eliminations, or to n-butane through two hydrogenolysis steps.

$-H_2S$

SH

$2 H_2$

S

S

H_2

$-H_2S$

SH

H_2

DFT calculations have been carried out of the two pathways in the HDS of thiophene on MoS_2 edge structures, the **hydrogenation (HYD)** pathway, which starts with hydrogenation of the S-containing molecule, and the **direct desulfurisation (DDS) pathway**, which starts with breaking a C–S bond. The calculations show that the HYD pathway involves hydrogenation of thiophene to 2,5-dihydrothiophene and can take place at the Mo($10\bar{1}0$) edge even in the absence of coordinatively unsaturated Mo edge sites [8.22]. This is ascribed to the metallic-like brim sites (cf. Section 8.2.1.1). The HYD reaction pathway can also occur at the S($\bar{1}010$) edge, but at this edge sulfur vacancies must be created for the reaction to proceed. The sulfur vacancies at the S($\bar{1}010$) edge are also active sites for the DDS pathway, which involves mono-hydrogenation to 2-hydrothiophene, followed by S–C scission. As hydrogenation is most facile on the Mo edge and S–C scission can only occur on the S edge, thiophene may be hydrogenated at the Mo edge, then moves to the S edge and undergoes S–C scission. Also on the Co–Mo–S edge, both hydrogenation and C–S bond scission can take place without the initial creation of vacancies [8.23]. The S atom that is removed from thiophene is bound to the Co–Mo–S edge and later removed in a catalyst regeneration step. The equilibrium Co–Mo–S($\bar{1}010$) edge contains bridged S atoms and if these shift to a position in the sulfur plane then vacancies are created at which HDS can take place; no S atoms have to be removed before the reaction can start. Co-promotion decreases the barrier of hydrogenation reactions and active site regeneration but increases the barrier of S–C scission reactions. As a result, and in agreement with experimental information, the hydrogenation activity and the relative importance of the DDS pathway are increased.

Dibenzothiophene can react by the HYD and DDS pathways [8.1]. It reacts mainly to biphenyl (70–90% for CoMo and NiMo catalysts), most probably through the DDS pathway, by a twofold hydrogenolysis.

$$\text{dibenzothiophene} \xrightarrow[-H_2S]{2\,H_2} \text{biphenyl}$$

Alkyl groups next to the sulfur atom, as in 4,6-dialkyldibenzothiophene, sterically hinder the adsorption of the molecule perpendicular to the catalyst surface (Fig. 8.6, left). A strong bond between the S atom of 4,6-dialkyldibenzothiophene and a Mo atom at the catalyst surface

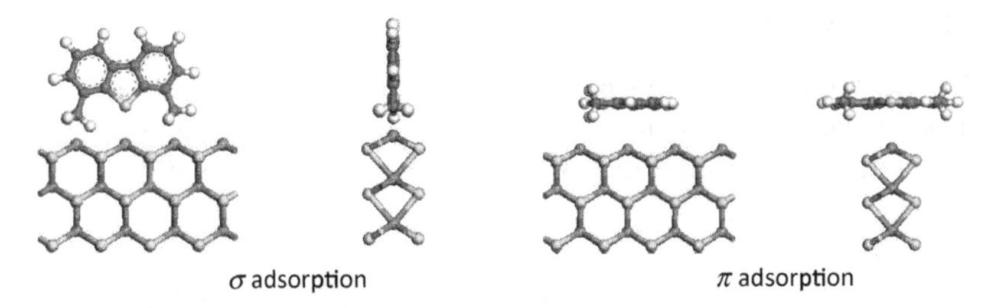

Fig. 8.6 Perpendicular (σ) and parallel (π) adsorption of 4,6-dimethyldibenzothiophene on MoS_2 edge.

cannot be established and the removal of the S atom through the DDS route is strongly suppressed. Furthermore, H_2S adsorbs strongly on the metal atoms at the catalyst surface and this suppresses the DDS route even more.

Dibenzothiophene can also react by the HYD pathway, by hydrogenation of one of the phenyl rings, followed by C–S bond scission by hydrogenolysis or by elimination and hydrogenation of the resulting double bond. The resulting arylthiol (2-thiophenylcyclohexane) is desulfurised to phenylcyclohexane (cyclohexylbenzene), in the same way as thiophenol is desulfurised to benzene. For dibenzothiophene the HYD pathway is a minor route (10–30% for CoMo and NiMo catalysts).

S $\xrightarrow{3\,H_2}$ S $\longrightarrow$ SH

SH $\xrightarrow{H_2}$ SH $\xrightarrow[H_2]{-H_2S}$

The HYD pathway is about equally fast for 4,6-dialkyldibenzothiophene and dibenzothiophene, which is explained by π adsorption and reaction of the (4,6-dialkyl)dibenzothiophene parallel to the catalyst surface (Fig. 8.6, right). Because the DDS rate is fast for dibenzothiophene but slow for 4,6-dialkyldibenzothiophene, the HYD route is the minority route for dibenzothiophene and the major (but equally slow) route for 4,6-dialkyldibenzothiophene. For these reasons, 4,6-dialkyldibenzothiophene molecules are among the molecules that are most difficult to desulfurise.

8.3.2 *Hydrodenitrogenation*

Polycyclic aromatic nitrogen compounds, such as quinoline, indole, acridine and carbazole, are the main nitrogen-containing compounds in oil products, while monocyclic pyridine and pyrrole as well as anilines are present in coal tar [8.1].

pyridine quinoline acridine

pyrrole indole carbazole

Aliphatic amines and saturated cyclic amines are hardly observed in oil and only at low concentrations during HDN, since their nitrogen atoms are easily removed. In the past it was assumed that the removal of NH_3 from alkylamines takes place by classic organic reactions, the Hofmann elimination to alkenes and the NH_2–SH substitution, followed by C–S hydrogenolysis, to alkanes [8.24]. Later studies showed that NH_2–SH substitution of alkylamines to alkylthiols and NH_2–NH_2 substitution to dialkylamines are the predominant reactions in the HDN of alkylamines [8.25]. Eventually all amines are transformed by reaction with H_2S into alkylthiols, which irreversibly react to alkenes and alkanes and H_2S (Fig. 8.7).

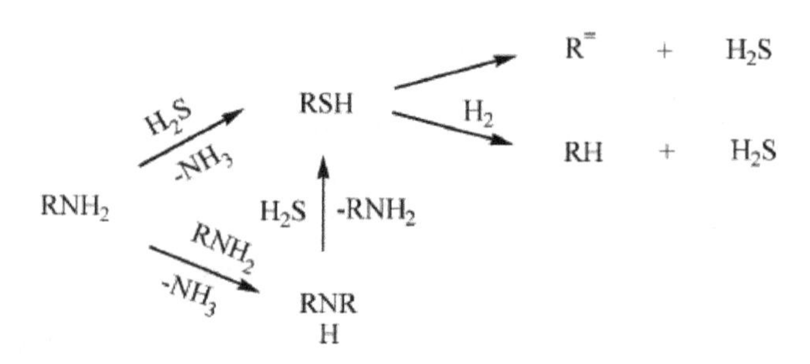

Fig. 8.7 Reaction network for the removal of nitrogen from alkylamines.

The substitution of the NH_2 group by an SH group does not take place directly by a classic S_N2 mechanism, but by an indirect sequence of reactions [8.26]. The amine is first dehydrogenated to an imine, then H_2S is added to the imine and ammonia is eliminated from the amino-thiol. Hydrogenation of the thiocarbonyl gives a thiol, which reacts further to an alkene and alkane as described in Section 8.3.1. This mechanism uses the metallic character of the surface of MoS_2.

$$RCH_2{-}NH_2 \xrightarrow{-H_2} RCH{=}NH \xrightarrow{+H_2S} HS{-}CHR{-}NH_2 \xrightarrow{-NH_3} S{=}CHR \xrightarrow{+H_2} HS{-}CH_2R \xrightarrow{+H_2} H_2S \rightarrow RCH$$

C–N bonds in aromatic rings are much stronger than those in aliphatic rings. Consequently, C–N bonds in pyridine and pyrrole can be broken only after hydrogenation of the ring to give the saturated endocyclic amines piperidine and pyrrolidine respectively (with the N atom in the ring). Direct cleavage of the exocyclic C–X bond in C_6H_5–X is the dominant reaction for X = Cl and SH, giving benzene and HX. For X = OH and NH_2, on the other hand, direct cleavage occurs only to a limited extent and only at high H_2/H_2S ratios and above 400 °C. The strength of the C–O bond in phenol and of the C–N bond in aniline are increased by conjugation with the aromatic ring. Under normal hydrotreating conditions, the C–N bond can, therefore, only be broken when it is aliphatic. Hydrogenation of the N-containing heterocycle (e.g. pyridine and pyrrole) or of the aromatic ring to which the amine group is attached (as in aniline) is necessary to remove a substantial portion of the nitrogen. Consequently, the main products of HDN and HDO are aliphatic, whereas those of HDS and hydrodechlorination are aromatic; correspondingly, HDN and HDO consume more hydrogen than HDS. This difference is a consequence of the lower energies of the C_6H_5–Cl and C_6H_5–SH bonds relative to the C_6H_5–OH and C_6H_5–NH_2 bonds.

The HDN of pyridine starts with hydrogenation of the ring and the resulting piperidine reacts to pentane and pentene by a twofold application of the imine mechanism [8.26]. Piperidine dehydrogenates to the imine tetrahydropyridine (or alternatively pyridine is hydrogenated to this imine), the first C–N bond is broken and the ring of the imine is opened by the addition of H_2S, elimination of the R–NH fragment and hydrogenation of the 1-amino-pentanethion, forming 1-amino-pentanethiol. This molecule reacts to pentylamine, and the second C–N bond is broken by a second round of imine formation, H_2S addition, NH_3 elimination, hydrogenation to pentanethiol and reaction to pentene and pentane [8.27].

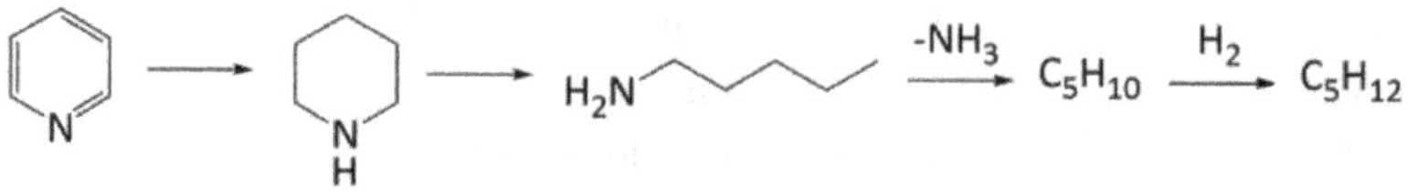

Depending on reaction conditions, the HDN of pyridine is thermodynamically or kinetically controlled [8.1]. The rate of hydrogenation of pyridine at 300 °C and 4 MPa is about as slow as the ring opening of piperidine on Co–Mo, Ni–Mo and Ni–W catalysts. As a consequence, H_2S may have a promoting as well as a poisoning effect on the HDN of pyridine under these conditions. At low H_2S/H_2 ratio, pyridine hydrogenation to piperidine is not poisoned by H_2S and proceeds smoothly but the ring opening of piperidine by nucleophilic substitution by H_2S is slow; this gives a low overall HDN conversion to C_5 hydrocarbons. At a higher H_2S/H_2 ratio, hydrogenation of pyridine to piperidine is slowed down, but ring opening is accelerated. Thus, although the conversion of pyridine decreases, the formation of hydrocarbons (the HDN conversion) increases. At high H_2S/H_2 ratio, the negative effect of H_2S on pyridine hydrogenation is responsible for low pyridine and low HDN conversion. At lower H_2 pressure and higher temperature, the thermodynamics of pyridine hydrogenation becomes rate-determining, and H_2S behaves more and more as a poison, whereas at high H_2 pressure and low temperature H_2S acts as a promoter because the ring opening reaction becomes rate-determining.

Aniline does not have a hydrogen atom in β position relative to the NH_2 group and removal of the NH_2 group by elimination is, therefore, not possible. Substitution of the amine group, attached to a $C(sp^2)$ atom, by a SH group is also highly unlikely. The main route for nitrogen removal from aniline (about 90%) is hydrogenation of the benzene ring to cyclohexylamine. This hydrogenation introduces β-hydrogen atoms which can be used in the reaction to cyclohexene, which reacts further to cyclohexane. A minor part of aniline (10%) undergoes direct removal of the N atom to benzene.

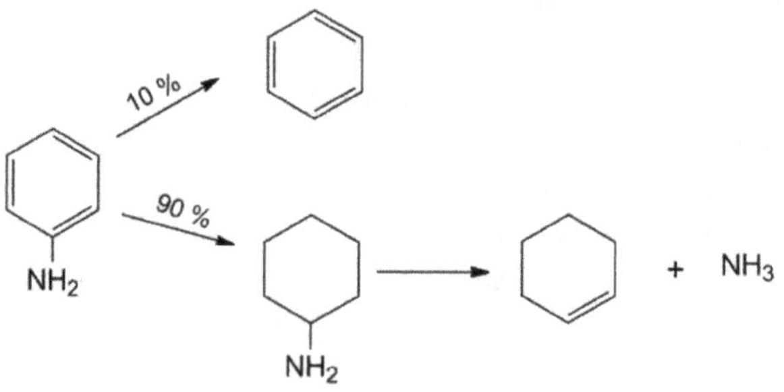

In the HDN of the bicyclic quinoline, the N-containing ring is hydrogenated preferentially, giving 1,2,3,4-tetrahydroquinoline. The first C–N bond of this molecule can be broken as shown for piperidine, giving 2-propylaniline. 2-Propylaniline reacts like aniline to give propylcyclohexane. However, the hydrogenation of the second aromatic ring of 1,2,3,4-tetrahydroquinoline to decahydroquinoline is faster than the ring opening of 1,2,3,4-tetrahydroquinoline to 2-propylaniline and the major part of the removal of nitrogen occurs through this route. Decahydroquinoline then loses the N atom and reacts to propylcyclohexane as described for piperidine.

N NH2 N H NH2 N H

8.3.3 *Hydrodeoxygenation*

Removal of oxygen from oxygen-containing molecules is at present only practised in the treatment of coal-derived liquids, since oxygen-containing molecules rarely occur in oil. The most abundant oxygen-containing molecules in coal liquids are phenols, aryl ethers and benzofurans (the oxygen equivalent of benzothiophene and indole, cf. Section 8.3.2). Catalytic hydrodeoxygenation (HDO) may play an important role in the future in the upgrading of biomass-derived liquids. Because the burning of fossil fuels to CO_2 has become a serious environmental problem, biomass has attracted attention as a renewable energy source. Triglycerides derived from animal fat and vegetable oil can be converted into biofuel (i.e. a mixture of methyl esters of fatty acids) by transesterification. However, this so-called first-generation biofuel has a high oxygen content and is incompatible with fossil fuels. Furthermore, it competes with the food industry and the creation of oil palm plantations causes deforestation. Second generation (green) oxygen-free biofuel can be prepared by HDO of first-generation biofuel [8.28] or by the direct HDO of waste vegetable oils and fats and of bio-oil formed in the pyrolysis of ligno-cellulosic biomass (which does not compete for food) [8.29]. Bio-oil contains many oxygen-containing compounds such as furans, pyrans, phenols and lactones, which are undesirable components in fuels. The hydrotreating of bio-oil can be done in two steps, starting with a stabilisation step, in which carbonyl and carboxyl groups are transformed into alcohols over supported noble metal catalysts. In a second step, cracking and HDO occur over Ru, Ni or sulfided Co–Mo catalysts. The resulting green biofuel is oxygen free, has a similar composition as petroleum fuel and can be employed directly in the fuel industry.

Transition metal sulfide catalysts (Co–Mo–S and Ni–Mo–S) have been widely tried in HDO due to their hydrogenation ability in refinery HDS and HDN processes, which are closely related to HDO processes. However, the surface of metal sulfides becomes oxidised in the presence of water or steam, a HDO byproduct. H_2S or other S-containing molecules must be co-fed to keep metal sulfide catalysts in the sulfided form, leading to increased interest in non-sulfide catalysts. Noble metals such as Pd, Pt, Rh and Ru have high activity in the HDO of bio-oil model compounds such as dibenzofuran (the oxygen equivalent of dibenzothiophene, cf. Section 8.3.1), phenol, guaiacol (2-methoxyphenol) and propanoic acid but their high cost and poor tolerance to sulfur and nitrogen create problems. Transition metal carbides, nitrides, phosphides and borides can be formed by the incorporation of C, N, P and B atoms into the lattices of transition metals. They have metallic character and catalyse hydrogenation and hydrogenolysis reactions and, therefore, have attracted attention as HDO catalysts [8.14, 8.30]. Metal phosphides are much more stable than metal sulfides in the presence of water and co-feeding PH_3 is not necessary in HDO. In an HDO study of phenol in the presence of excess steam, however, it was observed that Ni_2P was not stable and transformed into Ni_3P, but this catalyst actually had high activity [8.31].

Oxygen atoms can be removed from compounds that contain a carboxyl, carbonyl or alcohol group (as in fatty acids and fatty alcohols) in three ways, by hydrodeoxygenation (also called **direct deoxygenation DDO**), by **decarbonylation** or **decarboxylation** (DCO) and by **esterification** (EST). The DDO pathway (Eq. (8.1)) involves hydrogenation of the oxygen-containing compound to an alcohol, followed by water removal by elimination to an alkene and subsequent hydrogenation to an alkane, or by direct hydrogenolysis of the C–O bond. Catalysts that can remove CO or CO_2 can react by the DCO pathway (Eq. (8.2)). Metals that bind CO strongly, such as Ni, Co and Ru, can follow this pathway. Whether CO or CO_2 is removed is still an open question because even if CO is removed, CO_2 is always quickly formed by the water–gas shift reaction ($CO + H_2O$

$\rightarrow CO_2 + H_2$). A third way to remove oxygen is esterification (Eq. (8.3)), but only the DDO and DCO pathways lead to full removal of oxygen.

$$R{-}CH_2{-}COOH \xrightarrow[-H_2O]{+H_2} R{-}CH_2{-}CHO \xrightarrow{+H_2} R{-}CH_2{-}CH_2{-}OH \xrightarrow{-H_2O} R{-}CH{=}CH_2 \xrightarrow{+H_2} R{-}CH_2{-}CH_3 \quad (8.1)$$

$$R{-}CH_2{-}COOH \xrightarrow[-H_2O]{+H_2} R{-}CH_2{-}CHO \xrightarrow{-CO} R{-}CH_3 \quad (8.2)$$

$$R{-}CH_2{-}COOH + R{-}CH_2{-}CH_2{-}OH \xrightarrow{-H_2O} R{-}CH_2{-}COO{-}CH_2{-}CH_2{-}R \quad (8.3)$$

The HDO of cyclic molecules which contain an oxygen atom in the ring takes place through the DDO and/or DCO pathways. Before the oxygen can be removed, the O-containing ring must first be opened (and if needed be hydrogenated). O-containing rings are usually ring ethers, such as furan and tetrahydrofuran, and ring esters (officially called lactones), such as valerolactone.

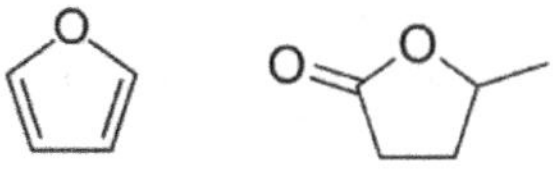

furan $\quad$ γ-valerolactone

Ring ethers and ring esters can be opened by acid catalysis, giving alkanediols and ω-hydroxycarboxylic acids (a carboxylic acid with an OH group at the other end of the molecule) respectively. The ring can also be opened by metal-catalysed hydrogenolysis. The HDO of furan occurs as follows. Furan is first hydrogenated to tetrahydrofuran ($C_4H_4O + 2\,H_2 \rightarrow C_4H_8O$), followed by opening of the ring, either by hydrogenolysis to butanol ($C_4H_8O + H_2 \rightarrow C_4H_9OH$) or by acid-catalysed ring opening to 1,4-butanediol ($C_4H_8O + H_2O \rightarrow HOC_4H_8OH$). Thereafter, C_4 hydrocarbons can be formed by C–O bond breaking (the DDO pathway), or by removing CO to form C_3 hydrocarbons by the DCO pathway. A Ru_2P/SiO_2 catalyst favoured C_4 products while a Ru/SiO_2 catalyst favoured C_3 products [8.32]. The Ru_2P/SiO_2 catalyst was more active than a Ru/SiO_2 catalyst and much more active than a sulfided Co–Mo/Al_2O_3 catalyst. In the same way, after ring opening of γ-valerolactone to form pentanoic acid ($C_5H_8O_2 + H_2 \rightarrow C_4H_9$–COOH) and subsequent hydrogenation to pentanal (C_4H_9–COOH + $H_2 \rightarrow C_4H_9$–CHO + H_2O), the oxygen removal proceeded mainly by the decarbonylation route (C_4H_9–CHO → C_4H_{10} + CO) to give CO and C_4 hydrocarbons on Ni_2P/MCM-41, CoP/MCM-41 and Pd/Al_2O_3, while on MoP/MCM-41 and WP/MCM-41 the DDO pathway produced mainly unsaturated C_5 hydrocarbons [8.33].

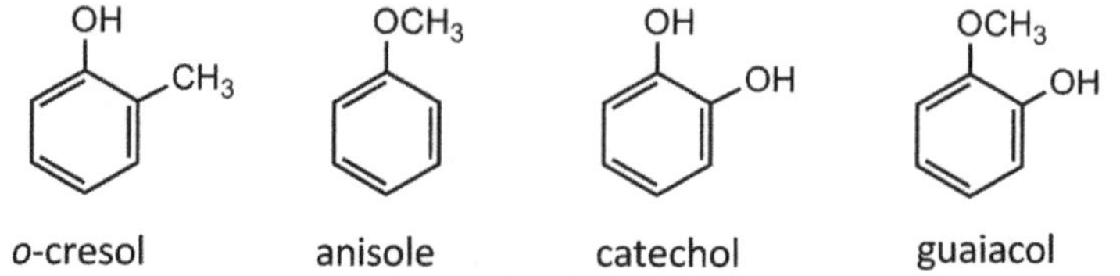

o-cresol anisole catechol guaiacol

Many studies of metal phosphides have been carried out with aryl alcohol model compounds (phenolics) such as phenol, cresol, anisole, catechol and guaiacol. Phenolic molecules can lose the oxygen atom in two ways, similar to the HYD and DDS pathways in the HDN of aniline. In the HYD pathway, phenol is hydrogenated to cyclohexanol, which forms cyclohexene and eventually cyclohexane [8.21]. In the second DDO pathway, benzene is formed by direct hydogenolysis of the C–O bond.

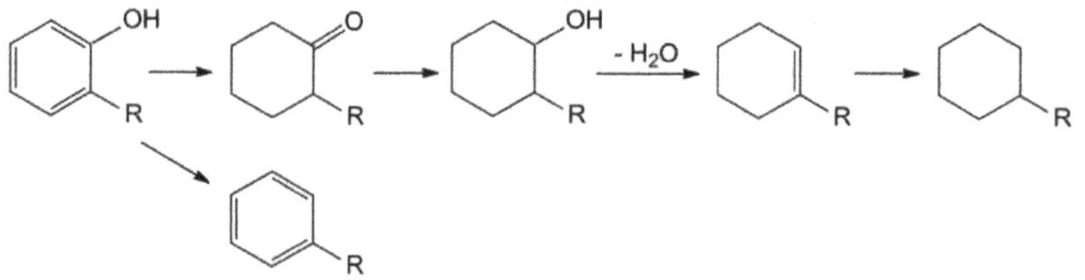

The HDO of benzofuran proceeds similar to the HDN of indole, via hydrogenation to dihydrobenzofuran, ring opening to 2-ethylphenol and removal of the oxygen atom as in phenol [8.34].

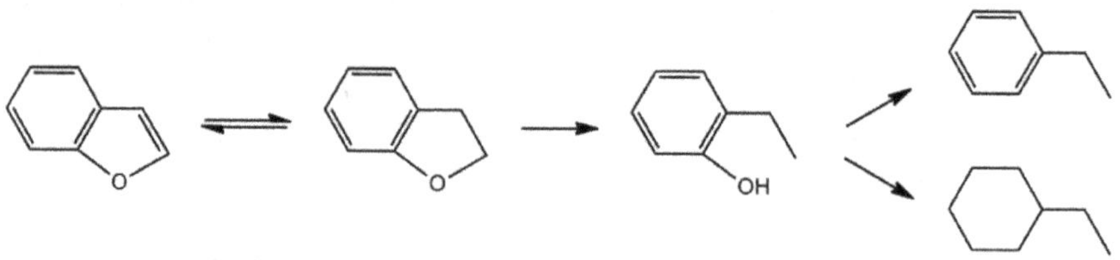

8.3.4 *Hydrotreating of Mixtures*

Competitive adsorption is very important in hydrotreating. Compound A may react faster than compound B, when each reacts separately, but may react slower when both react together because of stronger adsorption of compound B. Inhibition effects may be caused not only by the reacting molecules themselves, but also by the hydrotreating products and intermediates and by end products, e.g. H_2S, NH_3 and H_2O. A Langmuir–Hinshelwood-type rate expression (cf. Section 4.1.1) is often used to describe the effect of the competitive adsorption of all m components in the reaction mixture on the rate of hydrotreating of compound i:

$$r_i = \frac{k_i c_i}{1 + \sum_1^m K_j c_j}. \tag{8.4}$$

In this equation, r_i, k_i and c_i are the rate, rate constant and concentration of component i, respectively, and K_j and c_j are the adsorption constant and concentration, respectively, of component j in the mixture. All components are supposed to adsorb on the same site, while at constant H_2 pressure the influence of H_2 adsorption is lumped into k. The adsorption constants K increase quite strongly in the order aromatic hydrocarbons $<$ sulfur compounds $<$ oxygen compounds $<$ nitrogen compounds. The contents of N-containing molecules in an oil fraction may be one or more orders of magnitude lower than that of S-containing molecules, but they adsorb very strongly on the catalyst and inhibit HDS. As a consequence, HDS is strongly inhibited by the presence of nitrogen compounds under most conditions.

In the hydrotreating of a mixture of thiophene, phenol, aniline and benzene the reactivity order is thiophene $>$ phenol $>$ aniline $\gg$ benzene [8.21]. The conversion of all constituents decreased when the pressure of aniline was increased, demonstrating that aniline is the most strongly adsorbed component in the mixture. Although thiophene has the lowest adsorption constant of these four constituents, an increase in the thiophene pressure had a negative influence on the conversion of all components. This is because H_2S is a product, which can adsorb on the catalyst surface and diminish the hydrogenation activity of the catalyst. As a consequence, the hydrotreating rates of thiophene, phenol and aniline are decreased, because for these molecules hydrogenation is the rate-determining step. Molecules with the same heteroatom can also mutually influence their hydrotreating rates. For instance, aniline has one of the smallest adsorption constants of all N-containing compounds, and thus the HDN of aniline is strongly inhibited by indole and pyrrole but the HDN of indole and pyrrole is only weakly influenced by aniline [8.35]. An STM study substantiated that N-containing molecules such as pyridine and quinoline adsorb strongly on the S-edges of reduced MoS_2 and Co–Mo–S nanoparticles, where they can compete with the adsorption of S-containing molecules [8.36].

8.4 Hydrotreating Processes

Oil fractions are produced in a refinery by distillation of crude oil. They are mixed with high pressure H_2, preheated and fed to fixed-bed adiabatic reactors, where hydrotreating reactions take place over a catalyst (Fig. 8.8). The S-, N-, O- and M-containing hydrocarbons react to hydrocarbons and H_2S, NH_3, H_2O and M_xS_y respectively. The reaction products are cooled and separated in liquid and gas in high- and low-pressure L/G separators. The H_2-rich gas from the high-pressure separator is recycled and combined with fresh feed. The exit gas from the low-pressure separator is sent to a treating unit, where H_2S, NH_3 and H_2O are removed by absorption in a basic solution (e.g. ethanolamine). H_2S, NH_3 and H_2O are removed from the resulting liquid product by steam-stripping. The metal components in the oil fractions end up as metal sulfides on the catalyst surface.

The hydrotreating operating conditions depend on feed quality. The higher the boiling point range of the oil fraction is, the higher temperature and pressure have to be. Naphtha (b.p. 40–180 °C) contains mercaptans, sulfides and alkylthiophenes, while gasoil (b.p. 350–550 °C) contains alkylthiophenes, alkylbenzothiophenes, alkyldibenzothiophenes and higher boiling S-containing molecules as well as nitrogen-containing molecules such as pyridine, quinoline and carbazole. Not only the dialkyldibenzothiophenes and higher boiling S-containing molecules are difficult to remove, but the N-containing molecules are even more refractory. The catalyst deactivates during operation and the loss in activity is compensated by an increase of reactor temperature. When temperature reaches about 400 °C the HDS

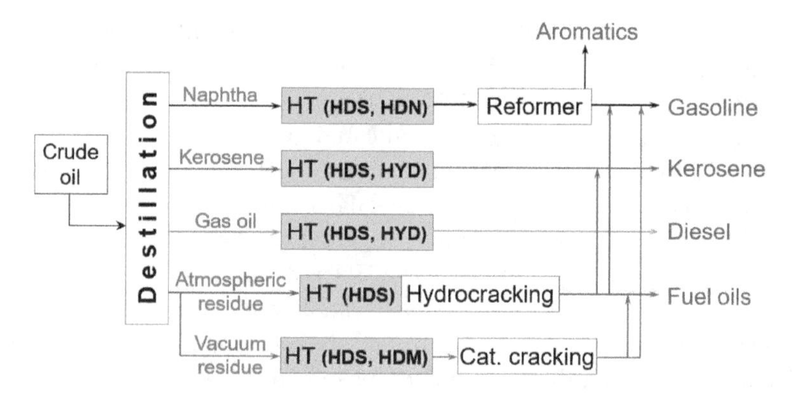

Fig. 8.8 Scheme of a refinery with in red the oil fractions that have to be treated in hydrotreating units (ochre) and the final products in blue.

process is stopped to avoid dehydrogenation of naphthenes to aromatics. The catalyst is then regenerated by coke burn-off and resulfidation off site by an external company. Times between regenerations are typically 2–3 years and depend on feed quality.

8.4.1 *Hydrodesulfurisation of Naphtha*

Gasoline is made from naphtha by reforming the hydrocarbon molecules, so that the product gets a higher octane number (Section 7.5.3). Naphtha is produced at several places in the refinery: in the primary distillation of crude oil, in the thermal cracking of gasoil and in the FCC of gasoil (Section 7.5.1). In the USA and Europe, naphtha produced in the FCC process contributes about one third to the total amount of naphtha in a refinery, but this FCC naphtha is responsible for more than 90% of the sulfur in the total naphtha. In China FCC naphtha contributes even much more to the naphtha pool. Because in the FCC process larger hydrocarbon molecules are cracked into smaller molecules (to produce more gasoline and diesel from crude oil), alkenes and even dienes (formed by too extensive cracking in the FCC process) are produced. Alkenes have a higher octane number than alkanes and, thus, are valuable naphtha components.

Naphtha contains S-containing molecules that are easy to desulfurise (mercaptans, sulfides, thiophenes) and hardly contains N-containing molecules. Therefore, a Co–Mo–S/γ-Al_2O_3 catalyst is used and operating conditions are relatively mild (370 °C, 1.5 MPa, LSV 2 h^{-1}). Because of the stringent sulfur specifications for gasoline ($<$ 10 ppm), sulfur must be removed from the naphtha and especially from the FCC naphtha. The problem of the treatment of FCC naphtha is that sulfur compounds must be removed without hydrogenating the alkenes (they must be kept, because they have a high octane number). The desulfurisation can either be done by treatment of the gasoil that is fed to the FCC unit, or by treating the naphtha product that comes from the FCC unit. Most refineries do the latter and combine the fractionating of the naphtha with catalytic HDS in order to optimise the overall HDS/hydrogenation selectivity and to minimise the loss in octane number. To avoid a loss in octane number in the deep HDS of FCC naphtha, one utilises the fact that the alkenes are mainly concentrated in the low-boiling fraction of the FCC naphtha, whereas most of the sulfur compounds are present in the high-boiling fraction. The FCC naphtha is fractionated by distillation before HDS and each fraction is desulfurised at appropriate conditions. The sulfur compounds in the light fraction (mainly mercaptans) are easy to desulfurise and alkenes are preserved. The intermediate fraction contains essentially thiophene and light alkylthiophenes as sulfur impurities, while the heavy fraction contains a relatively low concentration of alkenes but most of the sulfur impurities, including benzothiophene and alkylbenzothiophenes [8.37]. These sulfur compounds require more severe conditions but, because of the low amount of alkenes in this heavy fraction, the overall loss in alkenes is small.

Processes that treat FCC naphtha contain several units: a selective hydrogenation reactor, an isomerisation reactor, a splitter and an HDS reactor [8.1, 8.37, 8.38]. The selective hydrogenation reactor converts dienes to alkenes (with minimum saturation of alkenes) and, thus, protects the HDS catalyst in the HDS reactor from deactivation. At the same time, light mercaptans and light sulfides are converted into heavier sulfur compounds through thioetherification, a reaction of thiols with alkenes [8.39]. The resulting heavier sulfur compounds can be easily separated from a light sulfur-free fraction by distillation in the splitter. The double bonds of the alkenes are isomerised over a catalyst, which converts terminal into internal alkenes. This increases the octane number and limits alkene saturation in the HDS reactor, because internal alkenes are hydrogenated more slowly than terminal alkenes. The sulfur-containing medium and heavy FCC naphthas are hydrotreated in the HDS unit (Fig. 8.9). The selective hydrogenation reactor and splitter produce a mercaptan-free, low-sulfur light fraction, which does not need to be hydroprocessed and, thus, preserves the most valuable and reactive alkenes in the feed. The HDS reactor achieves a deep HDS of the middle and heavy fractions with a minimum of alkene saturation and no cracking reactions, which means nearly 100% liquid yield and limited hydrogen consumption.

8.4.2 *Hydrotreating of Diesel*

Diesel is a higher boiling-point fraction than naphtha and also the sulfur compounds have a higher boiling point. The sulfur compounds in diesel, especially the dialkyldibenzothiophenes, are much more difficult to desulfurise. As a consequence, catalysts with higher hydrogenating properties and more severe process conditions (higher temperature and pressure) are needed than in naphtha HDS. In the refinery, diesel fuel is produced from gas oil, which is obtained by primary distillation of crude oil (so-called straight-run gas oil), and light cycle oil, which is formed in the FCC of heavy oils. Light cycle oil has a high sulfur content, a high

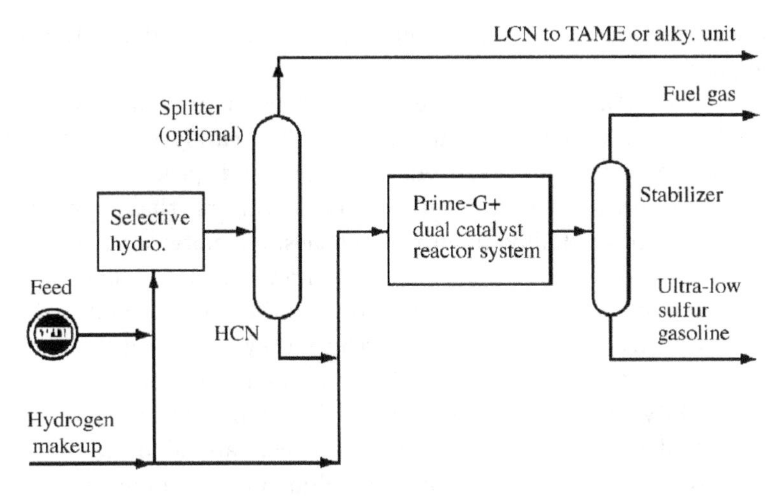

Fig. 8.9 The Prime G+ process developed by IFP. LCN and HCN stand for light and heavy FCC naphtha, respectively. Reprinted with permission from [8.1]. Copyright (2008) Wiley.

content of 4,6-dialkyldibenzothiophenes and a higher aromatics content. The problem of deep HDS of diesel fuel is caused by the much lower reactivity of 4,6-dialkyldibenzothiophenes than other sulfur compounds in fractions which are mixed into the diesel feed stock. As discussed in Section 8.3.1, the alkyl groups at the 4- and 6-positions hinder the interaction between the sulfur atom and the active site on the catalysts when the S-containing molecule is adsorbed perpendicular to the catalyst surface. This means that the most reactive HDS pathway is essentially blocked and that only the other less reactive pathway, through the chemisorption of 4,6-dialkyldibenzothiophene parallel to the surface of the hydrotreating catalyst, contributes to HDS. The problem is aggravated by the presence of polyaromatics and nitrogen compounds because these compounds adsorb more strongly than 4,6-dialkyldibenzothiophenes, thereby hindering the hydrogenation and subsequent HDS.

In diesel hydrotreating, nitrogen has to be removed to about the same level as sulfur to reach the ultra-low sulfur target (< 10 ppm) and conditions must be more severe (425 °C, 2 MPa) than for the HDS of naphtha. For feeds that have a significant aromatics and/or nitrogen content, and thus need a higher hydrogenation capability, a Ni–Mo–S catalyst is used. Gasoil hydrotreating is often carried out in two reactors and Ni–Mo–S is used as catalyst. H_2S produced in the first reactor is removed, to reduce poisoning of the catalyst by H_2S in the second reactor; otherwise deep HDS would not be possible.

Ultra-deep HDS can be achieved by using active catalysts and/or improved reactors. The advantage of active catalysts is that they can be implemented in traditional reactors and processes. Extensive investigations have been carried out on catalysts that enhance the hydrogenation of 4,6-dialkyldibenzothiophenes, but notwithstanding many investigations industry still uses the conventional metal sulfide catalysts. Ni–Mo catalysts have a high ability to saturate the aromatic rings of 4,6-dialkyldibenzothiophenes and therefore they are used in the deep desulfurisation of diesel fuels. Ni–W catalysts have even better hydrogenation ability but are more expensive. New and improved Ni–Mo (and Co–Mo) catalysts have been developed, which are able to desulfurise refinery streams down to a few ppm of sulfur and to significantly reduce the content of polyaromatics; they also improve the cetane number and density of diesel fuels under usual HDS operating conditions [8.37]. These improvements have come about through better dispersion of the active metal on the catalyst substrate. Bulk hydrotreating catalysts (Nebula, NEw BULk Activity [8.40], and Celestia [8.41]) do not contain a support. They are several times more active than conventional supported catalysts but, because of the high metal loading, are more expensive.

8.4.3 *Residue Hydroconversion*

In the primary distillation of crude oil in a refinery not only lighter boiling fractions are obtained, which after treatment are used as fuels (naphtha, kerosene and diesel), but also a large amount of atmospheric residue with high boiling point. This heavy fraction can be separated by vacuum distillation, giving additional fuels and vacuum residue. To make fuel out of residue, the residue must be cracked to molecules with the right boiling point in hydrocracking or catalytic cracking processes. Unfortunately, the availability of high-quality crude oil (with a low sulfur content) diminishes and low-quality crudes must be processed to meet the demand for fuels. These crude oils contain a high amount of sulfur-containing molecules (>3 wt.%), nitrogen-containing molecules (>0.2 wt.%), metal-containing molecules (>60 ppm V, >20 ppm Ni), which poison hydrocracking catalysts, and **asphaltene** molecules (>3 wt.%) (Fig. 8.10), which lead to coke deposition on the catalyst.

Treating heavy oils in fixed bed reactors leads to fast deactivation by catalyst poisoning and coke formation. Moving bed reactors, in which fresh

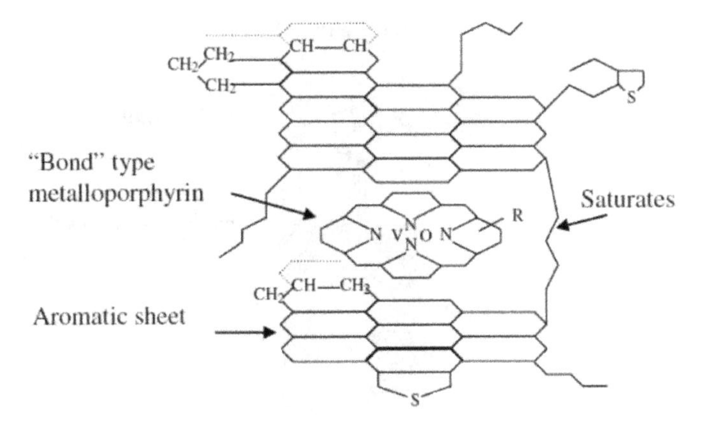

Fig. 8.10 Hypothetical asphaltene molecule and its interaction with a metalloporphyrin. The dotted lines indicate connections to sidechains. Reprinted with permission from [8.42]. Copyright (2007) Elsevier.

catalyst is added at the top and used catalyst is removed at the bottom, prevent fast deactivation. The process is carried out with a sequence of HDM–HDS–HDN-hydrocracking reactors [8.43]. The catalyst in the first moving-bed reactor has wide pores to allow a large uptake of Ni and V sulfide (formed from H_2S and the asphaltene compounds in the residue) and coke precursors and is regenerated after withdrawal from the moving-bed reactor. The subsequent reactors are normal fixed bed reactors loaded with classic HDS and HDN catalysts with decreasing pore size and increasing surface area to allow high hydrotreating efficiency.

When the impurity content of heavy oil is high ($> 100\,\text{ppm}$), even moving bed reactors become deactivated fast and these oils must be processed in ebullated bed or slurry bed reactors. In an ebullated-bed reactor, a liquid feed is mixed with hydrogen and reacted within an expanded catalyst bed that is maintained in turbulence by liquid up-flow in order to achieve efficient isothermal operation (Fig. 8.11). The reactor operates nearly isothermally, which leads to good heat transfer, minimum overheating of the catalyst bed and less coke formation. Catalyst is added and withdrawn on-stream, eliminating the need to shut down for catalyst replacement. The catalyst must be mechanically stable and resist attrition. Co–Mo/Al_2O_3 or Ni–Mo/Al_2O_3 catalysts are used, in the form of small beads or extrudates (1–1.5 mm), to facilitate suspension by the liquid phase in the reactor. The H-Oil$_{RC}$ process (RC = Resid Cracking) (IFP, Axens) and the LC-FINING process (Chevron Inc. and Lummus Technology) use **ebullated-bed reactors** to hydrocrack heavy feedstock residues (also

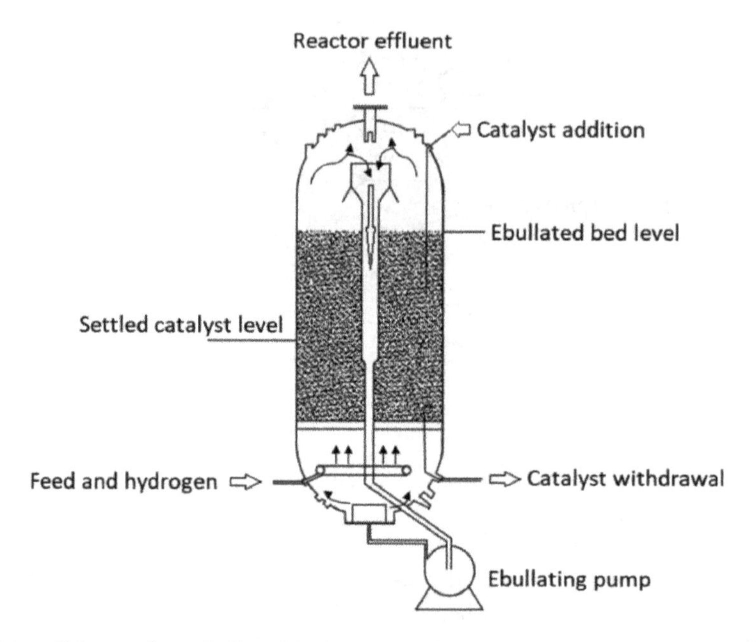

Fig. 8.11 Scheme of an ebullated bed reactor. Reprinted with permission from [8.44]. Copyright (2019).

Athabasca tar sand) that contain high amounts of metals, sulfur, nitrogen and asphaltenes at 410–450 °C and 10–20 MPa.

Ebullated bed processes are flexible with respect to the feedstock. They can handle substantial amounts of metals and coke, but the conversion is limited to about 80% [8.45, 8.46]. Slurry reactors can have a higher conversion. They operate with catalyst particles finely dispersed in the liquid feed and the reactants are kept in suspension with hydrogen flowing upward in the reactor. Most slurry hydroconversion processes are once-through processes, which limits the use of the catalyst to inexpensive and less active materials (Fe sulfide) or to low concentrations of a more active material (i.e. few hundreds ppmw Mo sulfide) [8.46]. The products are separated by fractionation and the unconverted heavy fraction is recycled to the reactor. The catalyst remains in the residue of the vacuum distillation unit together with the Ni and V sulfides derived from the metal-containing compounds in the feed. The slurry process is related to the Bergius process for coal liquefaction (Section 6.3), of which 12 commercial units were in operation in Germany between 1927 and 1945. Typical operating conditions are 440–460 °C and 10–15 MPa.

References

8.1 R. Prins, Hydrotreating, in *Handbook of Heterogeneous Catalysis*, Eds. G. Ertl, H. Knözinger, F. Schüth, J. Weitkamp, 2nd Edition, Vol. 6. Wiley-VCH, Weinheim, 2008, pp. 2695–2718.

8.2 H. Topsøe, B. S. Clausen, F. E. Massoth, *Hydrotreating Catalysis*, Springer Verlag, Berlin, 1996.

8.3 P. Arnoldy, J. A. M. van den Heijkant, G. D. de Bok, J. A. Moulijn, Temperature-programmed sulfiding of MoO_3/Al_2O_3 catalysts, *J. Catal.* 92, 35–55, 1985.

8.4 In the **Extended X-ray Absorption Fine Structure** (**EXAFS**) and **X-ray Absorption Near Edge Structure** (**XANES**) techniques (together called **XAS** techniques) a sample is irradiated with X-rays and the absorption of the X-rays is measured by recording the incident and transmitted X-ray intensities as a function of the incident X-ray energy. When the incident X-ray energy matches the binding energy of a core electron of an atom in the sample, the X-ray absorption increases strongly. The resulting absorption edge (Fig. 8.12) corresponds to excitation of a 1s, 2s, 2p or another electron. The binding energies of these electrons are element specific, making EXAFS element selective.

After each edge a series of oscillations is observed (Fig. 8.12). These are due to the fact that the emitting atom is surrounded by other atoms. Electrons also have wave character and the waves (electrons), which are emitted from the absorbing atom, can scatter back against the neighbouring atoms. The resulting interference pattern shows up as oscillations in the EXAFS spectrum. The oscillations are characteristic of the atoms that surround the central atom and by Fourier transformation of the EXAFS spectrum one can determine the type of surrounding atoms, their distance to the emitting atom and their coordination numbers. The XANES parts of

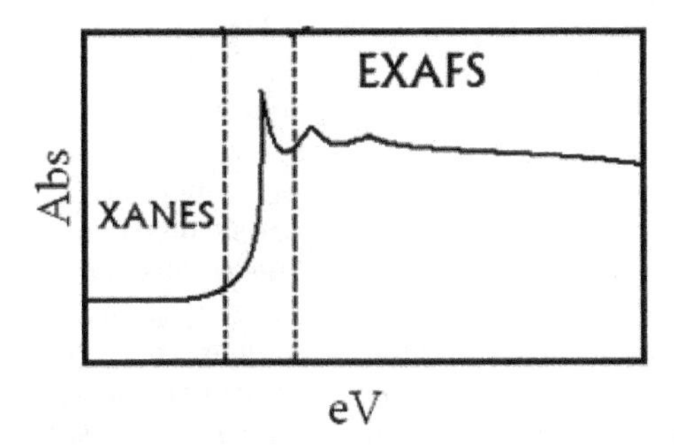

Fig. 8.12 Schematic X-ray absorption spectrum of a specific electron (e.g. 1s electron) from an atom in a sample. Electrons that are just excited are responsible for the XANES part of the spectrum, while ejected electrons with a larger kinetic energy scatter against the electron shells of neighbouring atoms and, through interference, create undulations in the EXAFS part of the spectrum.

the XAS spectra are also sensitive to the coordination environment of the absorbing atom in the sample. XAS spectra need continuous X-ray radiation and, therefore, have to be measured at synchrotrons. The high intensity of synchrotron X-ray sources is another reason for using these facilities. Because X-rays are highly penetrating, XAS spectra can be measured of gases, solids and liquids.

8.5 S. M. A. M. Bouwens, R. Prins, V. H. J. de Beer, D. C. Koningsberger, Structure of the molybdenum sulfide phase in carbon-supported Mo and Co–Mo sulfide catalysts as studied by extended X-ray absorption fine structure spectroscopy, *J. Phys. Chem.* 94, 3711–3718, 1990.

8.6 J. V. Lauritsen, M. V. Bollinger, E. Laegsgaard, K. W. Jacobsen, J. K. Nørskov, I. Stensgaard, B. S. Clausen, H. Topsøe, F. Besenbacher, Atomic-scale insight into structure and morphology changes of MoS_2 nanoclusters in hydrotreating catalysts, *J. Catal.* 221, 510–522, 2004.

8.7 J. V. Lauritsen, J. Kibsgaard, G. H. Olesen, P. G. Moses, B. Hinnemann, S. Helveg, J. K. Nørskov, B. S. Clausen, H. Topsøe, E. Lægsgaard, F. Besenbacher, Location and coordination of promoter atoms in Co- and Ni-promoted MoS_2-based hydrotreating catalysts, *J. Catal.* 249, 220–233, 2007.

8.8 N. Y. Topsøe, H. Topsøe, Characterisation of the structures and active sites in sulfided Co-Mo/Al_2O_3 and Ni-Mo/Al_2O_3 catalysts by NO chemisorption, *J. Catal.* 84, 386–401, 1983.

8.9 P. Raybaud, Understanding and predicting improved sulfide catalysts: Insights from first principles modeling, *Appl. Catal. A* 322, 76–91, 2007.

8.10 **Scanning tunneling microscopy** (STM) produces atomic-scale images of surfaces. It was developed by Binnig and Rohrer, who in 1986 received the Nobel Prize in Physics for the development of this technique, together with Ruska for his development of the TEM. STM can reach 0.1 nm lateral resolution and 0.01 nm depth resolution and can routinely image individual atoms at surfaces. The STM can be used in ultra-high vacuum as well as in air, gas, water and other liquids at ambient pressure and at temperatures from near 0 to 300 °C.

STM is based on the concept of quantum tunneling. When a conducting tip is brought close to a surface, a voltage difference between the tip and the surface enables electrons to tunnel through the vacuum between them (Fig. 8.13). The resulting tunneling current depends on the position of the tip, on the voltage and on the local density of states of the sample. When the tip is moved across the sample in the $x-y$ plane, changes in the surface height and the density of states cause changes in current. These changes can be mapped in images in two ways: either the change in the current with respect to position can be measured at constant height of the tip, or the height of the tip can be varied and measured while keeping the current constant. In the constant current mode, feedback electronics adjust the height by sending a voltage to the piezoelectric height control mechanism. This leads to a height variation and, thus, the image comes from the tip topography across the sample and gives a constant charge density surface;

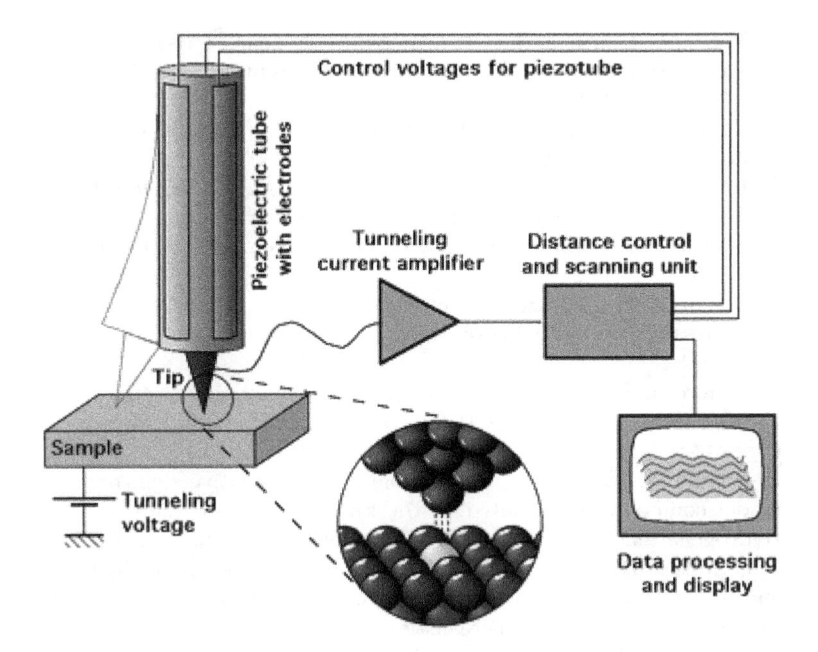

Fig. 8.13 Scheme of an STM instrument.

the contrast on the image is due to variations in charge density. In constant height mode, the voltage and height are kept constant, while the current changes. This leads to an image based on changes in the current over the surface, which can be related to charge density.

8.11 M. V. Bollinger, J. V. Lauritsen, K. W. Jacobsen, J. K. Nørskov, S. Helveg, F. Besenbacher, One-dimensional metallic edge states in MoS_2, *Phys. Rev. Lett.* 87, 196803, 2001.

8.12 M. H. Farias, A. J. Gellman, G. A. Somorjai, R. R. Chianelli, K. S. Liang, The coadsorption and reactions of sulfur, hydrogen and oxygen on clean and sulfided Mo(100) and on MoS_2(0001) crystal faces, *Surf. Sci.* 140, 181–196, 1984.

8.13 P.-Y. Prodhomme, P. Raybaud, H. Toulhoat, Free-energy profiles along reduction pathways of MoS_2 M-edge and S-edge by dihydrogen: A first-principles study, *J. Catal.* 280, 178–195, 2011.

8.14 K. J. Smith, Introduction to the application of nitrides, carbides, phosphides and amorphous boron alloys in catalysis, Catalysis Series No. 34, Alternative catalytic materials: Carbides, nitrides, phosphides and amorphous boron alloys, in *Royal Soc. Chem.*, Eds. J. S. J. Hargreaves, A. R. McFarlane, S. Laassiri, Chapter 1, pp. 1–26, 2018.

8.15 P. A. Clark, S. T. Oyama, Alumina-supported molybdenum phosphide hydroprocessing catalysts, *J. Catal.* 218, 78–87, 2003.

8.16 X. Wang, P. Clark, S. T. Oyama, Synthesis, characterisation, and hydrotreating activity of several iron group transition metal phosphides, *J. Catal.* 208, 321–331, 2002.

8.17 M. E. Bussell, R. Prins, Metal phosphides: Preparation, characterisation and catalytic reactivity, *J. Catal.* 219, 85–96, 2003.

8.18 S. J. Sawhill, K. A. Layman, D. R. Van Wyk, M. H. Engelhard, C. Wang, M. E. Bussell, Thiophene hydrodesulfurisation over nickel phosphide catalysts: Effect of the precursor composition and support, *J. Catal.* 231, 300–313, 2005.

8.19 M. J. Girgis, B. C. Gates, Reactivities, Reaction networks, and kinetics in high-pressure catalytic hydroprocessing, *Ind. Eng. Chem. Res.* 30, 2021–2058, 1991.

8.20 T. Todorova, Th. Weber, R. Prins, A density functional theory study of the hydrogenolysis reaction of CH_3SH to CH_4 on the catalytically active (100) edge of 2H-MoS_2, *J. Catal.* 236, 190–204, 2005.

8.21 H. Schulz, M. Schon, N. M. Rahman, Hydrogenative denitrogenation of model compounds as related to the refining of liquid fuels, in: Catalytic Hydrogenation, L. Červený (Ed.), *Stud. Surf. Sci. Catal.* 27, 201–255, 1986.

8.22 P. G. Moses, B. Hinnemann, H. Topsøe, J. K. Nørskov, The hydrogenation and direct desulfurisation reaction pathway in thiophene hydrodesulfurisation over MoS_2 catalysts at realistic conditions: A density functional study, *J. Catal.* 248, 188–203, 2007; corrigendum *J. Catal.* 260, 202–203, 2008.

8.23 P. G. Moses, B. Hinnemann, H. Topsøe, J. K. Nørskov, The effect of Co promotion on MoS_2 catalysts for hydrodesulfurisation of thiophene: A density functional study, *J. Catal.* 268, 201–208, 2009.

8.24 N. Nelson, R. B. Levy, The organic chemistry of hydrodenitrogenation, *J. Catal.* 58, 485–499, 1979.

8.25 Y. Zhao, R. Prins, Mechanisms of hydrodenitrogenation of alkylamines and hydrodesulfurisation of alkanethiols on NiMo/Al_2O_3, CoMo/Al_2O_3, and Mo/Al_2O_3, *J. Catal.* 229, 213–226, 2005.

8.26 P. Kukula, A. Dutly, N. Sivasankar, R. Prins, Investigation of the steric course of the C–N bond breaking in the hydrodenitrogenation of alkylamines, *J. Catal.* 236, 14–20, 2005.

8.27 H. Wang. C. Liang, R. Prins, Hydrodenitrogenation of 2-methylpyridine and its intermediates 2-methylpiperidine and tetrahydro-methylpyridine over sulfided NiMo/γ-Al_2O_3, *J. Catal.* 251, 295–306, 2007.

8.28 X. Li, X. Luo, Y. Jin, J. Li, H. Zhang, A. Zhang, J. Xie, Heterogeneous sulfur-free hydrodeoxygenation catalysts for selectively upgrading the renewable bio-oils to second generation biofuels, *Renew. Sustain. Energy Rev.* 82, 3782–3797, 2018.

8.29 Y. Han, M. Gholizadeh, C.-C. Tran, S. Kaliaguine, C.-Z. Li, M. Olarte, M. Garcia-Perez, Hydrotreatment of pyrolysis bio-oil: A review, *Fuel Process. Technol.* 195, 106140, 2019.

8.30 S. Boullosa-Eiras, R. Lødeng, H. Bergem, M. Stöcker, L. Hannevold, E. A. Blekkan, Potential for metal-carbide, -nitride, and -phosphide as future hydrotreating catalysts for processing of bio-oils, *Catalysis* 26, 29–71, 2014.

8.31 Z. Yu, Y. Wang, Z. Sun, X. Li, A. Wang, D. M. Camaioni, J. A. Lercher, Ni_3P as a high-performance catalytic phase for the hydrodeoxygenation of phenolic compounds, *Green Chem.* 20, 609–619, 2018.
8.32 R. H. Bowker, M. C. Smith, M. L. Pease, K. M. Slenkamp, L. Kovarik, M. E. Bussell, Synthesis and hydrodeoxygenation properties of ruthenium phosphide catalysts, *ACS Catal.* 1, 917–922, 2011.
8.33 G.-N. Yun, A. Takagaki, R. Kikuchi, S. T. Oyama, Hydrodeoxygenation of gamma-valerolactone on transition metal phosphide catalysts, *Catal. Sci. Technol.* 7, 281–292, 2017.
8.34 Y. Romero, F. Richard, Y. Renème, S. Brunet, Hydrodeoxygenation of benzofuran and its oxygenated derivatives 2,3-dihydrobenzofuran and 2-ethylphenol over $NiMoP/Al_2O_3$ catalyst, *Appl. Catal. A* 353, 46–53, 2009.
8.35 M. Callant, P. Grange, K. A. Holder, B. Delmon, Competitive hydrodenitrogenation of aniline and indole. Interaction with the hydrogenation function of a $NiMoP/\gamma$-Al_2O_3 catalyst, *Bull. Soc. Chim. Belg.* 100, 823–830, 1991.
8.36 N. Salazar, S. B. Schmidt, J. V. Lauritsen, Adsorption of nitrogenous inhibitor molecules on MoS_2 and CoMoS hydrodesulfurisation catalysts particles investigated by scanning tunneling microscopy, *J. Catal.* 370, 232–240, 2019.
8.37 C. Song, An overview of new approaches to deep desulfurisation for ultra-clean gasoline, diesel fuel and jet fuel, *Catal. Today* 86, 211–263, 2003.
8.38 I. V. Babich, J. A. Moulijn, Science and technology of novel processes for deep desulfurisation of oil refinery streams: A review, *Fuel* 82, 607–631, 2003.
8.39 Z. Shen, M. Ke, L. Lan, P. He, S. Liang, J. Zhang, H. Song, Active phases and reaction performance of Mo improved Ni/Al_2O_3 catalysts for thioetherification, *Fuel* 236, 525–534, 2019.
8.40 F. L. Plantenga, R. Cerfontain, S. Eijsbouts, F. van Houten, G. H. Anderson, S. Miseo, S. Soled, K. Riley, K. Fujita, Y. Inoue, "Nebula": A hydroprocessing catalyst with breakthrough activity, *Stud. Surf. Sci. Catal.* 145, 407–410, 2003.
8.41 K. Wilson, L. Burns, P. Lingaraju, R. Cerfontain, B. Leliveld, B. Slettenhaar, Increased flexibility and profitability from new hydroprocessing catalyst, www.digitalrefining.com/article/1000133, 2019.
8.42 M. S. Rana, V. Sámano, J. Ancheyta, J. A. I. Diaz, A review of recent advances on process technologies for upgrading of heavy oils and residua, *Fuel* 86, 1216–1231, 2007.
8.43 B. Scheffer, M. A. van Koten, K. W. Robschlager, F. C. de Boks, The Shell residue hydroconversion process: Development and achievements, *Catal. Today* 43, 217–224, 1998.
8.44 E. Manek, J. Haydary, Investigation of the liquid recycle in the reactor cascade of an industrial scale ebullated bed hydrocracking unit, *Chin. J. Chem. Eng.* 27, 298–304, 2019.
8.45 Y. Liu, L. Gao, L. Wen, B. Zong, Recent advances in heavy oil hydroprocessing technologies, *Recent Patents Chem. Eng.* 2, 22, 2009.

8.46 N. Panariti, A. Del Bianco, G. Del Piero, M. Marchionna, P. Carniti, Petroleum residue upgrading with dispersed catalysts Part 2. Effect of operating conditions, *Appl. Catal. A* 204, 215–222, 2000.

Questions

8.1 What are the M and S edges of MoS_2? Where are the Co atoms of Co–Mo–S located?

8.2 Explain how an STM instrument operates.

8.3 Metal-phosphides are good hydrotreating catalysts. Other Mo–M or Ni–M compounds could also be good catalysts. What is necessary to make them good hydrotreating catalysts and what could be their disadvantage?

8.4 Sulfur can be removed from organic molecules by an elimination or a hydrogenolysis reaction. What is the technical difference?

8.5 What is the main product of the HDS of benzothiophene and what of the HDN of indole? Explain the difference.

8.6 Why is a higher H_2 pressure needed in HDN than in HDS?

8.7 Organic esters and acids can undergo HDO by CO or CO_2 removal. Why is it difficult to distinguish between these mechanisms?

8.8 Explain how the hydrogenation of alkenes is avoided in the HDS of FCC naphtha.

8.9 What is the coordination of the Ni atoms in Ni_2P?

8.10 Naphtha and diesel fractions that are obtained in a refinery by distillation of crude oil must be desulfurised. Why is it not necessary to desulfurise the naphtha and diesel fractions that are produced in Fischer–Tropsch and methanol-to-hydrocarbon processes?

Chapter 9

Oxidation Catalysis

9.1 CO Oxidation

Catalytic oxidation plays an important role in the combustion of fuels, the abatement of noxious gases and in the production of industrial inorganic and organic chemicals. Section 9.1 of this chapter discusses the oxidation of carbon monoxide, one of the most important reactions in the cleaning of exhaust gas. In Section 9.2 we give examples of inorganic oxidation reactions and in Section 9.3 we discuss the oxidation of hydrocarbons. The emphasis is on the principles of oxidation reactions in relation to their industrial applications.

9.1.1 *Mechanism*

The oxidation of CO with oxygen to CO_2 is of practical importance in combustion and in the cleaning of exhaust gas. The CO oxidation reaction is one of the best-studied reactions in catalysis due, in part, to the relative ease with which small molecules such as CO and O_2 can be studied. A large number of studies have proven that CO oxidation proceeds according to a Langmuir–Hinshelwood (LH) mechanism over noble metals [9.1], i.e. not an Eley–Rideal (ER) mechanism, as assumed in earlier studies [9.2, 9.3]. In a LH mechanism both reactants in a bimolecular reaction chemisorb on the surface, whereas in an ER mechanism it is assumed that only one of the reactants chemisorbs. The other reactant either collides from the gas phase with the chemisorbed reactant or is weakly bonded to the surface in a physisorbed state.

In heterogeneous catalysis LH mechanisms are more probable than ER mechanisms (Section 4.1.2), but the mechanisms of very few reactions have been verified experimentally. The determination of the mechanism (LH or ER) is not always possible based on kinetics alone. Surface science studies of the mean surface residence times of the reacting species may provide the answer, but they are not easy to perform. If the adsorption energy of a molecule is low ($< 40\,\text{kJ}\cdot\text{mol}^{-1}$, ER mechanism), the molecule has a short residence time and, thus, must react fast ($< 10^{-6}\,\text{s}$). If the reaction takes place by a direct collision from the gas phase, then the duration of the reaction is even shorter (close to a vibrational period of $\approx 10^{-13}\,\text{s}$). On the other hand, if the adsorption of the reacting species on the surface and the formation of the reaction product take a considerably longer time, then this is clearly indicative of a LH mechanism, provided that the surface residence time of the product molecule is negligible.

Engel and Ertl (Nobel Prize in Chemistry, 2007) studied the surface residence time of the CO_2 product of the reaction of CO and O_2 by modulating the flux of the CO and of the O_2 [9.4]. The temperature-dependent phase lag for the CO_2 signal was in perfect agreement with the LH mechanism but was not in agreement with the ER mechanism. The surface residence time for CO on Pd(111) was 10^{-2}–$10^{-5}\,\text{s}$, which also rules out an ER mechanism.

G. Ertl

In the reaction of CO with O_2, both reactants adsorb, but dissociation of CO does not occur, whereas O_2 dissociates above $-100\,^{\circ}\text{C}$. Therefore, the following elementary steps were proposed [9.4]:

$$\text{CO} + * \rightleftharpoons \text{CO}* \tag{9.1}$$

$$\text{O}_2 + 2* \rightarrow 2\,\text{O}* \tag{9.2}$$

$$\text{CO}* + \text{O}* \rightarrow \text{CO}_2 + 2*, \tag{9.3}$$

where $*$ indicates a surface Pt atom, on which a CO molecule or an O atom can adsorb. In LH kinetics, the variations in the surface coverage θ_{CO} and θ_O are:

$$d\theta_{CO}/dt = k_1 P_{CO}(1 - \theta_{CO} - \theta_O) - k_{-1}\theta_{CO} - k_3\theta_O\theta_{CO} \quad (9.4)$$

$$d\theta_O/dt = k_2 P_{O2}(1 - \theta_{CO} - \theta_O)^2 - k_3\theta_{CO}\theta_O. \quad (9.5)$$

The first term of $d\theta_{CO}/dt$ describes the rate of adsorption, the second term the rate of desorption and the third term the reaction of chemisorbed CO. The first term of $d\theta_O/dt$ describes the dissociative adsorption of oxygen and the second term the reaction of chemisorbed O atoms, while oxygen desorption is neglected.

The ordinary differential Eqs. (9.4) and (9.5) are coupled non-linearly by the reaction term $k_3\theta_O\theta_{CO}$. When the reaction has significant effect on the surface coverage and $k_3\theta_O\theta_{CO}$ cannot be neglected, this set of equations has three solutions for the reaction rate $r = k_3\theta_O\theta_{CO}$: an unstable solution and two stable solutions, exhibiting bistability at fixed pressures of CO and O_2 [9.5]. The first stable solution corresponds to a surface, covered mainly by CO ($\theta_{CO} \gg \theta_O$). The high CO coverage effectively blocks the adsorption of oxygen, for which two neighbouring adsorption sites are necessary, and leads to a low θ_O and a low reaction rate $r_1 = k_3\theta_O\theta_{CO}$. The other stable solution corresponds to relatively high coverage of the surface by oxygen ($\theta_O > \theta_{CO}$). Only one adsorption site is required for CO adsorption, which is not inhibited by the high oxygen coverage. Thus, the O-dominated surface has a higher $\theta_O\theta_{CO}$ and, thus, a higher reaction rate than the CO-dominated surface. The asymmetry in the adsorption of CO and O_2 with increasing or decreasing CO pressure leads to hysteresis in the reaction rate, as demonstrated by simulations based on Eqs. (9.4) and (9.5) [9.6], and experimental results [9.7](Fig. 9.1).

Equations (9.4) and (9.5) explain why the CO oxidation reaction is bistable, but they do not explain why oscillations in the reaction occur, over time and depending on the position on the surface [9.5]. Careful analysis showed that a feedback mechanism is required to switch spontaneously between the two stable states and to regulate the observed self-sustained oscillations [9.5, 9.8]. For Pt(100) and Pt(110) single crystal surfaces at low pressures ($\sim 10^{-2}$ Pa), oscillations occurred between a CO-dominated, low-reactivity LH branch and an O-dominated, high-reactivity LH branch, caused by reconstruction of the surface. The Pt(100) and Pt(110) surfaces are unreconstructed when they are covered mainly by CO but, when they are covered mainly by oxygen, Pt(100) has a quasihexagonal structure

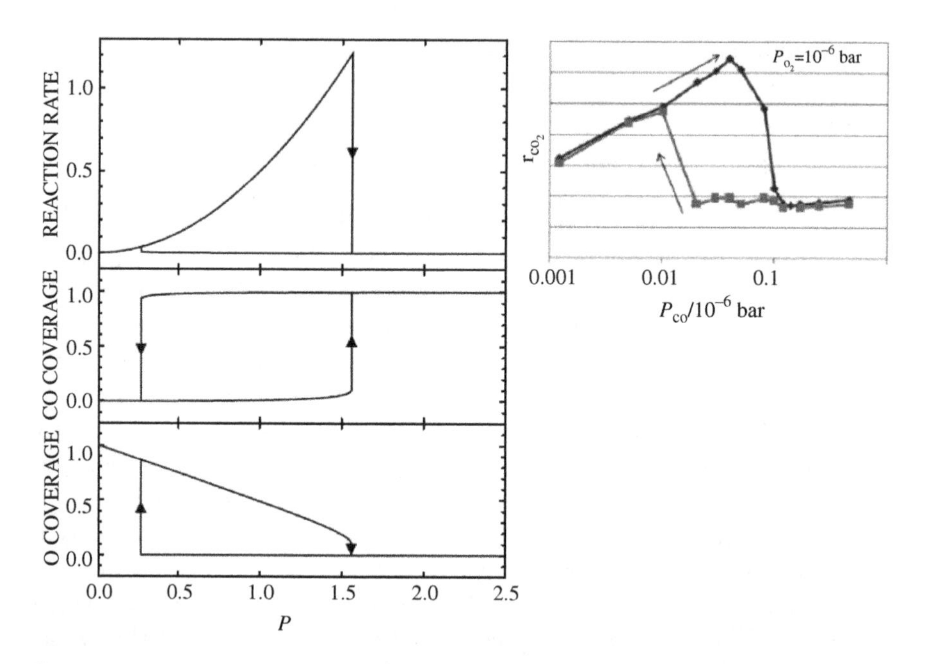

Fig. 9.1 Bistability in the oxidation of CO. Left: Simulation of the reaction rate and CO and O coverages as a function of $p = k_1 P_{CO}$ with $k_{-1} = 0.01$, $k_2 P_{O2} = 1$, and $k_3 = 100$. The reaction rate and the rate constants are normalised to $k_2 P_{O2}$ [9.6]. Right: Measured hysteresis in the rate of CO oxidation over Pt(111) [9.7].

(as in a (111) overlayer) and Pt(110) has a missing-row structure (Fig. 9.2). The reconstructions are reverted as a result of CO adsorption, when the coverage exceeds a critical value, and reconstruction occurs again when the CO coverage is below another critical value.

Oscillations in catalytic CO oxidation were also observed on Pd surfaces, which do not reconstruct, and on Pt (111) surfaces at higher pressures (> 100 Pa). These oscillations cannot be due to surface reconstruction. An IR study of the CO coverage of the surface of small Pt particles supported on SiO_2 at 150 °C and 100 kPa (O_2:CO:Ar = 6:0.1:94) and mass spectroscopy of the product concentration in the gas phase showed transitions between high (almost complete) and low (almost zero) CO coverage, demonstrating two distinctly different reaction regimes (Fig. 9.3). At high CO coverage the reaction proceeds in the CO self-poisoned regime at a relatively low rate and low CO_2 production; at low CO coverage the reaction rate is appreciably higher [9.9]. The frequency of the IR band indicates that CO adsorbs on oxidised Pt, indicating that islands of Pt oxide form, on which the reaction takes place during the reactive phases.

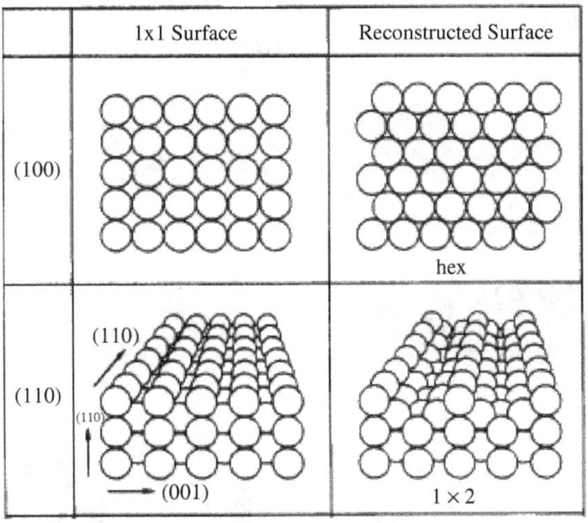

	1x1 Surface	Reconstructed Surface
(100)		hex
(110)	(110) (110) (001)	1 × 2

Fig. 9.2 Normal and reconstructed surfaces of Pt(100) and Pt(110) [9.4].

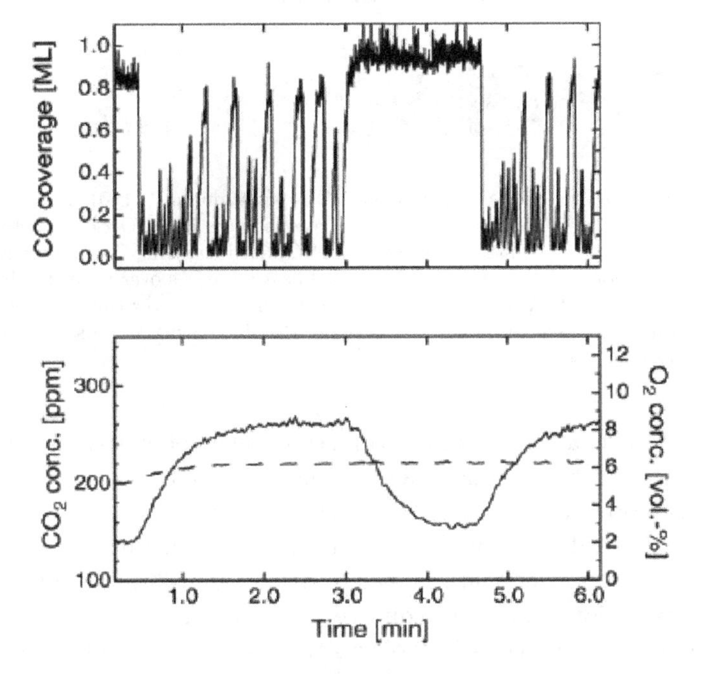

Fig. 9.3 Simultaneous FTIR and MS results of isothermal kinetic oscillations in the oxidation of CO over 1% Pt/SiO_2 at 150 °C and 100 kPa (O_2:CO:Ar = 6:0.1:94) [9.9]. Top: CO coverage obtained by integrating the IR band of CO bonded linearly to Pt. Bottom: CO_2 (solid line) and O_2 (dashed line) concentrations in the product stream.

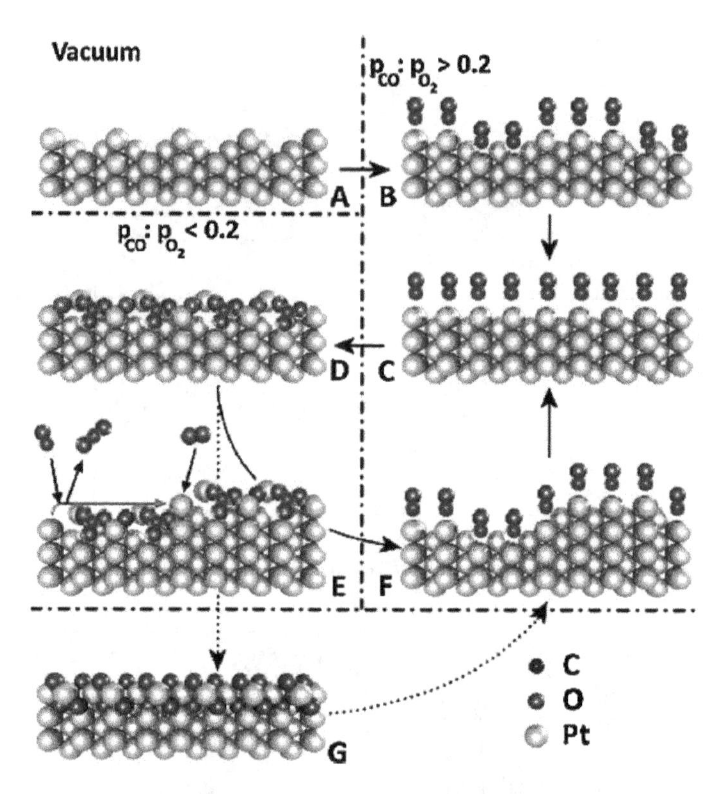

Fig. 9.4 Model of the different transitions in CO oxidation on Pt(110). The missing-row (110) surface (A), which is stable in vacuum, reconstructs in a CO-rich flow to a rough (1 × 1) surface (B), which flattens out over time (C). At increasing P_{O2}, a commensurate lifted-row oxide forms (D), and its surface roughness increases over time (E). Increasing P_{CO} results in a rough (1 × 1) surface (F), which smoothens over time, giving (C). (G) Incommensurate α-PtO_2 can form at high P_{O2}. Reprinted with permission from [9.11]. Copyright (2015) Elsevier.

A STM study at 150 °C and 100 kPa confirmed not only that a thin oxide layer formed on Pt(110) under actual conditions, but also showed that this oxide was more reactive than the metal surface (Fig. 9.4) [9.10, 9.11]. The thin, strained oxide layer on Pt(110) has properties that differ from those of bulk Pt oxide. It seems that strained or defective oxides are quite reactive.

The oxidation-reduction sequence suggests that the CO oxidation reaction on the surface oxide follows a Mars–Van Krevelen mechanism [9.11]. In a Mars–van Krevelen oxidation mechanism the reactant does not react directly with oxygen gas but with the metal oxide, consuming lattice oxygen (Section 1.1). In the particular case it means that CO reacts with an oxygen atom from the metal oxide layer. Thereafter, the reduced metal oxide layer is re-oxidised by oxygen. For CO oxidation this means that CO molecules adsorb from the gas phase on the Pt oxide layer and react with its oxygen atoms to form CO_2 (Fig. 9.4). The oxygen vacancies are refilled quickly with oxygen from the gas phase or from other parts of the oxide. In this way, the catalytic reaction proceeds by a continuous and local reduction and by re-oxidation of the oxide surface. As shown by STM, this process can lead to roughness, if occasionally one of the Pt atoms is so strongly under-coordinated that it diffuses out of the oxide. The diffusing Pt atom will oxidise with time, on top of the oxide. Since the rate, at which roughness builds up, is much lower than the CO oxidation rate, roughness is considered to be a byproduct of the reaction.

In agreement with the STM data, a combined XAS and IR study of a Pt/γ-Al_2O_3 catalyst showed that, at about 100 °C and 100 kPa (O_2:CO = 19), the structure of small Pt particles (1–2 nm) varied as a function of time and position in the catalytic reactor [9.12]. A highly disordered surface oxide, with many oxygen defects, acted as the catalytically active phase under technologically relevant conditions. Oxygen of the Pt surface oxide reacted with CO during an oscillation and was replenished by gas phase O_2 in the latter part of the oscillation.

The behaviour of large Pt nanoparticles ($\sim$ 50 nm) supported on a silicon nitride membrane in the CO oxidation reaction was monitored *in situ* in a nanoreactor at 380–450 °C and 25 kPa (O_2:CO:He = 21:4:75), conditions suitable for simultaneous time-resolved high-resolution transmission electron microscopy (TEM), mass spectrometry and calorimetry [9.13]. Mass spectrometry demonstrated periodic oscillations in O_2 and CO pressure in anti-phase with similar variations in the CO_2 pressure, synchronous with periodic reconstruction of the Pt nanoparticles from an almost spherical to a faceted morphology. Higher conversion led to more extended close-packed facet terminations and lower conversion to a higher number of higher-index stepped terminations of the Pt nanoparticles. The individual nanoparticles underwent oscillatory and reversible shape changes with a temporal frequency matching the oscillations in CO conversion, indicative of coupling of the oscillatory CO conversion and the dynamic shape change in the Pt nanoparticles. The discrepancy between the conclusion that oscillations occur on the metallic Pt surface (TEM) and that the oscillations occur on the Pt oxide surface (STM [9.11] and XAS-IR [9.12]) may be due to the fact that the temperature was much higher and oxygen pressure much lower in the TEM study.

9.1.2 *Three-way Catalysis*

The catalytic oxidation of CO is a prototypic reaction for understanding fundamental concepts in heterogeneous catalysis and is highly relevant for the control of automotive emissions. The exhaust gas contains unburnt hydrocarbon molecules, CO from incomplete combustion of fuel and NO_x that formed from N_2 and O_2 in air under the high temperatures in the vehicle's engine. Worldwide legislation demands that concentrations of hydrocarbons, CO and NO_x be severely reduced before these molecules are released into the air. This is accomplished by three-way catalytic converters on gasoline-powered vehicles. A three-way catalytic converter has three simultaneous tasks:

Oxidation of CO to CO_2: $2\,CO + O_2 \rightarrow 2\,CO_2$

Oxidation of unburnt hydrocarbons: $C_nH_{2n+2} + 0.5(3\,n + 1)\,O_2 \rightarrow n\,CO_2 + (n + 1)\,H_2O$

Reduction of NO_x to N_2 by CO and hydrocarbons: $2\,NO + 2\,CO \rightarrow 2\,CO_2 + N_2$

$(4n + m)\,NO + 2\,C_nH_m \rightarrow 0.5(4n + m)\,N_2 + 2n\,CO_2 + m\,H_2O$

The first two reactions occur under oxidative conditions and the third reaction under reductive conditions. All three reactions have high conversion when the engine is operated within a narrow range of air-fuel ratios close to stoichiometry (weight ratio 14.7), but the conversion decreases rapidly when the engine operates at a higher or lower ratio. With excess oxygen (lean fuel conditions), the exhaust gas contains little CO and/or HC but high concentrations of NO_x. With excess fuel (rich fuel conditions), all of the available oxygen is consumed in the engine by the time the exhaust gas reaches the catalyst. So there is no oxygen for oxidation over the catalyst; the concentrations of CO and HC are, thus, high and the concentration of NO_x is low.

Three-way catalysts usually contain three precious metals (Pt, Pd and Rh) as well as ceria. Pt is a reduction and oxidation catalyst, Pd an oxidation catalyst, Rh a reduction catalyst and ceria promotes the oxidation reactions. Ceria has a continuous range of compositions between Ce_2O_3 and CeO_2 and releases or takes up oxygen from the exhaust stream of a combustion engine without decomposing, thus acting as an oxygen buffer. When there is too little oxygen in the exhaust stream (e.g. when fast acceleration enriches the mixture), the oxygen stored in CeO_2 is involved in the oxidation reactions. During subsequent normal driving, the ceria is re-oxidised by air. Engines fitted with three-way catalytic converters are also equipped with a computerised closed-loop feedback fuel injection system with oxygen sensors. Precise control of the air-to-fuel ratio is necessary to achieve the satisfactory functioning of the catalyst to meet the conflicting requirements of effective NO_x reduction and CO and hydrocarbon oxidation. The control system must prevent complete oxidation of the NO_x reduction catalyst but still regenerate the oxygen storage material so that it continues to function as an oxidation catalyst.

A modern three-way catalyst consists of a monolithic ceramic flow-through support [9.14], usually made of cordierite (Mg–Al silicate, $Mg_2Al_4Si_5O_{18}$), and has a honeycomb structure (Fig. 9.5). Cordierite has a low thermal expansion coefficient, excellent heat stability, good resistance to thermal shock and chemicals and can be produced with large surface area. The surface of the monolithic flow-through support is coated with an aqueous slurry of inorganic oxides (Al_2O_3, TiO_2, SiO_2), by dipping the monolith in the slurry (called wash coating). Drying and calcination at high temperature ensures good bonding of the refractory oxide layer to the monolith support. Catalytic precious metals (Pt, Pd and Rh) are deposited on the washcoat surface by impregnation or by dissolving the metal salts in the washcoat slurry. The purpose of the washcoat is to provide a high surface area, to ensure a high dispersion of the precious metals, retention of the surface area and prevention of sintering of the metal particles, even at high temperature (1,000 °C). In the combustion of large gas streams, metal honeycomb monoliths, wash-coated with alumina and impregnated with precious metals, as described above, convert CO, NO_x and hydrocarbons.

Fig. 9.5 Cordierite ceramic honeycomb monolith.

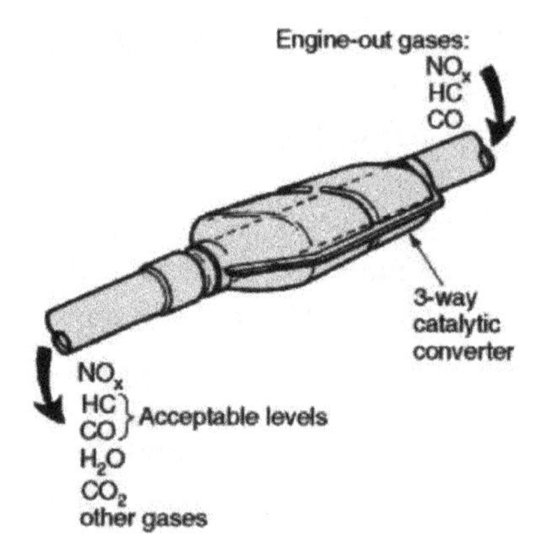

Fig. 9.6 Three-way catalytic converter.

The honeycomb monolith is contained in a metal can and wrapped in foil for thermal insulation and to dampen vibrations (Fig. 9.6).

9.2 Production of Sulfuric and Nitric Acid

9.2.1 *Sulfuric Acid*

After ammonia, sulfuric acid is the second most important base chemical; worldwide production in 2020 was 260 MT, about 60% of which is used in the production of ammonium phosphate and ammonium sulfate fertilisers. It is also important in the production of many chemicals (e.g. hydrochloric acid, nitric acid, sulfate salts, synthetic detergents, dyes and pigments, explosives and drugs) and in refineries to remove impurities from gasoline and other refinery products. Sulfuric acid is used to process metals (e.g. cleaning iron and steel before plating them with tin or zinc) and in making rayon and also serves as the electrolyte in the lead-acid storage battery.

Sulfuric acid is produced by the contact process [9.15]. In a first step, sulfur is burned to produce sulfur dioxide ($S + O_2 \rightarrow SO_2$). Most sulfur comes from refineries, where it is produced by the Claus process, which converts H_2S to sulfur by partly oxidising it to SO_2 followed by reaction of the remaining H_2S with SO_2 to give sulfur ($2\,H_2S + SO_2 \rightarrow 3\,S + 2\,H_2O$). Sulfur dioxide is oxidised to sulfur trioxide with oxygen in the presence of a vanadium potassium sulfate on silica catalyst [$K_3(VO)_2(SO_4)_4/SiO_2$].

$$2\,SO_2 + O_2 \leftrightarrow 2\,SO_3 \qquad \Delta H = -197\,\text{kJ}.$$

SO_2 oxidation proceeds by a Mars–van Krevelen mechanism (Section 1.1), by which SO_2 is oxidised by vanadium oxide (consuming lattice oxygen), not by oxygen, and the reduced vanadium oxide is re-oxidised by oxygen:

$$2\,SO_2 + 4\,V^{5+} + 2\,O^{2-} \rightarrow 2\,SO_3 + 4\,V^{4+},$$

$$4\,V^{4+} + O_2 \rightarrow 4\,V^{5+} + 2\,O^{2-}.$$

That the reactions occur in independent steps is supported by the fact that oxidation reactions of aromatic hydrocarbons, such as benzene, toluene, naphthalene and anthracene, can be carried out in the absence of oxygen [9.15], proving that the oxygen atom added to the substrate originates from the catalyst surface and not from oxygen gas. Oxidation experiments with isotopically labelled O_2 and unlabelled metal oxide catalyst confirmed that the initial product contained only unlabelled oxygen. The fact that the operating temperature of vanadium oxide catalysts did not depend on the substrate and that a catalyst in interaction with an oxidisable substrate and air is partly in the reduced state indicates that the rate is of re-oxidation of the catalyst is slower than the rate of reduction (i.e. the rate of oxidation of the substrate) [9.16]. Note that the reactions of naphthalene to phthalic anhydride, benzaldehyde to benzoic acid and toluene to benzoic acid all had a good yield at about the same temperature.

Mars and van Krevelen described the derivation of the kinetics of a two-step mechanism [9.16]. When the oxidation reaction is first order in the reactant S, the rate of oxidation of S and of the simultaneous reduction of the catalyst surface are

$$r_{\text{red}} = k_{\text{red}} P_S f,$$

where f is the degree of occupation of active oxygen groups on the catalyst surface. If the rate of re-oxidation of the surface is proportional to the partial pressure of oxygen and to the fraction of the catalyst surface area, not covered by oxygen but which can undergo re-oxidation, the rate of re-oxidation of the catalyst surface is:

$$r_{\text{ox}} = k_{\text{ox}} P_{O2}(1 - f).$$

In steady state, r_{ox} is equal to n times r_{red}, where n is the number of O_2 molecules required for the oxidation of one substrate molecule:

$$k_{\text{ox}} P_{O2}(1 - f) = n k_{\text{red}} P_S f.$$

Thus,

$$f = \frac{k_{\text{ox}} P_{\text{O2}}}{n k_{\text{red}} P_S + k_{\text{ox}} P_{\text{O2}}} \quad \text{and} \quad r = \frac{1}{\frac{1}{k_{\text{red}} P_S} + \frac{n}{k_{\text{ox}} P_{\text{O2}}}}.$$

The first step in the oxidation of aromatic hydrocarbons is substantially faster than the second step ($n k_{\text{red}} P_S > k_{\text{ox}} P_{\text{O2}}$) and to a good approximation $r = k_{\text{ox}} P_{\text{O2}}$.

Because the oxidation of SO_2 is exothermic, high temperature decreases the yield of SO_3, while low temperature results in a high yield but a low reaction rate. The reaction conditions (reactor inlet temperature 420 °C and 0.2 MPa) are a compromise between a faster reaction rate and a slightly lower yield. Because of the high exothermicity, an adiabatic reactor with four catalyst beds and intermediate cooling is the reactor of choice. In modern plants, the gas leaving the third bed is fed to an adsorption tower where the SO_3 in the product mixture of SO_3 and SO_2 is absorbed in 97–98% H_2SO_4 to form oleum ($H_2S_2O_7$), also known as fuming sulfuric acid. The oleum is then diluted with water to form concentrated sulfuric acid:

$$H_2SO_4(l) + SO_3(g) \rightarrow H_2S_2O_7(l),$$

$$H_2S_2O_7(l) + H_2O(l) \rightarrow 2\,H_2SO_4(l).$$

Direct reaction of SO_3 with water is impossible because the highly exothermic reaction leads to an aerosol, which is very difficult to condense to a liquid. The SO_2 remaining after absorption of SO_3 in the oleum is fed to the fourth catalyst bed, where further reaction to SO_3 can occur and a higher than 99.7% conversion of SO_2 can be reached. This is necessary for environmental reasons.

9.2.2 *Nitric Acid*

Nitric acid is another very important inorganic base chemical. Its main use (75%) is as ammonium nitrate, an agricultural fertiliser. Other applications are explosives, nylon precursors and speciality organic compounds. Commercial grade nitric acid solutions contain 52% to 68% nitric acid. Production of nitric acid takes place by the Ostwald process, in which ammonia oxidises to nitric oxide over a Pt–Rh catalyst at about 900 °C and 1 MPa [9.15].

$$4\,NH_3 + 5\,O_2 \rightarrow 4\,NO + 6\,H_2O \qquad \Delta H = -905\,\text{kJ}.$$

The mechanism of ammonia oxidation is simple. The NH_3 and O_2 reactants dissociate completely to N, H and O atoms on the metal surface and these recombine to the NO and H_2O product molecules. The rate of ammonia oxidation is determined by mass transport to the Pt wire and, thus, by the stagnant layer on the outer surface of the catalyst (Section 4.2). The industrial catalyst consists of a Pt–Rh (9:1) wire gauze, through which the ammonia-air mixture flows and 98% of the NH_3 is converted to NO. Only an increase in the gas velocity has an effect on this process, resulting in a thinner stagnant gas film and, thus, an increase in the mass transfer coefficient.

The resulting NO reacts with air in a non-catalytic reaction to NO_2:

$$2\,NO + O_2 \rightarrow 2\,NO_2 \qquad \Delta H = -114\,\text{kJ·mol}^{-1}.$$

NO_2 absorbs in water to form nitric acid and nitric oxide:

$$3\,NO_2(g) + H_2O(l) \rightarrow 2\,HNO_3(aq) + NO(g) \qquad \Delta H = -117\,\text{kJ·mol}^{-1}.$$

The nitric oxide is recycled for reoxidation. Alternatively, if the last step is carried out in air, then NO reacts *in situ* to NO_2:

$$4\,NO_2(g) + O_2(g) + 2\,H_2O(l) \rightarrow 4\,HNO_3(aq).$$

The aqueous HNO_3 product is concentrated by distillation up to about 68%. Further concentration to 98% can be achieved by dehydration with concentrated sulfuric acid. By using ammonia derived from the Haber process, HNO_3 is produced from nitrogen, hydrogen and oxygen, which are derived from air and natural gas (or coal) as the sole feedstocks.

9.2.3 *Selective Catalytic Reduction*

NO and NO_2 are intermediates in the production of nitric acid and also a small amount of N_2O is formed. For environmental reasons, these gases must be removed from the vent gas of the nitric acid plant. NO_x is removed by **selective catalytic reduction (SCR)** with ammonia or urea (NH_2–CO–NH_2). This process is also used to clean the vent gas in boilers in power plants (Fig. 9.7), gas turbines and diesel engines. The SCR reactor is filled with honeycomb structures (cf. Section 9.1.2), the walls of which are coated with active catalytic components [9.17]. If thermal stability is important, as in automotive SCR applications, then zeolite catalysts exchanged with Fe^{3+} and Cu^{2+} cations are the catalysts for urea SCR of the exhaust gas of diesel engines [9.18]. These catalysts withstand prolonged operation at 600 °C and transient conditions of up to 850 °C, and SO_2 damages zeolites to a lesser extent.

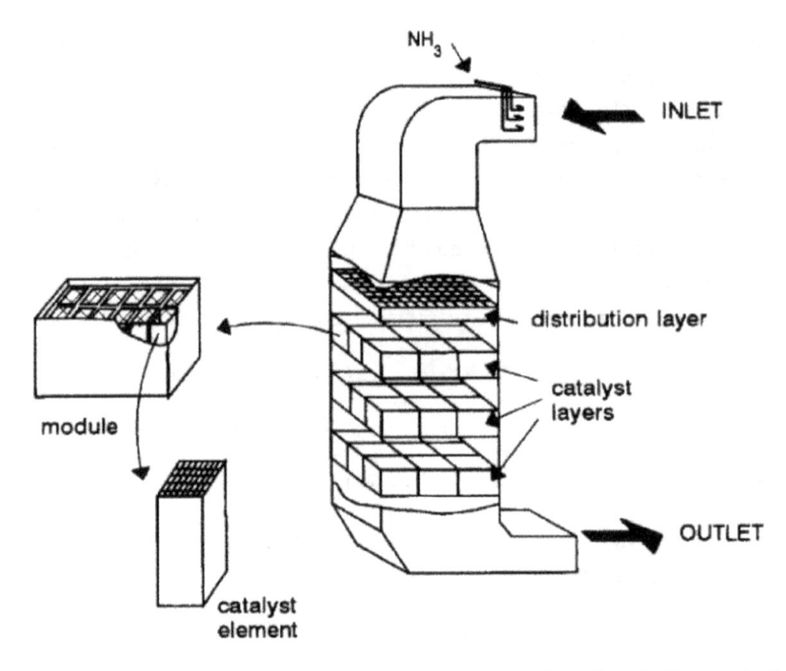

Fig. 9.7 Denox SCR reactor. Reprinted with permission from [9.17]. Copyright (2001) Elsevier.

In stationary reactors, such as boilers in power plants, the catalyst is V_2O_5–WO_3/TiO_2. V_2O_5 ($\sim$1 wt.%) catalyses the SCR, while TiO_2 (anatase) is used as the support, because it is only weakly and reversibly sulfated in the presence of SO_2 and oxygen and also promotes the catalyst activity. WO_3 ($\sim$10 wt.%) increases the acidity of the catalyst and, thus, limits the oxidation of SO_2. Furthermore, it acts as a structural promoter that hinders anatase TiO_2 from transforming to rutile TiO_2 and then sintering.

In the SCR reactor, ammonia (or ammonia formed *in situ* from urea) reduces NO_x, while NO_x oxidises NH_3.

$$4/x\,NO_x + 4\,NH_3 + O_2 \rightarrow (2 + 2/x)\,N_2 + 6\,H_2O \qquad (x = 1 \text{ or } 2),$$

$$2\,NO_x + 2\,NH_2\text{–}CO\text{–}NH_2 + (3 - x)\,O_2 \rightarrow 3\,N_2 + 4\,H_2O + 2\,CO_2 \qquad (x = 1 \text{ or } 2).$$

IR experiments showed that NH_3 chemisorbs on the catalyst surface but that NO does not chemisorb. The NH_3 either adsorbs on a $V^{5+}O^{2-}$ Lewis site and reacts to an amide [9.19] or adsorbs on a $V^{5+}OH^-$ Brønsted site and reacts to an amide group on a neighbouring $V^{5+}{=}O$ group [9.20].

In both mechanisms, the amide species reacts with gas-phase NO by radical coupling to a nitrosamide intermediate, which decomposes easily to N_2 and H_2O as a result of acid catalysis. The catalytic cycle is closed by re-oxidation of the reduced catalyst by gaseous oxygen. The mechanism on the Lewis sites [9.19] is:

$$\begin{aligned}
&2\,NH_3 + 2\,O^{2-}V^{5+} \rightarrow 2\,O^{2-}V^{5+}{:}\,NH_3 \rightarrow 2\,HO^-V^{4+} - NH_2\\
&2\,HO^-V^{4+}{-}NH_2 + 2\,NO \rightarrow 2\,HO^-V^{4+}{-}NH_2NO\\
&2\,HO^-V^{4+}{-}NH_2NO \rightarrow 2\,HO^-V^{4+} + 2\,N_2 + 2\,H_2O\\
&\underline{2\,HO^-V^{4+} + 0.5\,O_2 \rightarrow H_2O + 2\,O^{2-}V^{5+}}\\
&2\,NH_3 + 2\,NO + 0.5\,O_2 \rightarrow 2\,N_2 + 3\,H_2O.
\end{aligned}$$

Figure 9.8 gives the mechanism that takes place on the $V^{5+}OH^-$ Brønsted sites [9.20]. This and related mechanisms have also been studied by DFT [9.21, 9.22].

A slightly different mechanism is based on the assumption that the formation of nitrite is essential for the removal of NH_3 [9.23]. The precursor of nitrite, a mixture of NO and NO_2, forms by oxidation of NO:

$$NO + 0.5\,O_2 \rightarrow NO_2,$$

$$NO + NO_2 + H_2O + 2\,NH_3 \rightarrow 2\,NH_4NO_2.$$

The oxidation of NO to NO_2 occurs in air even in the absence of a catalyst, but a catalyst accelerates the reaction. This explains why SCR catalysts always contain a redox catalyst component such as V_2O_5, TiO_2 or Cu^{2+} or Fe^{3+} cations adsorbed on a zeolite. The formation of N_2 from

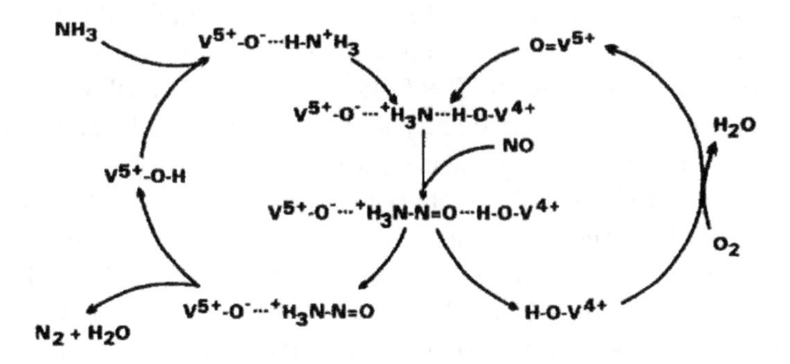

Fig. 9.8 Acid and redox cycles in the SCR reaction over V_2O_5/TiO_2. Reprinted with permission from [9.20]. Copyright (1995) Elsevier.

ammonium nitrite is assumed to occur by dehydration of nitrosamine (NH_2–N=O) [9.23]. This reaction is catalysed by acids, also in the absence of a redox catalyst.

$$NH_4^+NO_2^- + H_3O^+ \rightarrow H_2O + [(NH_4)^+NOOH \rightleftharpoons H_3N^+{-}N(OH)_2]$$
$$\rightarrow 2\,H_2O + H_2N{-}N^+(OH),$$
$$H_2N{-}N^+(OH) \rightarrow N_2 + H_3O^+.$$

The equilibrium between $(NH_4)^+NOOH$ and H_3N^+–$N(OH)_2$ is far to the left but is pulled to the right by the irreversible decomposition of the protonated nitrosamine, which is acid-catalysed.

All mechanisms are based on the assumption that a bifunctional catalyst, with acid and redox sites, is required for SCR. It is unclear, however, whether a LH or ER mechanism applies. Without a doubt, NH_3 chemisorbs on the catalyst surface but NO does not chemisorb, suggesting an ER mechanism. On the other hand, if the reaction does not occur by NO but by nitrite, then an LH mechanism might be feasible.

9.3 Oxidation of Hydrocarbons

The processes discussed thus far rely on hydrocarbons (from methane to heavy molecules) as the feedstock to produce fuel (gasoline, diesel) and base chemicals (methanol, ammonia, alkenes, aromatics). There are also many base chemicals containing oxygen or nitrogen atoms, and they can be produced by catalysis, as described in this section. Oxygen-containing molecules are prepared by adding an oxygen atom to a hydrocarbon molecule or by removing oxygen atoms from a molecule containing several oxygen atoms. Adding an oxygen atom to a hydrocarbon molecule is relatively easy when the hydrocarbon is an alkene. The addition of water to alkenes (hydration) gives alcohols ($RCH{=}CH_2 + H_2O \rightarrow RCH(OH)CH_3$), which can be oxidised to ketones or aldehydes (RCH_2COCH_3 or RCH_2CH_2CHO). Addition of synthesis gas to alkenes gives aldehydes by hydroformylation ($RCH{=}CH_2 + CO + H_2 \rightarrow RCH_2CH_2CHO$) (Section 8.4). Another relatively simple method is to add oxygen atoms to activated carbon atoms, as in the reaction of propene to acrolein ($CH_2{=}CH{-}CH{=}O$) and toluene to benzaldehyde C_6H_5–CHO). The C–H bonds of the methyl groups of propene and toluene are relatively weak, and stable allyl ($CH_2{=}CH{-}CH_2\cdot$) and toluyl ($C_6H_5{-}CH_2\cdot$) radicals can be produced, the starting point for the oxidation reaction.

Using small alkanes for the preparation of molecules containing heteroatoms would be a large advantage over the use of alkenes. Methane, ethane, propane and butane are much more widely available than ethene, propene and butene, and they are also much cheaper. Current research is, therefore, devoted to the exploration of possibilities of small alkanes. Success thus far is relatively modest, first because the C–H bonds of alkanes are strong and H atoms cannot be removed easily by oxygen or other oxidising agents. Second, and more important, however, is that the C–H bonds of the resulting oxygen-containing molecule are weaker than those of the original hydrocarbon molecule and, thus, oxidise more easily. As a consequence, the oxidation reaction tends to continue at the intermediate stage, CO_2 is produced and the selectivity-conversion curves of alkanes to oxygen-containing molecules are hyperbolic: either selectivity to the oxygen-containing molecule is high but the conversion of alkane is low, or conversion is high and selectivity low. High selectivity can be achieved at low conversion by working with a low oxygen concentration, so that after the oxidation of the first C–H bond, no oxygen remains for further oxidation to CO_2; this means intensive recycling of alkane. This explains why, in industry, methanol is not produced directly from methane but indirectly by synthesis gas (Section 6.4.2). Attempts have been made to oxidise methane to methanol [9.24] and acetic acid [9.25, 9.26], but the required experimental conditions (catalysis by Pd^{2+} and Cu^{2+} cations in strong acids) have prevented commercialisation.

The activation of ethane is also difficult and once it starts it easily leads to full oxidation. Therefore, ethane is used only in the chemical industry for thermal cracking to ethene. Propane would be an excellent feedstock for C_3 chemicals because it is abundant and inexpensive, but the C–H bond is strong and the formation of radicals by reaction with O_2 is unselective, leading to primary ($CH_3CH_2CH_2\cdot$) as well as secondary ($CH_3CH{\cdot}CH_3$) propyl radicals and eventually to the introduction of oxygen at the C1 and C2 atoms. Nevertheless, considerable research continues to develop a method to selectively transform propane to propene and heteroatom-containing molecules such as acrolein and acrylonitrile with reasonable conversion.

9.3.1 *Oxidation by Oxygen*

The introduction of an oxygen atom to a molecule, changing, for example, an alkane, alkene or aromatic molecule into an oxygen-containing molecule, brings substantial added value. The holy grail of oxidation in organic chemistry is the use of (inexpensive) oxygen (air) as the oxidant. This can be achieved for some small molecules; for example, oxygen or air is used in the oxidation of methanol to formaldehyde and of ethene to acetaldehyde and ethylene oxide. Oxidation of **methanol to formaldehyde** is performed with a silver or an iron molybdate catalyst [9.27]. The silver catalyst (in the form of a shallow bed of needles or several layers of fine gauze) operates at about 650 °C to induce the endothermic dehydrogenation of methanol to formaldehyde. Excess methanol is used (> 36.5%) to avoid the explosion region (6.7–36.5% methanol in air). Conversion is above 80% but methanol must be recycled. An admixture of air establishes an autothermic process. The addition of steam hinders sintering of the silver catalyst particles.

$$CH_3OH \rightarrow H_2CO + H_2 \qquad \Delta H = 84\,\text{kJ} \quad \text{dehydrogenation}$$

$$H_2 + 0.5\,O_2 \rightarrow H_2O \qquad \Delta H = -243\,\text{kJ} \quad \text{oxidation}$$

$$CH_3OH + 0.5\,O_2 \rightarrow CH_2O + H_2O \qquad \Delta H = -159\,\text{kJ} \quad \text{oxidative dehydrogenation.}$$

In the second process, methanol and excess air react at about 400 °C in presence of Fe_2MoO_6 to produce formaldehyde.

$$CH_3OH + 0.5\,O_2 \rightarrow 2\,CH_2O + 2\,H_2O.$$

This is a partial oxidation reaction (Section 9.3.3) that occurs by a Mars–van Krevelen mechanism, in which the oxygen atoms of the Fe_2MoO_6 oxide lattice oxidise the methanol and air re-oxidises the catalyst. Thermodynamics are favourable to high conversion (> 95%) per pass, and the recycling of methanol is unnecessary, but as a result of exothermicity a tubular fixed-bed reactor is necessary for cooling.

The formaldehyde process uses 25% of all produced methanol and about half of the formaldehyde produced (8 MT/y worldwide) is required for the production of resins [9.27]. When treated with phenol or urea, formaldehyde produces phenol–formaldehyde or urea–formaldehyde resin, respectively. These polymers are common adhesives used in plywood and carpeting. Formaldehyde-based materials are also important in automobile components.

Acetaldehyde is produced from ethene and air or oxygen in the **Wacker–Hoechst process**, in which a palladium chloride solution functions as a homogeneous catalyst, cupric ions re-oxidise the resulting Pd atoms and oxygen re-oxidises the cuprous ions [9.28].

$$C_2H_4 + H_2O + PdCl_2 \rightarrow CH_3CHO + Pd + 2\,HCl$$

$$Pd + 2\,CuCl_2 \rightarrow PdCl_2 + 2\,CuCl$$

$$2\,CuCl + 2\,HCl + 0.5\,O_2 \rightarrow 2\,CuCl_2 + H_2O$$

$$C_2H_4 + 0.5\,O_2 \rightarrow CH_3CHO \qquad \Delta H = -243\,\text{kJ}.$$

The mechanism starts with the exchange of two Cl^- anions in the square-planar tetrachloropalladate anion by a water molecule and an ethene molecule to form the square-planar dichloro-aquopalladate that binds ethene, in π fashion, perpendicular to the chloropalladate plane.

$$[PdCl_4]^{2-} + H_2O \rightarrow [PdCl_3(OH_2)]^- + Cl^-,$$

$$[PdCl_3(OH_2)]^- + C_2H_4 \rightarrow PdCl_2(OH_2)(\pi - CH_2{=}CH_2) + Cl^-.$$

A nucleophilic attack of a water molecule on the π-complexed ethene occurs from outside the palladium complex, resulting in a σ-bonded 2-hydroxyethyl group.

$$PdCl_2(OH_2)(\pi{-}CH_2{=}CH_2) + H_2O$$
$$\rightarrow [PdCl_2(OH_2)(CH_2{-}CH_2{-}OH)]^- + H^+.$$

The σ-bonded 2-hydroxyethyl group rearranges by an intramolecular H shift to π-bonded $CH_2{=}CHOH$ and further to σ-bonded 1-hydroxymethyl-methine ($HO{-}CH{-}CH_3$), which forms acetaldehyde.

$$[PdCl_2(OH_2)(CH_2{-}CH_2{-}OH)]^- \rightarrow [PdHCl_2(OH_2)(\pi{-}CH_2{=}CH{-}OH)]^-$$
$$\rightarrow [PdCl_2(OH_2)\{CH(CH_3)(OH)\}]^-$$
$$[PdCl_2(OH_2)\{CH(CH_3)(OH)\}]^- \rightarrow Pd + 2\,Cl^- + H_3O^+ + CH_3{-}CHO.$$

Re-oxidation of the resulting Pd atoms by O_2 occurs indirectly. First cupric ions oxidise Pd, and the resulting cuprous ions are re-oxidised by oxygen.

$$Pd + 2\,Cu^{2+} + 3\,Cl^- + H_2O \rightarrow [PdCl_3(OH_2)]^- + 2\,Cu^+$$
$$2\,Cu^+ + 2\,H^+ + 0.5\,O_2 \rightarrow 2\,Cu^{2+} + H_2O.$$

The Wacker process is carried out in one or in two reactors. In the two-reactor process the formation of acetaldehyde and re-oxidation of palladium take place in the same reactor at about 130 °C and 10 atm; the re-oxidation of the cupro ions with air takes place in the other reactor. In the single-reactor process pure oxygen is used at about 4 atm. The importance of acetaldehyde has decreased in recent decades. Acetic acid is the main product of acetaldehyde, but the preferred route for producing acetic acid is no longer the oxidation of acetaldehyde but rather the carbonylation of methanol (Monsanto and Cativa processes) (Section 1.2).

Ethylene oxide (**EO**, also referred to as oxirane) can also be made by oxidation of ethene with oxygen. It is produced at about 250 °C and 1.5 MPa over a Ag/α-Al_2O_3 catalyst [9.15, 9.29]. The main reaction is

$$CH_2{=}CH_2 + 0.5\ O_2 \rightarrow \text{(ethylene oxide)} \qquad \Delta H = -105\,\text{kJ}\cdot\text{mol}^{-1}.$$

but isomerisation to acetaldehyde may occur, followed by total oxidation:

$$\text{(ethylene oxide)} \rightarrow CH_3CHO$$

$$CH_3CHO + 2.5\,O_2 \rightarrow 2\,CO_2 + 2\,H_2O.$$

Another reaction to avoid is direct total oxidation:

$$CH_2{=}CH_2 + 3\,O_2 \rightarrow 2\,CO_2 + 2\,H_2O \qquad \Delta H = -1327\,\text{kJ}\cdot\text{mol}^{-1}.$$

Addition of alkylchloride (ppm) in the gas phase ensures that chlorine atoms are present on the catalyst surface, increasing the EO selectivity from ~50% to ~75%. The presence of Cs increases selectivity to ~80%, and further promotion by high-valent oxyanions of Re, Mo or S leads to an increase of up to 90% [9.30]. This is important not only for efficient use of the feedstock but also for process control. The combustion reaction is very exothermic and requires careful control of temperature in a tubular reactor. The feed contains excess ethene to stay outside the explosion limits and to maintain low conversion and, thus, low production of heat. Methane is a diluent gas because of its high thermal conductivity. The support, α-Al_2O_3, is chosen to diminish combustion. α-Al_2O_3 has wide pores, which enable the EO molecules to quickly migrate from the catalyst particles and, thus, avoid further oxidation. Another advantage is that α-Al_2O_3 is neutral and does not isomerise EO to acetaldehyde.

It was believed originally that ethene reacts with a chemisorbed O_2 molecule to EO and with a chemisorbed O atom to acetaldehyde, which combusts quickly to CO_2. Today, however, CO_2 and EO formation are ascribed to atomic oxygen. DFT calculations indicate that, for silver surfaces with low oxygen coverage, an Ag–CH_2–CH_2–O–Ag oxometallacycle is the surface intermediate for the formation of both EO and acetaldehyde

[9.31]. The calculations suggest that Cs cations on the Ag surface and Cl and O anions below the surface (subsurface) have an electronic effect by lowering the transition state for EO formation to a greater extent than the transition state for acetaldehyde formation [9.32, 9.33].

A different mechanism is based on the assumption that the catalyst surface consists of Ag_2O and that the oxygen atoms on the Ag_2O surface interact directly with the C=C double bond, resulting in the formation of EO [9.34]. In this mechanism, EO selectivity is ascribed to a higher activation barrier for the formation of acetaldehyde than for the formation of EO. When oxygen atoms are removed from the Ag_2O surface, ethene adsorbs on the Ag atoms around the vacancy and forms an oxometallocycle. The oxometallocycle reacts to acetaldehyde with lower activation energy than to EO. The positive influence of O_2 pressure and Cl on EO selectivity is explained in the Ag_2O model by the blocking of vacant O sites to prevent oxometallocycle formation.

EO is one of the most important raw materials in the large-scale production of chemicals. Worldwide production was 26 MT in 2014 [9.35]. Most EO is used for the synthesis of ethylene glycols, including diethylene glycol and triethylene glycol, which account for up to 75% of global consumption [9.27]. Other important products are ethylene glycol ethers, ethanolamines and ethoxylates. Ethylene glycol is used as a diol in the production of polyester (for fibres, films and resins) and polyethylene terephthalate (PET, a raw material for plastic bottles) and as an antifreeze and solvent.

Ethylene glycol is produced by hydrolysis of EO:

$$\text{(EO)} + H_2O \rightarrow CH_2OH{-}CH_2OH \qquad \Delta H = -80\,\text{kJ·mol}^{-1}.$$

A large excess of water is used in the reaction because EO reacts faster with ethylene glycol than with water and forms di- and triethylene glycols, which would be detrimental in the manufacture of polyesters, where a monomer purity of 99.9% is required. In the Omega process (Shell) the formation of di- and triethylene glycol is prevented by first reacting EO with CO_2 to ethylene carbonate, which is hydrolysed to ethylene glycol and CO_2 [9.35]. The CO_2 is recycled.

$$CO_2 + \text{(epoxide)–R} \longrightarrow \text{(cyclic carbonate)} \qquad R = H$$

$$C_2H_4O_2C{=}O + H_2O \rightarrow CH_2OH{-}CH_2OH + CO_2.$$

9.3.2 *Oxidation by Hydroperoxide*

While some hydrocarbons can be oxidised by oxygen to valuable chemicals, other molecules must be oxidised by more selective oxidation agents, such as H_2O_2 and organic peroxides (ROOH). Thus, in contrast to EO, it is not possible to produce propene oxide (PO) by the oxidation of propene by oxygen because the main reaction would be complete combustion, not the desired epoxidation. Due to the relatively weak C–H bonds in the CH_3 group of propene, the combustion of propene is much stronger than that of ethene. Abstraction of an H atom gives the allyl radical $CH_2{=}CH{-}CH_2\cdot$ which is stabilised by conjugation. Thus, both the C=C double bond as well as the CH_3 group are attacked by oxygen. Therefore, PO can only be produced indirectly with high selectivity.

Propene oxide (**PO**, propylene oxide) is a significant organic chemical product used ($\sim$60%) as a reaction intermediate in the production of polyether polyols ($HO{-}CH(CH_3){-}CH_2{-}[O{-}CH_2CH(CH_3){-}]_nOH$). Polyether polyols are coupled with diisocyanates to give polyurethanes resins. Other uses of PO are propylene glycol, alkanolamines and glycol ethers [9.35]. PO is produced industrially by the chlorohydrin process and the hydroperoxide process. In 2010, the global PO capacity was about 6 MT, about 50% of which was produced in 36 chlorohydrin plants (one large DOW plant and many small plants in Asia). The other 50% was produced in 13 large plants, four working with *t*-butane and nine with ethylbenzene as the hydroperoxide transfer agent [9.27].

The oldest PO process is the **chlorohydrin process** [9.27, 9.36], in which propene and chlorine react in the presence of water to the two isomers of chloropropanol (propene chlorohydrin).

$$CH_3{-}CH{=}CH_2 + Cl_2 + H_2O \rightarrow CH_3{-}CHOH{-}CH_2Cl$$
$$(+CH_3{-}CHCl{-}CH_2OH) + HCl.$$

The reaction can be considered as the addition of hypochlorite (HOCl) to the double bond. Hypochlorite forms *in situ* from chlorine and water ($H_2O + Cl_2 \rightarrow HCl + HOCl$). The mixture of chloropropanols is then dehydrochlorinated to PO ($CH_3CH(O)CH_2$) with a base (lime, calcium hydroxide):

$$CH_3CH(OH)CH_2Cl + HCl + Ca(OH)_2 \rightarrow CH_3\overset{\overbrace{\quad O \quad}}{CH{-}{-}CH_2} + CaCl_2 + 2\,H_2O.$$

The chlorohydrin process produces large amounts of inorganic and organic byproducts. A useless aqueous calcium chloride solution ($40\,L{\cdot}kg^{-1}$ PO, containing, e.g. 5–6 wt.% $CaCl_2$ and 0.1 wt.% $Ca(OH)_2$), as well as organic waste (1,2-dichloropropane, 2,2′-dichloro-diisopropylether and propylene glycol) are produced [9.27].

PO production by oxidation of propene by an organic hydroperoxide or hydrogen peroxide is a much more cleanly process.

$$CH_3{-}CH{=}CH_2 + ROOH \rightarrow CH_3\overset{\;}{CH}\overset{O}{\overline{\quad\;}}CH_2 + ROH.$$

In the concerted reaction between a peroxide and an alkene the peroxide acts as an electrophile and the alkene as a nucleophile (Fig. 9.9).

Epoxidation reactions can also be carried out in the inner sphere of coordination complexes, where the alkene and peroxide act as ligands and are, thus, close together. For instance, PO is produced by the **hydroperoxide process**, which involves oxidation of propene to PO by an organic hydroperoxide in the presence of a titanium catalyst (Fig. 9.10).

In the **hydroperoxide process** for the oxidation of propene to PO, an organic hydroperoxide forms *in situ* from a hydrocarbon that can easily react to a hydroperoxide ($RH + O_2 \rightarrow ROOH$) [9.14, 9.26, 9.37]. In industry, isobutane and ethylbenzene are used. In both cases, an alcohol (tert-butanol and 1-phenylethanol, respectively) is produced as a co-product. The hydroperoxide is produced in a separate reactor in the liquid phase at about 130 °C and 3.5 MPa, and the product solution is transferred to a second reactor for the epoxidation of propene in the presence of a

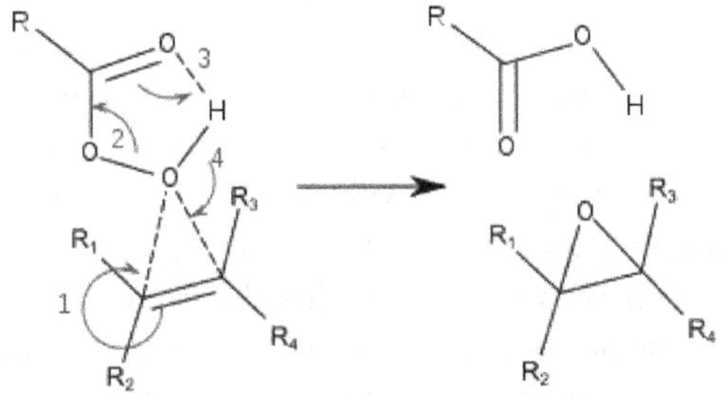

Fig. 9.9 Peroxidation of an alkene by a peroxy acid to an oxirane and carboxylic acid.

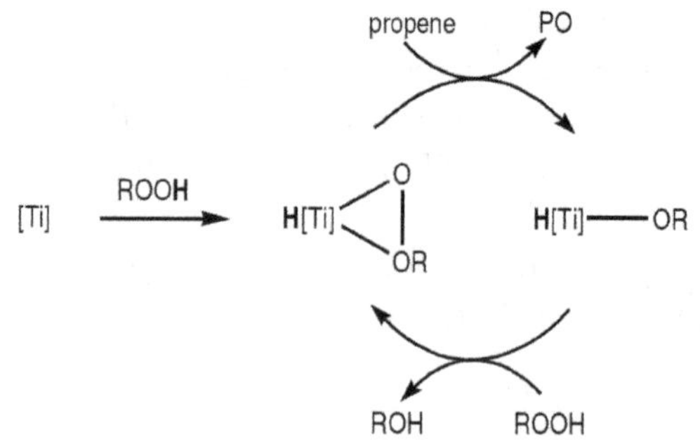

propene PO
[Ti] ROOH H[Ti](O)(OR) H[Ti]—OR
ROH ROOH

Fig. 9.10 Propene oxidation with hydroperoxide over a heterogeneous Ti complex. Reprinted with permission from [9.37]. Copyright (2004) Elsevier.

soluble metal organic catalyst (Mo, Ti) or of a heterogeneous Ti catalyst.

$$\mathrm{CH_3CH{=}CH_2 + (CH_3)_3CH + O_2 \rightarrow CH_3\overset{O}{\overbrace{CH{-}CH_2}} + (CH_3)_3COH,}$$

$$\mathrm{CH_3CH{=}CH_2 + C_6H_5{-}CH_2CH_3 + O_2 \rightarrow CH_3\overset{O}{\overbrace{CH{-}CH_2}} + C_6H_5{-}CH(CH_3)OH.}$$

The tert-butanol and 1-phenylethanol co-products are valuable feedstocks for other products. For example, tert-butanol can react with methanol to give methyl-t-butylether (MTBE), which has a high octane number and was added to gasoline for many years in the USA. However, contamination of drinking water due to leakage from storage tanks led to a ban of MTBE. Styrene is the dehydration product of 1-phenylethanol and is a bulk chemical. Styrene and PO are, therefore, coproduced in several very large plants. Figure 9.11 gives the separate steps for producing PO and styrene.

A disadvantage of co-production is the dependence of the process economics on the market price of two bulk products. It is possible to avoid co-production by recycling the co-product. For instance, in the co-oxidation process, *t*-butanol, cumylalcohol and styrene, which form from isobutane, cumene and ethylbenzene, respectively, can be recycled after dehydration of the alcohols and hydrogenation of the resulting alkenes. Thus, Sumitomo Chemical commercialised a co-product-free hydroperoxidation route to PO [9.38]. They use cumene as the intermediate for the hydroperoxidation of propene, dehydrate the resulting 2-phenyl-2-propanol and hydrogenate the

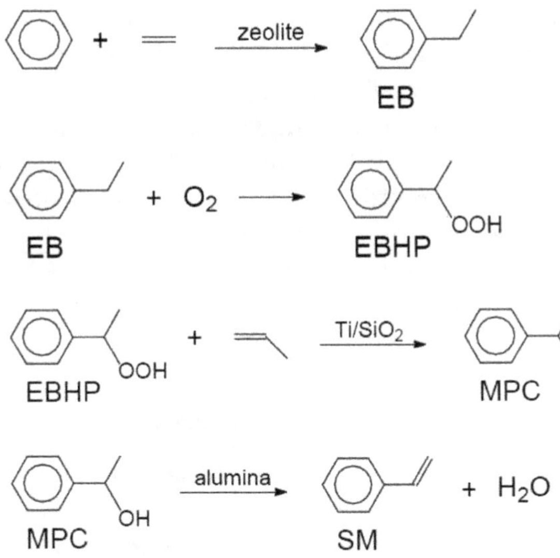

Fig. 9.11 Reaction steps in the co-production of propene oxide (PO) and styrene (SM) from propene and ethylbenzene (EB). Reprinted with permission from [9.37]. Copyright (2004) Elsevier.

resulting α-methylstyrene back to cumene.

$$CH_3CH{=}CH_2 + C_6H_5{-}CH(CH_3)_2 + O_2 \rightarrow CH_3\overset{O}{\overbrace{CH{-\!-}CH_2}} + C_6H_5{-}C(CH_3)_2OH$$

$$C_6H_5{-}C(CH_3)_2OH \rightarrow C_6H_5{-}C(CH_3){=}CH_2 + H_2 \rightarrow C_6H_5{-}CH(CH_3)_2.$$

Several companies developed technologies for the production of PO from propene and hydrogen peroxide (H_2O_2) [9.36]. The advantage of H_2O_2 is that only water is generated as side product.

$$CH_3CH{=}CH_2 + H_2O_2 \rightarrow CH_3\overset{O}{\overbrace{CH{-\!-}CH_2}} + H_2O.$$

In 2008, SK Corp started up a plant in Ulsan, South Korea, based on Evonik–Uhde technology; DOW Chemical and BASF started up a plant in Antwerp in 2009 [9.38].

Current global PO production is carried out for the most part in hydroperoxidation processes that co-produce styrene and t-butyl alcohol. These are capital intensive production processes and a balance must be struck between cost and the size of the markets for PO and the co-product. Significant amounts of PO are still produced by the chlorohydrin process, but this route is not environmentally friendly. Furthermore, inexpensive electric power must be available to provide chlorine. PO production by the chlorohydrin route has recently been restricted in China, and new projects must follow alternative routes. The cumene hydroperoxide process and the H_2O_2–PO process require a lower capital investment than the other established routes, are environmentally friendly and do not produce significant amounts of by-products. The first H_2O_2–PO plant (Evonik) in China started up in 2014 (Jilin Shenhua Group Co.) [9.38].

9.3.3 *Selective Partial Oxidation of Hydrocarbons*

Hydrocarbons such as propene, butane and methanol can be selectively oxidised with oxygen to valuable products over transition-metal oxides. The reactions occur by Mars–van Krevelen mechanisms, in which an oxygen atom from the transition-metal oxide lattice oxidises the hydrocarbon, and pure oxygen or air re-oxidises the catalyst.

9.3.3.1 *Oxidation of propene to acrylic acid and acrylonitrile*

Acrylic acid ($CH_2{=}CHCOOH$) is used as a chemical intermediate in the production of alkyl acrylate esters and resins. The esters are polymerised and are used in the production of paints, coatings, textiles, adhesives, polishes and plastics. The second major application of acrylic acid is in the production of polyacrylic acid polymers. These polymers are cross-linked polyacrylates and are super absorbent; they can absorb and retain more than 100 times their own weight and are used in the production of diapers and personal care products. Acrylic acid is also utilised in the production of detergent polymers. The worldwide production of acrylic acid is about 5 MT/y [9.15].

The main process for producing acrylic acid is the vapour phase oxidation of propene. It is carried out in two steps with two reactors in series and different catalysts. In the first reactor the propene reacts to **acrolein** (2-propenal), while in the second reactor the acrolein reacts further to acrylic acid [9.36].

$$CH_2{=}CHCH_3 + O_2 \rightarrow CH_2{=}CHCHO + H_2O \quad \Delta H = -368\,\text{kJ·mol}^{-1},$$

$$CH_2{=}CHCHO + 0.5\,O_2 \rightarrow CH_2{=}CHCOOH \quad \Delta H = -266\,\text{kJ·mol}^{-1}.$$

A two-step process is implemented because, as often in oxidation reactions, the first reaction (propene to acrolein) is more difficult than the second reaction (acrolein to acrylic acid). While the optimal temperature in the first step is around 320 °C, it is only 230 °C in the second reaction. Consequently, in a one-step process a temperature of 320 °C would be required to achieve adequate propene conversion in the first reaction, but that would lead to substantial overoxidation in the second reaction and, thus, to low selectivity.

In the partial oxidation of propene with air to acrolein, a multi-component catalyst (e.g. $Fe_4BiW_2Mo_{10}Si_{1.3}K_{0.6}$) is used with bismuth molybdate providing the main active sites. The reaction follows the Mars–van Krevelen mechanism and is initiated by coordination of the propene double bond to a Bi–Mo site. Abstraction of a hydrogen atom by a surface O atom gives a Mo-bound allyl intermediate surface species [9.39]. The C–O bond forms through insertion of an oxygen atom from the catalyst surface, followed by a second hydrogen abstraction. The lattice oxygen atoms on the catalyst surface contribute to the specific selectivity. Their mobility is determined by the structure of the catalyst and plays a critical role in both propene conversion and acrolein selectivity in the Mars–van Krevelen mechanism. The mobility of the lattice oxygen is promoted by the presence of divalent (Co^{2+}, Ni^{2+}, Mg^{2+}) and trivalent (e.g. Fe^{3+}, Cr^{3+}) cations, by alkali metals and by one of the elements Sb, Te, W, V or Nb and P or B. Gaseous molecular oxygen re-oxidises the catalyst and ensures sufficient availability of lattice oxygen atoms.

The oxidation in the second step of acrolein to acrylic acid is catalysed by V–Mo–(W) mixed oxides (e.g. $V_{4.6}Cu_{2.2}Cr_6W_{2.4}Mo_{12}/Al_2O_3$). Conversion and selectivity in the second step are very high. Multitubular fixed-bed reactors (up to 20,000 tubes per reactor, surrounded by molten salt for heat transfer) are in operation to remove the heat of reaction. The catalysts are highly efficient and can last at least three years.

Acrylonitrile is a large volume chemical intermediate (word wide production 6.5 MT/y) for the manufacture of polyacrylonitrile fibre, used in the production of apparel and home furnishings. Other major uses are in the manufacture of acrylonitrile–styrene–butadiene (ABS) resins for engineering (metal-replacing) polymers, adiponitrile for nylon polymers, and arylamide for homopolymers and copolymers. An emerging speciality application is in the production of carbon fibre as a high-strength, low-weight composite material for the transportation industry. Acrylonitrile is produced by the Sohio process, in which propene reacts with ammonia and oxygen over a bismuth molybdate catalyst at about 450 °C and 0.15 MPa [9.15, 9.36].

$$CH_2{=}CHCH_3 + 1.5\,O_2 + NH_3 \rightarrow CH_2{=}CH{-}CN + 3\,H_2O$$

$$\Delta H = -\,502\,\text{kJ}\cdot\text{mol}^{-1}.$$

The mechanism starts with the abstraction of a hydrogen atom of propene by a surface O atom to create an allyl group π-bonded to a metal cation and with adsorption of NH_3 to surface-bonded amide (replacement of an O atom by an NH group).

$$2\,CH_2{=}CHCH_3 + O* + * \rightarrow 2\,CH_2{=}CHCH_2* + H_2O,$$

$$2\,NH_3 + 2\,O* \rightarrow 2\,H_2O + 2\,NH*.$$

Reaction of the allyl group with surface O atoms leads to σ-bonded acroleine. This can react with a surface-bonded amide group to a surface-bonded imine $CH_2{=}CH{-}CH{=}NH$, which is oxidatively dehydrogenated to acrylonitrile ($CH_2{=}CH{-}C{\equiv}N$).

$$2\,CH_2{=}CHCH_2* + 3\,O* \rightarrow 2\,CH_2{=}CH{-}CH{=}O* + H_2O + 3\,*$$

$$2\,CH_2{=}CH{-}CH{=}O* + 2\,NH* \rightarrow 2\,CH_2{=}CH{-}CH{=}NH* + 2\,O*$$

$$2\,CH_2{=}CH{-}CH{=}NH* + 2\,O* \rightarrow 2\,CH_2{=}CH{-}C{\equiv}N + 2\,H_2O + 4\,*.$$

Alternatively, the allyl group can react directly with a surface-bonded amide group to a σ-bonded amine $CH_2{=}CH{-}CH_2{-}NH$. This can be oxidatively dehydrogenated to acrylonitrile.

$$2\,CH_2{=}CHCH_2* + 2\,NH* \rightarrow 2\,CH_2{=}CH{-}CH_2{-}NH* + 2\,*$$

$$2\,CH_2{=}CH{-}CH_2{-}NH* + 2\,O* \rightarrow 2\,CH_2{=}CH{-}CH{=}N* + 2\,H_2O + 2\,*$$

$$2\,CH_2{=}CH{-}CH{=}N* + O* \rightarrow 2\,CH_2{=}CH{-}C{\equiv}N + H_2O + 3\,*.$$

The organic reactant and intermediates react with surface O atoms (O*) or amide groups (NH*), and the resulting vacancies are re-oxidised by molecular oxygen to surface O atoms.

$$3\,O_2 + 6\,* \rightarrow 6\,O*.$$

The total reaction is highly exothermic, and a fluidised-bed reactor is required to remove the heat and achieve a propene conversion above 95% as well as a yield of acrylonitrile close to 80%. Steam coils inside the reactor keep the temperature as constant as possible to prevent thermal runaway. Byproducts are acetonitrile (CH_3CN) and HCN. Unreacted NH_3

is captured in sulfuric acid to form $(NH_4)_2SO_4$, used as a fertiliser. Two types of catalysts are used in this process: a multi-metal molybdate $(K,Cs)_{0.1}(Ni,Mg,Mn)_{7.5}(Fe,Cr)_{2.3}Bi_{0.5}Mo_{12}O_x$, made up of Mo, Bi, M^{2+} (Mg, Ni, Co, Mn) and M^{3+} (Fe, Cr) cations and additives (e.g. Cr, Mg, Rb, K, Cs, P, B, Ce, Sb and Mn) dispersed on silica, and an antimonate with rutile structure, made up of four metal antimonate cations and a redox couple of Fe, Ce, U and Cr [9.40].

In the near future there might be a shortage of propene. Ever-increasing amounts of propene are required for polypropene production and its production by thermal cracking of naphtha has declined in countries that have invested in fracking. Thus, considerable research aims at developing strategies for the use of propane, less expensive than propene as a feedstock, in the selective catalytic oxidation to propene, acrolein and acrylic acid [9.41]. Mixed metal oxides of Mo, V, Te and Nb activate propane for the selective formation of acrylic acid. Vanadia sites are responsible for selectively activating propane by oxidative dehydrogenation to propene during its oxidation over these bulk mixed-metal oxide catalysts. Thus far, the MoV(Nb,Ta)(Te,Sb)O catalyst, which consists of a mixture of two phases (M1/M2) and was discovered by Mitsubishi/Asahi, is the most promising catalyst [9.42]. The problem in the search for a material that can catalyse the reaction from propane to acroleine, acrylic acid or acrylonitrile is similar to that faced in finding a catalyst for the reaction of methane to methanol or formaldehyde. The first C–H bond to break is the strongest, meaning that subsequent reactions must be slowed down. Though considerable progress has been made in the selective oxidation of propane, much remains to be done.

In the future, acrolein and, thus, acrylic acid, may be produced from both propene and propane as well as from glycerol (by dehydration).

$$CH_2OH{-}CHOH{-}CH_2OH \rightarrow CH_2{=}CH{-}CHO + 2\,H_2O,$$

$$CH_2{=}CH{-}CHO + 0.5\,O_2 \rightarrow CH_2{=}CH{-}COOH.$$

Glycerol is cogenerated in the production of biodiesel from fatty acids, and large quantities are becoming available. However, at present, glycerol dehydration cannot yet compete with propene oxidation.

9.3.3.2 *Oxidation of C4 and C6 molecules*

Maleic anhydride (2,5-furandione, the anhydride of *cis*-butenedicarboxylic acid, HOOC–CH=CH–COOH) is produced by the oxidation of *n*-butane

[9.15]. The overall process converts the methyl groups to carboxylate groups and dehydrogenates the backbone. That maleic anhydride can be made from butane, whereas ethane and propane still cannot be selectively oxidised, is determined by the stability of its conjugated double-bond skeleton (O═C–C═C–C═O), which protects it against further oxidation. Nevertheless, to keep selectivity high (and minimise heat production) the operation is carried out at incomplete conversion. Butane and air are passed through a catalyst bed at high temperature with vanadyl pyrophosphate, $(VO)_2P_2O_7$ as the catalyst.

$$C_4H_{10} + 3.5\,O_2 \rightarrow C_2H_2(CO)_2O + 4\,H_2O \qquad \Delta H = -1236\,\text{kJ}\cdot\text{mol}^{-1}.$$

The strongly exothermic reaction requires very efficient cooling, which is obtained with a multi-tubular reactor with, e.g. 13,000 externally cooled tubes.

About 50% of the world's production of maleic anhydride (1.4 MT in 2012) is required for the manufacture of unsaturated polyester resins. Chopped glass fibres are added to these esters to reinforce the plastics, for a wide range of products such as boats, bathroom fixtures, automobiles, tanks and pipes. Some maleic anhydride is converted to 1,4-butanediol, which is one of the world's fastest growing chemicals used in the production of thermoplastic polyurethanes, polybutylene terephthalate resins and other products.

Phthalic anhydride (world production 4.5 MT in 2005) is produced by catalytic oxidation of *o*-xylene [9.27]:

$$C_6H_4(CH_3)_2 + 3\,O_2 \rightarrow C_6H_4(CO)_2O + 3\,H_2O \qquad \Delta H = -1110\,\text{kJ}\cdot\text{mol}^{-1}.$$

The reaction proceeds with about 70% selectivity at 320–400 °C over a catalyst containing vanadium pentoxide supported on titania; about 10% of maleic anhydride is coproduced:

$$C_6H_4(CH_3)_2 + 7.5\,O_2 \rightarrow C_2H_2(CO)_2O + 4\,H_2O + 4\,CO_2.$$

Phthalic anhydride and maleic anhydride are recovered by distillation. The primary use of phthalic anhydride is in phthalate esters, which function as plasticisers in flexible polyvinyl chloride products such as cables, pipes and hoses, leather cloth and film for packaging. The phthalate esters maintain the flexibility of these products; without them polyvinyl chloride would be hard and brittle. Phthalic anhydride is also used in the production of polyester resins.

9.4 Platform Chemicals

Base chemicals which contain oxygen atoms can be prepared by adding an oxygen atom to a hydrocarbon molecule (Section 9.3) or by removing oxygen atoms from molecules which contain several oxygen atoms. Biomass contains a high percentage of oxygen and oxygen-containing base chemicals could be prepared from biomass by removal of oxygen atoms. Such base chemicals are called **platform chemicals**. Because their production from biomass instead of from oil, gas or coal would be sustainable, a substantial amount of research has been undertaken worldwide to identify attractive chemical transformations of biomass into platform chemicals in economically feasible processes. Hemicellulose and cellulose are two of the three main constituents of biomass. They are carbohydrate-based polymers that can be depolymerised to low molecular weight sugars by hydrolysis using an acid catalyst. In the conversion of lignocellulosic biomass into platform chemicals the lignocellulosic biomass is first separated into hemicellulose, cellulose and lignin, and then each fraction is processed under different conditions to optimise the yields of the target products. The highly functionalised and reactive C_6 and C_5 sugars are processed at mild conditions to obtain platform molecules (Fig. 9.12), which still have enough

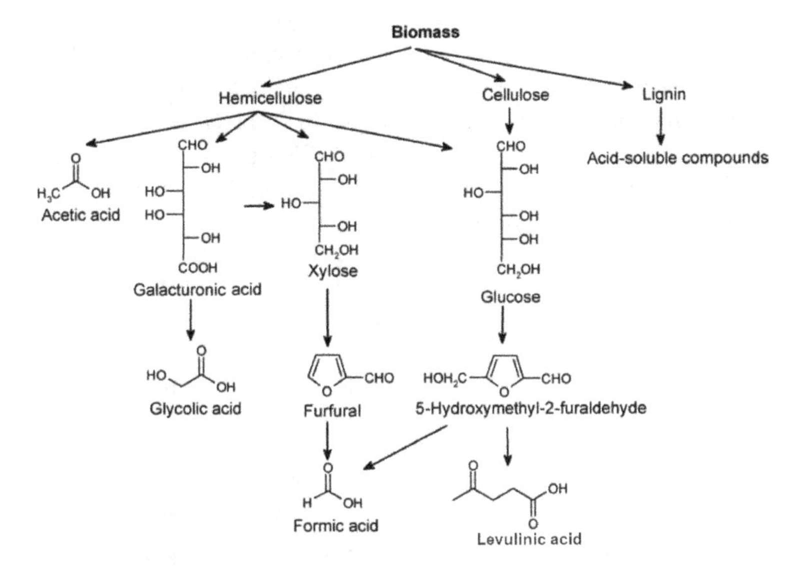

Fig. 9.12 Separation of lignocellulosic biomass into hemicellulose, cellulose and lignin and their conversion to platform molecules. Reprinted with permission from [9.43]. Copyright (2006) American Chemical Society.

functionality to be used as building blocks to produce a variety of chemicals. Platform chemicals that could be produced from sugars derived from biomass are ethanol, furans (5-hydroxymethylfurfural, HMF, and 2,5-furan dicarboxylic acid, 2,5-FDCA), glycerol, organic acids (e.g. lactic, succinic, malic, fumaric, glucaric and aspartic acid), hydroxypropionic acid, levulinic acid (LA), sorbitol, xylitol and γ-valerolactone (GVL). In the following, we will discuss the preparation and potential use of glycerol, HMF, LA and GVL.

At the end of the 20th century, when world energy consumption soared, the possibility that the world would run out of oil was seriously considered. Interest in the production of sustainable fuels, such as **biodiesel**, a mixture of fatty esters (esters of fatty acids and alcohols) and classic diesel increased. Fatty esters are made by trans esterification of fats (triglycerides, the esters of fatty acids and glycerol) as indicated in Fig. 9.13.

Originally, triglycerides were extracted from oil seeds, and transesterified with alcohols to fatty esters RCOOR'. Because biodiesel competes with the food and the oleochemical industry for oil seeds, second generation fatty esters are made from waste fats (cooking oil, animal fat). A large amount of **glycerol** is co-produced in the production of fatty esters and this led to efforts in finding new applications for glycerol. Many useful intermediates and speciality chemicals can be made from glycerol. A few catalytic reactions of industrial interest are presented in Fig. 9.14. There are plans to use glycerol for the synthesis of epichlorohydrine ($CH_2(O)CH_2$–Cl), an intermediate chemical used in the manufacture of epoxy resins), propylene glycol, acrolein and acrylic acid. These future outlets will add to the existing applications of glycerol in food ingredients, cosmetics, pharmaceuticals and explosives.

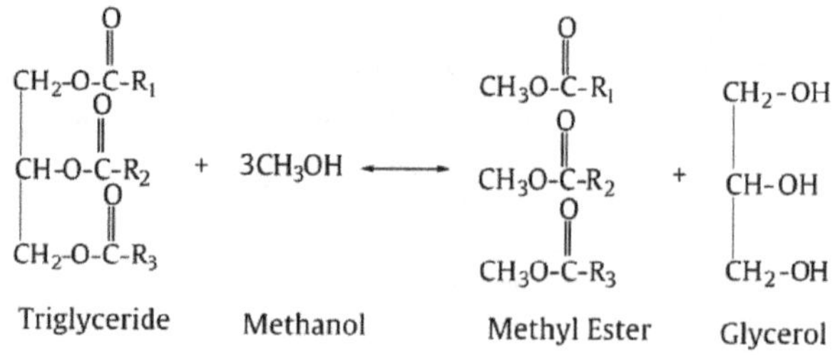

Triglyceride		Methanol		Methyl Ester		Glycerol
CH_2-O-C(=O)-R_1				CH_3O-C(=O)-R_1		CH_2-OH
CH-O-C(=O)-R_2	+	$3CH_3OH$	⟵⟶	CH_3O-C(=O)-R_2	+	CH-OH
CH_2-O-C(=O)-R_3				CH_3O-C(=O)-R_3		CH_2-OH

Fig. 9.13 Transesterification of triglyceride with methanol to the methyl esters of fatty acids and glycerol.

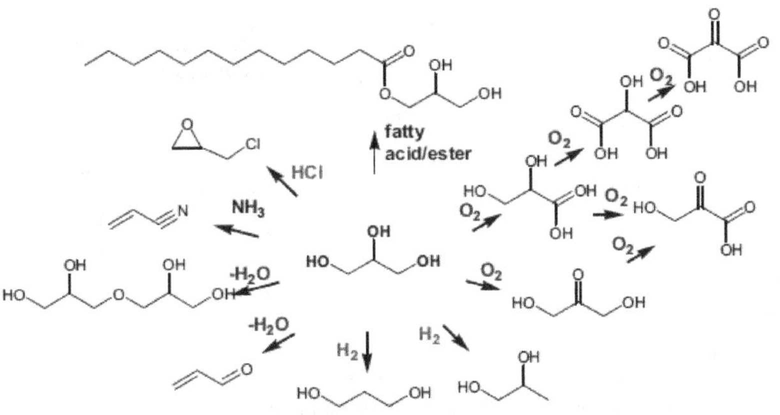

Fig. 9.14 Examples of conversion of glycerol to intermediate chemicals. Reprinted with permission from [9.44]. Copyright (2011) Elsevier.

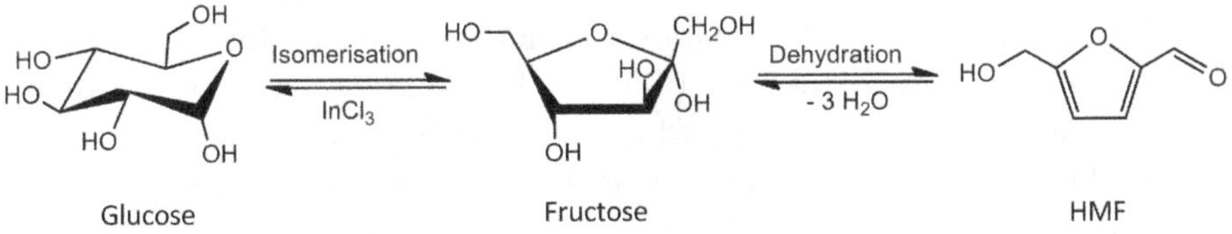

Fig. 9.15 Acid-catalysed transformation of glucose into HMF. Reprinted with permission from [9.45]. Copyright (2014) Elsevier.

A promising platform chemical is **5-hydroxymethylfurfural, HMF**. HMF can be made by acid-catalysed dehydration of C_6 sugars, such as glucose. Glucose, which in the cyclic form has a six-membered ring consisting of five C atoms and one O atom, is first isomerised with an acid catalyst (e.g. $InCl_3$, a Lewis acid metal-halide catalyst) to fructose, with a five-membered ring in the cyclic form [9.45]. Fructose is dehydrated in three steps to HMF (Fig. 9.15) [9.46]. HMF consists of a furan ring (with four C atoms and one O atom) that contains an aldehyde and an alcohol functional group and is a yellow, low-melting water-soluble solid. An economically viable synthesis of HMF has not been achieved yet, but still attracts a lot of attention [9.45].

HMF can be converted into several other chemicals (Fig. 9.16). For instance, oxidation of HMF gives 2,5-furandicarboxylic acid (FDCA),

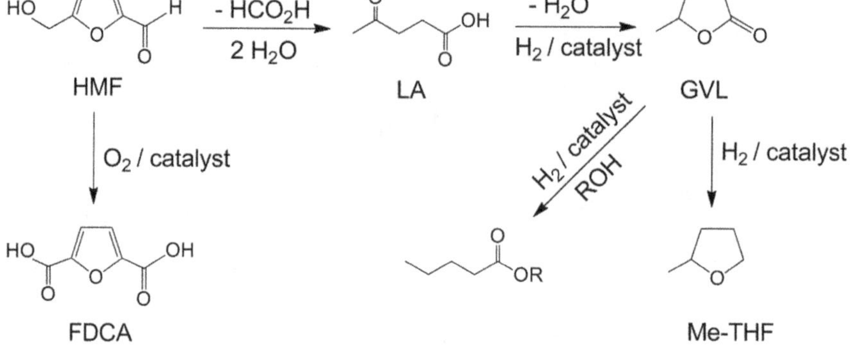

Fig. 9.16 Derivatives of HMF. Reprinted with permission from [9.47]. Copyright (2016) Elsevier.

which has two carboxylic groups and has been proposed as a replacement for terephthalic acid in the production of polyesters. HMF can also be converted to other platform chemicals, such as **levulinic acid (LA)** and **γ-valerolactone (GVL)**. LA contains a ketone group and a carboxylic acid group, which make it a very versatile building block for the synthesis of various organic (bulk)-chemicals as shown in Fig. 9.17.

LA can be hydrogenated to γ-valerolactone, GVL, a ring ester with 5 C atoms, three (γ) of which are between the O atom and the CO group (Fig. 9.17). GVL can be used as a fuel or fuel additive. It retains 97% of the energy content of glucose and performs comparably to ethanol when used as a blending agent in gasoline. One molecule of formic acid is produced per molecule of LA. This unavoidable by-product might be used to reduce LA to GVL. GVL can not only be used to produce renewable fuels, but also interesting monomers for the chemical and polymer industry [9.48] (Fig. 9.18).

Synthesising HMF, LA and GVL in an economic way is not easy. With their reactive groups, these molecules give several byproducts and, also because of hydration and dehydration steps, the selectivity or conversion are not very high in aqueous solution. Use of organic solvents gives better results but is hindered by separation of solvent and product.

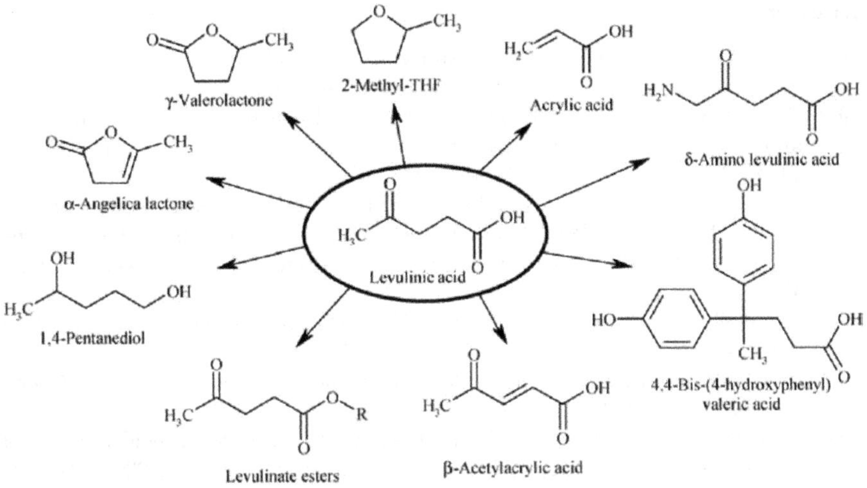

γ-Valerolactone
2-Methyl-THF
Acrylic acid
δ-Amino levulinic acid
α-Angelica lactone
Levulinic acid
1,4-Pentanediol
4,4-Bis-(4-hydroxyphenyl) valeric acid
Levulinate esters
β-Acetylacrylic acid

Fig. 9.17 Derivatives of levulinic acid, LA. Reprinted with permission from [9.43]. Copyright (2006) American Chemical Society.

Acrylic polymers
Green solvent
α-methylene-γ-valerolactone
Methyl pentenoate
Dimethyl adipate
Butene
Pentenoic acid
γ-valerolactone
1,4 pentanediol
C8+alkanes (jet fuel)
Pentanoic acid
Gasoline blender
Methyl tetrahydrofuran (gasoline)
5-nonanone
C9 alkanes (Diesel fuel/gasoline)
C18-C27 alkanes (Diesel fuel)

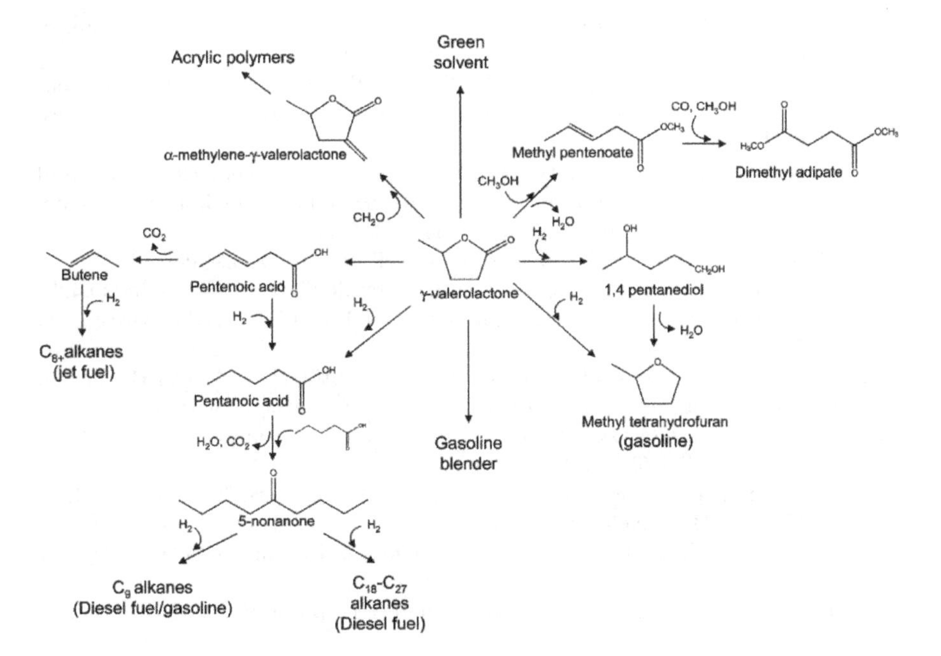

Fig. 9.18 Reaction pathways for the conversion of GVL to chemicals. Reprinted with permission from [9.48]. Copyright (2010) American Chemical Society.

References

9.1 T. Engel, G. Ertl, Elementary steps in the catalytic oxidation of carbon monoxide on platinum metals, *Adv. Catal.* 28, 1–77, 1979.

9.2 G. Ertl, P. Rau, Chemisorption und katalytische Reaktion von Sauerstoff und Kohlenmonoxid an einer Palladium (110)-Oberfläche, *Surf. Sci.* 15, 443–465, 1969.

9.3 W. H. Weinberg, R. P. Merrill, Crystal field surface orbital-bond energy bond order (CFSO-BEBO) calculations for surface reactions I. The reactions CO + O_2, NO + CO and H_2 + O_2 on a platinum (111) surface, *J. Catal.* 40, 268–280, 1975.

9.4 T. Engel, G. Ertl, A molecular beam investigation of the catalytic oxidation of CO on Pd (111), *J. Chem. Phys.* 69, 1267–1281, 1978.

9.5 R. Imbihl, G. Ertl, Oscillatory kinetics in heterogeneous catalysis, *Chem. Rev.* 95, 697–733, 1995.

9.6 P.-A. Carlsson, V. P. Zhdanov, B. Kasemo, Bistable mean-field kinetics of CO oxidation on Pt with oxide formation, *Appl. Surf. Sci.* 239, 424–431, 2005.

9.7 M. Kolodziejczyk, R. E. R. Colen, M. Berdau, B. Delmon, J. H. Block, CO oxidation on a copper-modified Pt(111) surface, *Surf. Sci.* 375, 235–249, 1997.

9.8 R. Imbihl, Nonlinear dynamics on catalytic surfaces: The contribution of surface science, *Surf. Sci.* 603, 1671–1679, 2009.

9.9 P.-A. Carlsson, V. P. Zhdanov, M. Skoglundh, Self-sustained kinetic oscillations in CO oxidation over silica-supported Pt, *Phys. Chem. Chem. Phys.* 8, 2703–2706, 2006.

9.10 B. L. M. Hendriksen, S. C. Bobaru, J. W. M. Frenken, Bistability and oscillations in CO oxidation studied with scanning tunnelling microscopy inside a reactor, *Catal. Today* 105, 234–243, 2005.

9.11 M. A. van Spronsen, G. J. C. Van Baarle, C. T. Herbschleb, J. W. M. Frenken, I. M. N. Groot, High-pressure operando STM studies giving insight in CO oxidation and NO reduction over Pt(1 1 0), *Catal. Today* 244, 85–95, 2015.

9.12 J. Singh, M. Nachtegaal, E. M. C. Alayon, J. Stötzel, J. A. van Bokhoven, Dynamic structure changes of a heterogeneous catalyst within a reactor: Oscillations in CO Oxidation over a Supported platinum catalyst, *ChemCatChem.* 2, 653–657, 2010.

9.13 S. B. Vendelbo, C. F. Elkjær, H. Falsig, I. Puspitasari, P. Dona, L. Mele, B. Morana, B. J. Nelissen, R. van Rijn, J. F. Creemer, P. J. Kooyman, S. Helveg, Visualisation of oscillatory behaviour of Pt nanoparticles catalysing CO oxidation, *Nature Mater.* 13, 884–891, 2014.

9.14 J. L. Williams, Monolith structures, materials, properties and uses, *Catal. Today* 69, 3–9, 2001.

9.15 G. Ertl, H. Knözinger, F. Schüth, J. Weitkamp, Eds., in *Handbook of Heterogeneous Catalysis.* Wiley-VCH, Weinheim, 2nd Edition, 2008.

9.16 P. Mars, D. W. van Krevelen, Oxidations carried out by means of vanadium oxide catalysts, *Chem. Eng. Sci. Special Suppl.* 3, 41–59, 1954.
9.17 P. Forzatti, Present status and perspectives in de-NOx SCR catalysis, *Appl. Catal. A* 222, 221–236, 2001.
9.18 D. W. Fickel, E. D'Addio, J. A. Lauterbach, R. F. Lobo, The ammonia selective catalytic reduction activity of copper-exchanged small-pore zeolites, *Appl. Catal. B* 102, 441–448, 2011.
9.19 L. Lietti, G. Ramis, F. Berti, G. Toledo, D. Robba, G. Busca, P. Forzatti, Chemical, structural and mechanistic aspects on NOx SCR over commercial and model oxide catalysts, *Catal. Today* 42, 101–116, 1998.
9.20 N. Y. Topsøe, J. H. Dumesic, H. Topsøe, Present status and perspectives in de-NOx SCR catalysis, *J. Catal.* 151, 241–252, 1995.
9.21 J. Li, S. Li, New insight into selective catalytic reduction of nitrogen oxides by ammonia over H-form zeolites: A theoretical study, *Phys. Chem. Chem. Phys.* 9, 3304–3311, 2007.
9.22 T. C. Brüggeman, F. J. Keil, Theoretical investigation of the mechanism of the selective catalytic reduction of nitrogen oxide with ammonia on Fe-form zeolites, *J. Phys. Chem. C* 115, 23854–23870, 2011.
9.23 M. Li, Y. Yeom, E. Weitz, W. M. H. Sachtler, An acid catalysed step in the catalytic reduction of NOx to N_2, *Catal. Lett.* 112, 129–132, 2006.
9.24 M. Lin, T. Hogan, A. Sen, A highly catalytic bimetallic system for the low-temperature selective oxidation of methane and lower alkanes with dioxygen as the oxidant, *J. Am. Chem. Soc.* 119, 6048–6053, 1997.
9.25 R. Periana, O. Mironov, D. Taube, G. Bhalla, C. Jones, Homogeneous, catalytic, oxidative coupling of methane to acetic acid in one step, *Top. Catal.* 32, 169–174, 2005.
9.26 M. Zerella, A. Kahros, A. T. Bell, Methane oxidation to acetic acid catalysed by Pd^{2+} cations in the presence of oxygen, *J. Catal.* 237, 111–117, 2006.
9.27 H.-J. Arpe, *Industrial Organic Chemistry*, 5th Edition, Wiley-VCH, Weinheim, 2010.
9.28 P. W. N. M. van Leeuwen, *Homogeneous Catalysis*, Kluwer, Dordrecht, 2004.
9.29 R. A. van Santen, H. P. C. E. Kuipers, The mechanism of ethylene epoxidation, *Adv. Catal.* 35, 265–321, 1987.
9.30 W. Diao, C. D. DiGiulio, M. T. Schaal, S. Ma, J. R. Monnier, An investigation on the role of Re as a promoter in Ag-Cs-Re/α-Al_2O_3 high-selectivity, ethylene epoxidation catalysts, *J. Catal.* 322, 14–23, 2014.
9.31 S. Linic, M. A. Barteau, Control of ethylene epoxidation selectivity by surface oxametallacycles, *J. Am. Chem. Soc.* 125, 4034–4035, 2003.
9.32 S. Linic, M. A. Barteau, On the mechanism of Cs promotion in ethylene epoxidation on Ag, *J. Am. Chem. Soc.* 126, 8086–8087, 2004.
9.33 P. J. van den Hoek, E. J. Baerends, R. A. van Santen, Ethylene epoxidation on Ag(110): The role of subsurface oxygen, *J. Phys. Chem.* 93, 6469–6475, 1989.
9.34 M. O. Özbek, I. Önal, R. A. van Santen, Ethylene epoxidation catalysed by silver oxide, *Chem Cat Chem* 3, 150–153, 2011.

9.35 www.shell.com/global/products-services.
9.36 H. A. Witcoff, B. G. Reuben, J. S. Plotkin, *Industrial Organic Chemicals*, 3rd Edition, Wiley, 2013.
9.37 J. K. F. Buijink, J. J. M. van Vlaanderen, M. Crocker, F. G. M. Niele, Propylene epoxidation over titanium-on-silica catalyst — The heart of the SMPO process, *Catal. Today* 93-95, 199–204, 2004.
9.38 F. Cavani, Catalytic selective oxidation: The forefront in the challenge for a more sustainable chemical industry, *Catal. Today* 157, 8–15, 2010.
9.39 C. Zhao, I. E. Wachs, Selective oxidation of propylene to acrolein over supported V_2O_5/Nb_2O_5 catalysts: An *in situ* Raman, IR, TPSR and kinetic study, *Catal. Today* 118, 332–343, 2006.
9.40 D. Cespi, F. Passarini, E. Neri, I. Vassura, L. Ciacci, F. Cavani, Life cycle assessment comparison of two ways for acrylonitrile production: The SOHIO process and an alternative route using propane, *J. Cleaner Prod.* 69, 17–25, 2014.
9.41 G. Mestl, J. L. Margitfalvi, L. Végvári, G. P. Szijjártó, A. Tompos, Combinatorial design and preparation of transition metal doped MoVTe catalysts for oxidation of propane to acrylic acid, *Appl. Catal. A* 474, 3–9, 2014.
9.42 R. K. Grasselli, C. G. Lugmair, A. F. Volpe Jr., Towards an understanding of the reaction pathways in propane ammoxidation based on the distribution of elements at the active centers of the M1 phase of the MoV(Nb,Ta)TeO system, *Top. Catal.* 54, 595–604, 2011.
9.43 B. Girisuta, L. P. B. M. Janssen, H. J. Heeres, Green chemicals. A kinetic study on the conversion of glucose to levulinic acid, *Chem. Eng. Res. Des.* 84, 339–349, 2006.
9.44 P. Gallezot, Direct routes from biomass to end-products, *Catal. Today* 167, 31–36, 2011.
9.45 Y. Shen, Y. Xu, J. Sun, B. Wang, F. Xu, R. Sun, Efficient conversion of monosaccharides into 5-hydroxymethylfurfural and levulinic acid in $InCl_3$–H_2O medium, *Catal. Commun.* 50, 17–20, 2014.
9.46 M. J. Climent, A. Corma, S. Iborra, Converting carbohydrates to bulk chemicals and fine chemicals over heterogeneous catalysts, *Green Chem.* 13, 520–540, 2011.
9.47 R. A. Sheldon, Green chemistry, catalysis and valorization of waste biomass, *J. Mol. Catal. AChem.* 422, 3–12, 2016.
9.48 J. Q. Bond, D. M. Alonso, R. M. West, J. A. Dumesic, γ-Valerolactone ring-opening and decarboxylation over SiO_2/Al_2O_3 in the presence of water, *Langmuir* 26, 16291–16298, 2010.

Questions

9.1 Give the elementary steps of an Eley–Rideal mechanism for the oxidation of CO.

What is the surface coverage of CO under these conditions? Is it possible to have oscillations of the CO oxidation in an ER mechanism?

9.2 Should Mars–van Krevelen-type oxidation catalysis be carried out in one or two reactors?

9.3 Why is it possible to prove a Mars–van Krevelen oxidation mechanism with isotope labelling only by analysing the initial products?

9.4 How do the following parameters influence the conversion of SO_2 to SO_3?

   (a) the O_2/SO_2 ratio,
   (b) the temperature,
   (c) the pressure,
   (d) the catalyst.

9.5 According to the amide-nitrosamide explanation of the SCR reaction, the amide species reacts with gas-phase NO. Is this a Langmuir–Hinshelwood or an Eley–Rideal mechanism? What does this suggest about the reliability of this mechanism?

9.6 Why does WO_3 increase the acidity of the V_2O_5/TiO_2 SCR catalyst?

9.7 What information do we gain from carrying out the Wacker reaction between ethene and labelled oxygen?

9.8 The use of acetaldehyde has shrunk because the preferred routes for making products has changed for both acetic acid and butanol. In the past butanol was made by aldol condensation of acetaldehyde, whereas today the preferred route is through propene. Write the chemical equations for these two routes.

9.9 Give the components present in the EO catalyst. What role does each component play?

9.10 What are the EO process conditions and why?

9.11 Write the chemical equations of the formation of 1,2-dichloropropane, 2,2′-dichloro-diisopropylether and propylene glycol, which are byproducts in the production of PO in the chlorohydrin process.

9.12 In the past benzene rather than butane was partially oxidised to maleic anhydride. What are the disadvantages of benzene?

9.13 Which reactor is used in the production of maleic anhydride from butane in industry?

Chapter 10

Electrocatalysis

10.1 Introduction

In electrocatalysis a substance (electrocatalyst) accelerates an electrochemical reaction while it remains unaltered. Whereas in heterogenous catalysis only molecules are involved in the reaction, in electrochemical reactions **molecules, electrons and ions** play a role [10.1]. Electrochemical reactions are either driven by electricity or produce electricity. They can, thus, play a crucial role in the transition towards sustainable energy and, for example, tackle the intermittency of renewable energy sources via the conversion of electrical energy to chemical energy and vice versa [10.2]. As with all chemical reactions, electrochemical reactions, too, obey the laws of thermodynamics. For instance, water splitting is a non-spontaneous (endergonic) reaction at 25 °C since its standard Gibbs free energy (ΔG) is positive ($+223\,\mathrm{kJ{\cdot}mol^{-1}}$). Endergonic reactions require input of energy to proceed and, thus, to drive this reaction electrochemically, electricity should be supplied to the system. Likewise, the opposite reaction (water formation from hydrogen and oxygen) is spontaneous (exergonic) at 25 °C, since $\Delta G < 0$, and, thus the reaction can produce electricity.

An electrocatalyst is either part of an electrode or it is the electrode; its role is to decrease the activation energy barrier of an electrochemical reaction. The events that take place during electrochemical reactions include adsorption/desorption of species and bond breakage/formation, as in heterogeneous catalysis, but in addition there is a transfer of charge (ions and electrons). A unique feature of electrochemical reactions is that they take place by two spatially separated half-reactions. For instance, the overall reaction of water splitting to hydrogen and oxygen ($H_2O \rightarrow H_2 + 1/2\,O_2$) in the presence of proton (H^+)-conducting media takes place by (i) $H_2O \rightarrow 1/2\ O_2 + 2\,H^+ + 2\,e^-$ (oxygen evolution reaction (OER)) and (ii) $2\,H^+ + 2\,e^- \rightarrow H_2$ (hydrogen evolution reaction, HER).

The first electrocatalytic devices were developed at the beginning of the 19th century, typical examples being the first battery, i.e. the Voltaic pile of Alessandro Volta and the first H_2/O_2 fuel cell, i.e. the gas voltaic battery of Sir William R. Grove. The core of theoretical electrocatalysis was established a century later and published in the revolutionary works of Tafel, Butler and Volmer [10.3, 10.4, 10.5]. The Tafel equation was introduced in the early 1900s [10.3] primarily to characterise the activity of various metals towards HER and it is still the main criterion for electrocatalyst characterisation. A few decades later [10.4], Butler's development of kinetic theory at equilibrium and the work of Erdey–Grúz and Volmer [10.5] on identifying the current-potential relationship led to the development of the Butler–Volmer equation, establishing the basis of electrode kinetics.

Despite the similarities between electrochemistry and heterogeneous catalysis, i.e. similar isotherms are used to describe adsorption, these fields were studied separately for decades. The term electrocatalysis was introduced in the German language by Kobosev and Monblanova in 1936 [10.6], but the close interconnection of the two fields was recognised internationally only after the work of Grubb was published in the early 1960s [10.7]. Today electrocatalysis finds several applications, such as in the conversion and storage of energy, organic electrosynthesis, corrosion science, electroanalytical sensors and wastewater treatment. Electrochemical processes that play a significant role in industrial manufacturing of chemicals and materials are the chloralkali and the Hall–Héroult processes [10.8]. In the chloralkali process, the electrolysis of sodium chloride takes place for the simultaneous production of chlorine and sodium hydroxide (caustic soda), two compounds, which are the precursors in the manufacture of a wide range of speciality chemicals. In the Hall–Héroult process, which is used in the metallurgy industrial sector, alumina (Al_2O_3, typically obtained from bauxite, cf. Section 2.3.1) is reduced to aluminium, while carbon (coke) is simultaneously oxidised.

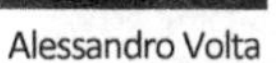

10.2 Fundamental Aspects [10.1, 10.9, 10.10]

10.2.1 *Electrochemical Cells*

An electrochemical reaction is accompanied by the passage of an electric current, thus involving the transfer of electrons. The overall electrochemical reaction consists of two half-reactions: an **oxidation** reaction (i.e. accompanied by release of electrons) and a **reduction** reaction (i.e. consumption of electrons). Typical examples of electrochemical reactions are:

$$H_2 \rightarrow 2\,H^+ + 2\,e^- \quad \text{(oxidation half-reaction)}$$

$$1/2\,O_2 + 2\,H^+ + 2\,e^- \rightarrow H_2O \quad \text{(reduction half-reaction).}$$

Electrochemical reactions take place in electrochemical cells (Fig. 10.1), devices that consist of two electrodes (anode and cathode) and an **electrolyte**. An electrolyte is a substance that can conduct ions. The four main types are electrolytic solutions, polymer electrolytes, solid oxides and molten salts. Electrochemical reactions occur at the electrode/electrolyte interface, with oxidation reactions taking place at the anode and reduction reactions at the cathode. When the overall reaction in an electrochemical cell is spontaneous ($\Delta G < 0$), chemical energy can be converted to electrical energy and the electrochemical cell operates as a galvanic cell (electricity is produced). In essence, any galvanic cell can be used as a battery. In the opposite situation, electrolytic cells use electricity to drive a non-spontaneous electrochemical cell reaction ($\Delta G > 0$). During charging, rechargeable batteries operate as electrolytic cells.

Fuel cells and water electrolysers operate similarly as galvanic and electrolytic cells, respectively, but their operation requires the constant supply of one or more reactants. Figure 10.1 is a schematic representation of their operation when a proton exchange membrane (PEM) electrolyte

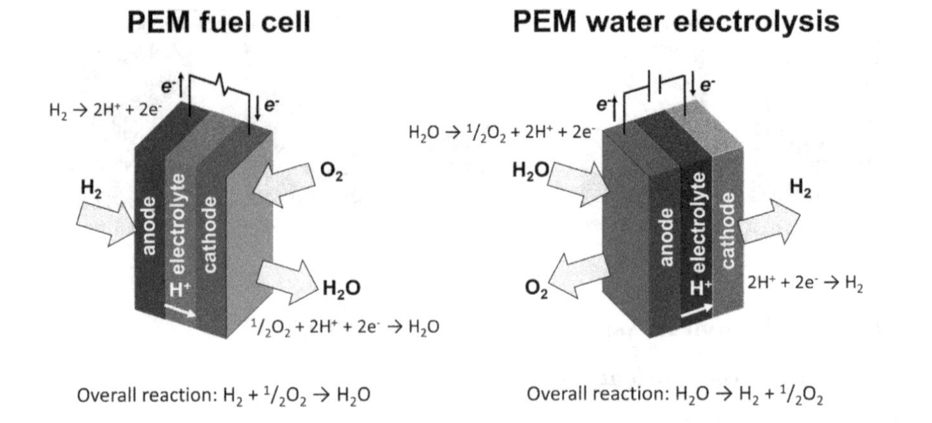

Fig. 10.1 Operation principles of fuel cells and water electrolysers with a PEM electrolyte.

Table 10.1 Main types of electrolytes used for fuel cells and water electrolysers.

Cell type	Electrolyte	Mobile ion	Operating temperature
Alkaline	KOH solution	OH^-	20–90 °C
Proton exchange membrane (PEM)	Polymeric membrane	H^+	20–80 °C
Solid oxide electrochemical cells (SOEC)	Ion conducting ceramic	O^{2-}	800–1,000 °C

(with H^+ conductivity) is used. In fuel cells, hydrogen oxidation takes place at the anode. The produced protons travel through the polymer electrolyte to the cathode, while the produced electrons flow in an external electrical circuit to produce electricity. At the cathode, H^+, electrons and oxygen react and produce water. The reverse operation takes place in water electrolysers, where electrical energy is supplied in order to split water into oxygen (released from the anode) and hydrogen (released from the cathode). There are different types of fuel cells and water electrolysers; their primary classification is based on the type of electrolyte (Table 10.1), which determines the operating temperature, the kind of electrochemical half-reactions and the required electrocatalysts.

The **rate** of electrochemical reactions is represented by the current density (i.e. current per unit area, typically expressed in $A \cdot cm^{-2}$) which flows out of an electrochemical cell, according to **Faraday's law**:

$$r = i/n\mathrm{F}, \tag{10.1}$$

where r is the reaction rate (in $mol \cdot s^{-1} \cdot cm^{-2}$), i is the current density, n is the number of electrons taking part in the reaction and F is the Faraday's constant (96,485 $C \cdot mol^{-1}$). It is worth noting that the conduction at the electrodes and through the external electrical circuit is electronic, while that in the electrolyte is ionic. Therefore, a constant direct current can pass through the electrolyte only when the electrochemical reactions take place at the boundaries between the electronic and ionic components of the circuit.

10.2.2 *Cell and Electrode Potentials*

The driving force of the electron transfer between anode and cathode is

$$E_{\text{cell}} = E_{\text{cathode}} - E_{\text{anode}}, \tag{10.2}$$

where E_{cathode} and E_{anode} are the absolute potentials of the cathode and the anode, respectively. E_{cell} is a measurable parameter (in Volts) that represents the potential difference between the anode and the cathode. The absolute electrode potentials, on the other hand, cannot be measured because the transfer of electrons requires the presence of both an electron donor and an electron acceptor. Electrode potentials have been tabulated based on the standard hydrogen electrode (SHE), the potential of which is assigned to zero. Typical in thermodynamics, the superscript "o" (e.g. E^o) in cell and electrode potentials denotes standard conditions (1 bar, 1 M, 25 °C). Table 10.2 gives the standard cell potentials (E^o_{cell}) of fuel cells and water electrolysers. E^o_{cell} depends only on the overall cell reaction, irrespective of the kind of electrolyte. The cell potential is related to the change in Gibbs free energy according to the equation:

$$\Delta G = -n\mathrm{F}\mathrm{E}_{\text{cell}}, \tag{10.3}$$

where n is the number of electrons participating in the reaction and F is Faraday's constant. Eq. (10.3) is valid both under standard and non-standard conditions. At constant temperature and pressure a chemical reaction occurs spontaneously when $\Delta G < 0$. Therefore, under standard conditions, an electrochemical reaction is spontaneous when $E^o_{\text{cell}} > 0$, while

Table 10.2 Standard potentials for fuel cells and water electrolysers with acidic and alkaline electrolytes.

Type of electrolyte	Half-reaction	E^o/V	E^ocell/V	E^ocell/V water electrolysis
H^+ conducting (i.e. acidic solution)	$2\,H^+ + 2\,e^- \leftrightarrow H_2$ (g)	0.00	$1.23 - 0.00 = 1.23$	$-1.23 = -1.23$
	$1/2\,O_2$ (g) $+ 2\,H^+ + 2\,e^- \leftrightarrow H_2O$	+1.23		
OH^- conducting (i.e. alkaline solution)	$2\,H_2O + 2\,e^- \leftrightarrow 2\,H_2$ (g) $+ 2\,OH^-$	−0.83	$0.40 + 0.83 = 1.23$	$-0.83 - 0.4 = -1.23$
	$1/2\,O_2$ (g) $+H_2O + 2\,e^- \leftrightarrow 2\,OH^-$	+0.40		

in the opposite case, energy is required for the reaction to take place when $E^o_{\text{cell}} < 0$.

10.2.3 *The Nernst Equation*

The Gibbs free energy under non-standard conditions is

$$\Delta G = \Delta G^o + RT \ln Q_{\text{R}}, \tag{10.4}$$

where R is the ideal gas constant ($8.314\,\text{J}\cdot\text{mol}^{-1}\cdot\text{K}^{-1}$), T is the temperature and Q_R is the reaction quotient (i.e. the mathematical product of the concentration of the products of the reaction divided by the mathematical product of the concentration of reactants). Combining Eqs. (10.3) and (10.4), E_{cell} is

$$E_{\text{cell}} = E^o_{\text{cell}} - (RT/n\text{F})\ln\, Q_R. \tag{10.5}$$

Equation (10.5) is the Nernst equation and is one of the most important relationships in electrochemistry. This equation enables the calculation of E_{cell} for all types of electrochemical cells under any set of operational conditions. At 25 °C, the Nernst equation is used in its simplified form:

$$E_{\text{cell}} = E^o_{\text{cell}} - (0.0592/n)\log_{10} Q_R. \tag{10.6}$$

At equilibrium, ΔG equals zero, the reaction quotient equals the equilibrium constant of the reaction ($Q_{\text{R}} = K_{\text{eq}}$) and there is no net charge transfer (thus $E_{\text{cell}} = 0$). It follows that:

$$E^o_{\text{cell}} = (RT/\text{nF})\ln K_{\text{eq}}. \tag{10.7}$$

Equation (10.7) indicates that for all electrochemical reactions, K_{eq} can be calculated based on the tables of standard electrode potentials.

10.2.4 *Overpotential*

Thus far, we have explained how, under standard conditions, thermodynamics predicts whether or not electrochemical reactions are spontaneous. Following is an explanation of the processes that control the rate of electrochemical reactions. In heterogeneous catalysis, when a closed system is at equilibrium no net reaction takes place at the catalyst surface, since the forward and backward reactions take place at equal rates. Similarly, when an electrochemical cell is at equilibrium (i.e. open circuit), forward and backward reactions take place at equal rates and the overall reaction rate and, thus, the current (Eq. (10.1)) is zero. While moving beyond equilibrium, the net reaction rate is different to zero, since the reaction proceeds in one direction and current is flowing through the electrochemical cell. In this case, the potential is different than the equilibrium potential (reversible potential). This potential difference is called overpotential and is typically represented by the symbol η. Overpotential can be experimentally quantified by measuring the cell potential at a given current density. A good electrocatalyst shows a high current density at a low overpotential.

Figure 10.2 shows a typical potential-current plot (polarisation plot) of a hydrogen fuel cell operating at 25 °C. The cell overpotential consists of three main components: **concentration** (η_{conc}), **ohmic** (η_{ohm}) and **activation overpotential** (η_{act}). In short, η_{conc} is caused by slow mass transport to/from the electrode, η_{ohm} by slow electronic/ionic conduction in the cell and η_{act} by slow charge transfer at the electrode/electrolyte interface. Concentration overpotentials can be minimised by tuning the morphological characteristics of electrodes (i.e. porosity, catalyst particle size) or by tuning the operation conditions (i.e. supply rate of reactants, stirring of the electrolyte solution). Ohmic overpotentials can be lowered by improving electrolyte conductivity (i.e. tuning the pH of aqueous electrolytes or minimising the thickness of the solid oxide and polymer electrolytes). Activation overpotentials can be optimised by correct selection of electrocatalysts and by the appropriate design of the electrode/electrolyte interface.

Another type of overpotential may also be present at currents close to zero due to the crossover (permeation) of the reactant or product species through the electrolyte. For instance, hydrogen can permeate the polymer electrolyte membrane (hydrogen crossover), depending on the nature and thickness of the membrane and the hydrogen concentration at the anode feed. Permeated hydrogen reaches the cathode, which is exposed to air, and a mixed potential is established. This, in turn, leads to open-circuit potentials (cell potential at zero current), which are lower than

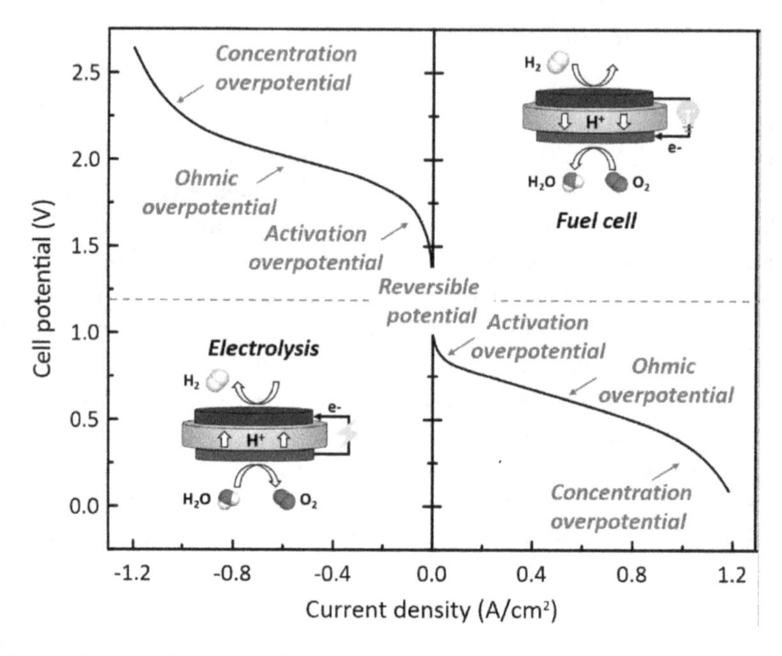

Fig. 10.2 Typical polarisation plots (current–potential dependence) of a PEM hydrogen fuel cell and a water electrolyser, where the different types of overpotential are indicated. Cell potentials on the vertical axis appear as absolute values.

the reversible cell potential under the given conditions. Experimentally observed open-circuit potentials of H_2/O_2 fuel cells are typically close to 1.0 V (Fig. 10.2), depending on the operating conditions.

The concentration overpotential dominates at high current densities and contains the contribution of both anode and cathode. It occurs when mass transport is slow. Similar to thermal catalysis, mass transport limitations are the result of slow diffusion of reactive species (Section 4.2). When the transport of reactants to the electrode surface is slow, their surface concentration decreases as the current increases and the catalyst surface is not fully used (similar to the situation presented in Fig. 4.3, Section 4.2). Once the reactants are depleted, the reaction is limited by mass transport and the current density reaches a critical value (limiting current density, i_L), which is independent of potential. The limiting current density ($A{\cdot}cm^{-2}$) depends linearly on the bulk concentration of electroactive species, C^0 ($mol{\cdot}cm^{-3}$), according to the equation:

$$i_L = n\mathrm{FD}(\mathrm{C}^0/\delta_N), \tag{10.8}$$

where D is the diffusion coefficient of the species ($cm^2{\cdot}s^{-1}$) and δ_N is the Nernst diffusion layer thickness (cm). According to Eq. (10.8), i_L largely depends on the experimental conditions, since, for instance, D increases with temperature and δ_N decreases when the electrolyte solution is stirred. Under the conditions where charge transfer happens fast, the concentration overpotential can be expressed as:

$$\eta_{\mathrm{conc}} = (RT/n\mathrm{F})\ln(1 - i/i_L). \tag{10.9}$$

Measurements of η_{conc} can reveal transport resistances in electrochemical cells and are important for the optimisation of the electrode and cell design. Bubble overpotential is a specific form of concentration overpotential, which is present in the specific case of gas evolving electrodes in aqueous electrolytes (e.g. in alkaline water electrolysis). It is related to the decrease in the effective area of the electrode due to the formation of bubbles from the evolved gases. Bubble overpotential can be eliminated by applying appropriate coatings to the electrode, by adding surfactants to the electrolyte solution or by stirring as discussed in Section 4.2.

The ohmic cell overpotential is predominant at medium current densities. It is related mainly to the ohmic resistance of the electrolyte, but it also includes the contribution of the resistance of the anode, cathode and electrode–electrolyte interface. Ohmic overpotential can be determined experimentally by electrochemical impedance spectroscopy (EIS) (Section 10.3.4).

10.2.5 *Electrode Kinetics*

At equilibrium, the forward (oxidation) and backward (reduction) reactions occur simultaneously at the electrode and at equal rates. The two partial currents are, thus, numerically equal, and both have the magnitude of the exchange current density. The exchange current density (i_o) is a kinetic parameter and depends on the kind of electrode, on the type of the electrochemical reaction and on the operating conditions.

When the electrochemical reaction is not in equilibrium, the forward and backward reactions both occur at the electrode but at different rates. For single-step electrochemical reactions ($\mathrm{R} \leftrightarrow \mathrm{O} + \mathrm{e}^-$) the forward and backward reaction rates are

$$i = i_{\mathrm{o,forward}} \exp(\beta\mathrm{F}\eta/RT), \tag{10.10a}$$

$$i = i_{\mathrm{o,backward}} \exp[-(1-\beta)\mathrm{F}\eta/RT]. \tag{10.10b}$$

Here, i_o is the exchange current density and β is the **symmetry factor**, which is a dimensionless kinetic parameter. The symmetry factor indicates the extent that the polarisation affects the forward and backward reactions at an electrode. If the symmetry factor is β for the forward (oxidation) reaction, then it must be $(1 - \beta)$ for the backward (reduction) reaction. Typically, the symmetry factor values are between 0.3 and 0.7, while it can never exceed unity.

Based on Eqs. (10.10a) and (10.10b), the net rate of a single-step electrochemical reaction, the difference between the forward and backward current densities, is given by:

$$i = i_o[\exp(\beta \mathrm{F}\eta/RT) - \exp(-(1-\beta)\mathrm{F}\eta/RT)]. \tag{10.11a}$$

Equation (10.11a) is known as the **Butler–Volmer equation** and is valid only for activation-controlled reactions. The symmetry factor is a fundamental parameter in electrode kinetics and is always related to a single one-electron step electrochemical reaction. However, electrochemical reactions are usually multi-electron reactions, which consist of several elementary steps (see Section 10.4). To describe a multistep reaction (e.g. $O_2 + 4H^+ + 4e^- \leftrightarrow 2H_2O$) the Butler–Volmer equation is re-written by replacing the symmetry factor by the **charge transfer coefficient**, α:

$$i = i_o[\exp(\alpha_{\mathrm{anodic}}\mathrm{F}\eta/RT) - \exp(-\alpha_{\mathrm{cathodic}}\mathrm{F}\eta/RT)]. \tag{10.11b}$$

where α_{anodic} and $\alpha_{\mathrm{cathodic}}$ correspond to the charge transfer coefficients of the anodic (forward) and cathodic (backward) reactions, respectively. The charge transfer coefficient is considered as phenomenological coefficient. It is a dimensionless experimental parameter, which derives from the current-potential relationship of multistep reactions. It includes parameters related to the number of electrons which precedes the rate-determining step and the number of times that this step should be repeated so that the overall reaction occurs once. Unlike the symmetry factor, α_{anodic} and $\alpha_{\mathrm{cathodic}}$ do not necessarily sum up to unity.

The Butler–Volmer equation has two limiting cases.

(i) When the anodic overpotential is high, $\eta > RT/F$ (but still not limited by mass transport), the reaction is far from equilibrium and the backward reaction rate can be neglected compared to the forward rate at the electrode. The total current in this case can be approximated as

$$i = i_o \exp(\alpha_{\mathrm{anodic}}\mathrm{F}\eta/RT). \tag{10.12a}$$

(ii) When the cathodic overpotential is high, $\eta > RT/F$, the reaction is far from equilibrium and the forward rate can be ignored, while the total current can be approximated as

$$i = i_o \exp(-\alpha_{\text{cathodic}} \text{F}\eta/RT). \tag{10.12b}$$

Equation (10.12) can be re-written in the logarithmic form as

$$2.3 \log(i/i_o) = (\alpha_{\text{anodic}} \text{F}/\text{RT})\eta, \tag{10.13a}$$

$$2.3 \log(i/i_o) = (-\alpha_{\text{cathodic}} \text{F}/\text{RT})\eta. \tag{10.13b}$$

According to these equations, plots of the overpotential, η, versus the $\log(i)$ yield a straight line (Fig. 10.3); these so-called **Tafel plots** are very useful for evaluating kinetic parameters. The Tafel slope (measured in mV per decade of current density and reported as an absolute value) equals $(2.3\, RT/\alpha F)$. In general, high Tafel slopes indicate slow electrode reaction kinetics. Extrapolation of the linear segments of Tafel plots to the intercept with the x axis allows the calculation of the exchange current density, as shown in Fig. 10.3.

The Butler–Volmer equation (Eq. (10.11a)) can also be found in literature written as

$$i = i_o[\exp(\beta \text{nF}\eta/RT) - \exp(-(1-\beta)\text{nF}\eta/RT)],$$

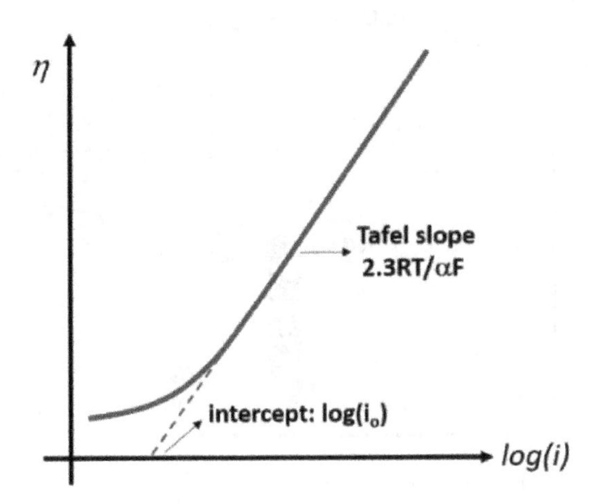

Fig. 10.3 Anodic branch of a typical Tafel plot.

where n is the number of electrons participating in the reaction. However, this equation is valid only when all n electrons are transferred simultaneously during the rate-determining step, which in reality is unlikely to happen.

10.3 Experimental Methods and Techniques

10.3.1 *Three-Electrode Cell Configuration* [10.11, 10.12]

A three-electrode cell is used to screen the performance of electrocatalysts towards specific half-reactions (Fig. 10.4). In this configuration, three electrodes are connected to two different electric circuits. The working electrode is the one where the half-reaction under investigation takes place. Current passes through the working and counter electrodes (first electric circuit). The counter electrode should, therefore, be highly non-polarisable to ensure that potential changes are much smaller than those at the working electrode. The working and reference electrodes are connected to the second circuit, through which no current flows. While current passes through the first circuit, the changes in the potential of the working electrode are measured against the reference electrode, the potential of which can be considered to be constant. Therefore, the measured changes in the working-reference potential difference correspond to changes in the working electrode

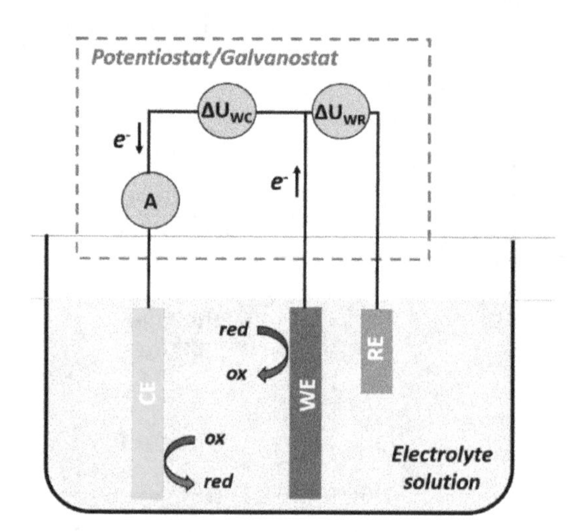

Fig. 10.4 Three-electrode cell configuration: Current passes through the working and counter electrodes, leading to a potential difference ΔU_{WC}, while the change in the potential of the working electrode is measured against a reference electrode (ΔU_{WR}).

potential. This procedure corresponds to the case of measuring changes in potential, while controlling the current, and is a galvanostatic measurement. However, the current–potential relationship can also be determined by measuring the changes in the current while controlling the potential, referred to as a potentiostatic measurement. Both types of measurement are used regularly and typically lead to the same result.

Although in potentiostatic and galvanostatic measurements the potential is applied and measured between the working and reference electrodes, the potential between working and counter electrodes is the one that drives the reaction. In practice, the working-reference potential difference includes some part of the potential drop in the electrolyte solution between these electrodes. This drop is the **residual potential drop** and is influenced by the electrolyte's conductivity, the distance between the working and reference electrodes, and the magnitude of the current. When the residual potential drop is large, it can lead to significant shifts in the polarisation plot and, thus, to misinterpretation of the polarisation data. Subtracting the potential drop from experimental measurements is of paramount importance, especially when current densities are high or when an electrolyte with low conductivity is used. To determine the ohmic drop, the related ohmic resistance should be measured. This is typically carried out by EIS (Section 10.3.4), but the current interruption technique can also be used.

10.3.2 *Techniques for Electrocatalyst Evaluation*

As in thermal catalysis, a complete evaluation of electrocatalytic performance includes the investigation of activity, selectivity and stability (Table 10.3). The electrocatalytic activity is typically assessed by voltametric techniques where the current is recorded as a function of applied potential. Stability is typically assessed by recording the potential at constant current or the current at constant applied potential as a function of polarisation time. Consecutive voltammetric scans can also be used for stability investigations, which give information about degradation mechanisms at specific potentials. To gain information about reaction kinetics, hydrodynamic techniques (i.e. rotating disc electrode, rotating ring disc electrode) are employed. Finally, EIS is a powerful technique, both for mechanistic investigations, since it separates the contribution of various processes that occur simultaneously, and for investigation of stability (alteration at electrode–electrolyte interfacial properties against time). Sections 10.3.3–10.3.5 describe the most common electrochemical techniques.

Table 10.3 Electrochemical, analytical and microscopy-based techniques in electrocatalysis.

Activity		Selectivity	
Technique	**Features**	**Technique**	**Features**
Voltammetry (i.e. CV, LSV)	Preliminary activity assessment	Hydrodynamic techniques (e.g. RDE, RRDE)	Detection of reaction intermediates
Hydrodynamic (i.e. RDE, RRDE)	Activity assessment at controllable mass transport. Kinetics investigation	Chromatography (e.g. GC, HPLC) and spectrometry (e.g. MS)	Quantification of reaction products and intermediates
Electrochemical impedance spectroscopy (EIS)	Separates the contribution of interfacial processes which occur simultaneously		

Stability	
Technique	**Features**
Amperometry/Potentiometry	Stability assessment
Voltammetry	Identification of degradation mechanisms related with electrochemical properties (i.e. changes occurring at specific potentials)
Electrochemical impedance spectroscopy	Changes in electrode/electrolyte interface over time
Microscopy (i.e. SEM, TEM)	Changes in morphology, particle size and distribution
Spectroscopic techniques (i.e. XPS, XRD, XAS etc.)	Changes in structure, electronic properties, etc.

Electrochemical methods cannot quantify the intermediate or final products of the reaction. Therefore, they are complemented by analytical and spectroscopic techniques, which provide mechanistic information [10.13]. Moreover, for specific electrocatalytic systems (e.g. CO_2 electroreduction, where multiple products can be formed), quantification of the products is essential for determining the product selectivity. Finally, a range of *ex situ* microscopic and spectroscopic techniques can be used to investigate degradation phenomena on electrocatalysts.

10.3.3 *Linear Sweep Voltammetry and Cyclic Voltammetry*

Linear sweep voltammetry is one of the most common techniques used in electrocatalysis [10.14, 10.15]. The potential is varied linearly with time between two values, while the response of the resulting current is followed. In an anodic sweep or forward scan the potential is swept from negative to positive values and in a cathodic sweep or backward scan the potential is swept in the opposite direction. When the sweep rate is low (i.e. $< 5\,\mathrm{mV{\cdot}s^{-1}}$) the measurement is conducted at quasi-steady-state conditions and the current-potential relationship is the same when the direction of the potential sweep is reversed. Cyclic voltammetry (CV) is the technique of subsequent anodic and cathodic sweeps.

An instructive example of CV is the case of a polycrystalline Pt film (Fig. 10.5). HER occurs at the extreme left and OER at the extreme right. Between the two extremes, there are three main regions. In the hydrogen region, adsorption/desorption of atomic hydrogen on/from the Pt surface occurs ($H^+ + e^- \leftrightarrow H_{ads}$), leading to two pairs of anodic and cathodic peaks in the CV. The presence of two symmetric pairs indicates the presence of strongly and weakly bonded hydrogen, the adsorption/desorption of which is electrochemically reversible.

In view of electroneutrality, as the electrode becomes further charged the solution side of the double layer is also charged with an equivalent charge

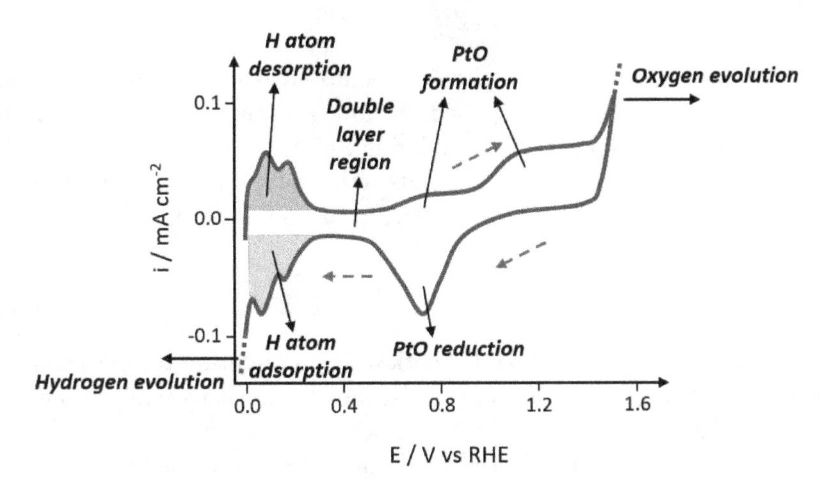

Fig. 10.5 Cyclic voltammogram of polycrystalline Pt in 0.5 M H_2SO_4 ($0.05\,\mathrm{V{\cdot}s^{-1}}$, 25 °C).

of the opposite sign. This charging is a non-Faradaic process (no charge transfer) and causes the **double layer current** (or capacitance current):

$$i_{\mathrm{dl}} = C_{\mathrm{dl}} v_{\mathrm{s}}, \tag{10.14}$$

where C_{dl} is the capacitance of the double layer and v_s is the sweep rate. When analysing CV measurements, i_{dl} should be subtracted from the measured current. According to Eq. (10.14), i_{dl} becomes more important at high sweep rates or on highly porous electrodes where C_{dl} is high.

In the oxygen region, wide anodic shoulders correspond to platinum oxidation to hydroxides/oxides. In the backward scan, platinum oxides are reduced, and a sharp cathodic peak is observed. The asymmetry between anodic and cathodic branches, and thus the irreversibility of oxide formation/reduction, can be lowered by reducing the upper potential limit. This indicates that easily reducible species form earlier during the anodic sweep.

10.3.4 *Electrochemical Impedance Spectroscopy*

EIS [10.16, 10.17] is an alternating current technique for characterising the time response of electrochemical systems over a range of frequencies. The total impedance, Z, consists of a real part (Z_R) and an imaginary part (Z_{im}):

$$Z = (Z_R^2 + Z_{im}^2)^{1/2}. \tag{10.15}$$

During EIS, the working electrode is polarised at V_{o} and a probing signal $V(t) = V_{\mathrm{o}} + V_{\mathrm{m}} \sin(2\pi \mathrm{ft})$ is applied, where V_{m} is the amplitude (typically 5–25 mV at room temperature) and f is the frequency (in the mHz–kHz range). The response (Fig. 10.6a) leads to a sinusoidal current with phase shift θ, $I(t) = I_{\mathrm{m}} \sin(2\pi \mathrm{ft} + \theta)$, with real and imaginary parts of the impedance:

$$Z_R = |Z| \cos\theta; \quad Z_{im} = |Z| \sin\theta. \tag{10.16}$$

EIS provides quantitative information by representing the electrochemical system (black box in Fig. 10.6a) by an equivalent electrical circuit. A typical half-cell contains three main impedance contributions, represented schematically in Fig. 10.6b as the impedance of the electrolyte (Z_{el}), of the electrode/electrolyte interface (Z_{int}) and of the electrochemical reaction (Z_{F}). The most frequently used circuit to describe such half-cells is the modified Randles equivalent circuit (Fig. 10.6c), which represents the

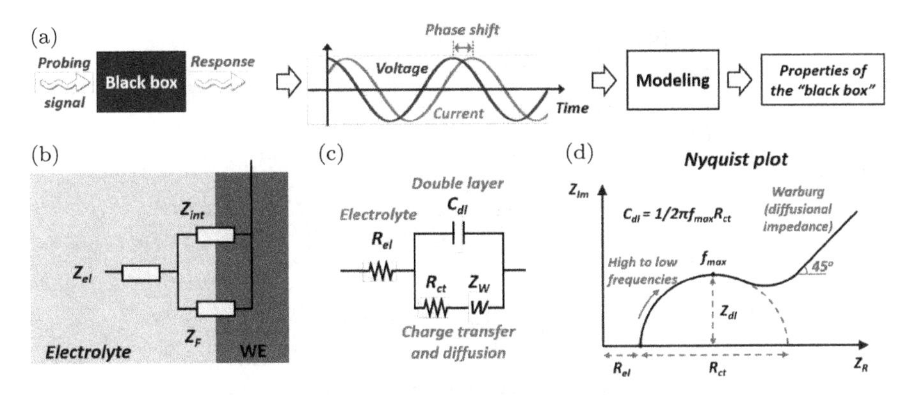

Fig. 10.6 (a) EIS investigates an electrochemical system by following the response to a probing signal and modelling the obtained impedance data, (b) the impedance contributions of a typical half-cell, (c) the modified Randles circuit, (d) Nyquist plot and extracted quantitative information. The figure is adapted from [10.17].

electrolyte by a resistor, the electrode/electrolyte interface by an ideal capacitor and the electrochemical reaction by a resistor (characterising the charge transfer) and by the so-called Warburg impedance (Z_{w}, characterising diffusion limitations). EIS data are usually presented as Nyquist plots. These typically show a semicircle at high frequencies, from which the charge-transfer resistance (R_{ct}), electrolyte resistance (R_{el}) and double layer capacitance (C_{dl}) are obtained. The phase shift by Z_{w} is independent of frequency and consequently a straight line with a slope of unity appears at the low-frequency end.

10.3.5 *Rotating Disc Electrode*

The rotating disc electrode is the most common technique for eliminating the influence of mass transport limitations and is, therefore, a very useful tool for studying the intrinsic kinetics of electrochemical reactions [10.9, 10.15, 10.18]. It consists of a disc (e.g. glassy carbon), which is mirror-polished on one side. The disc is placed in an insulating and chemically stable shroud (e.g. Teflon®) and is inserted into the electrochemical cell, so that only the polished surface is exposed to the electrolyte solution. The rotating disc electrode rotates around its vertical axis and a well-defined flow pattern is established (Fig. 10.7), while different rotation speeds yield different fluxes of the electrolyte across the electrode's surface. This forced convection leads to a decrease in the diffusion layer thickness, thus affecting the limiting current (Fig. 10.7b).

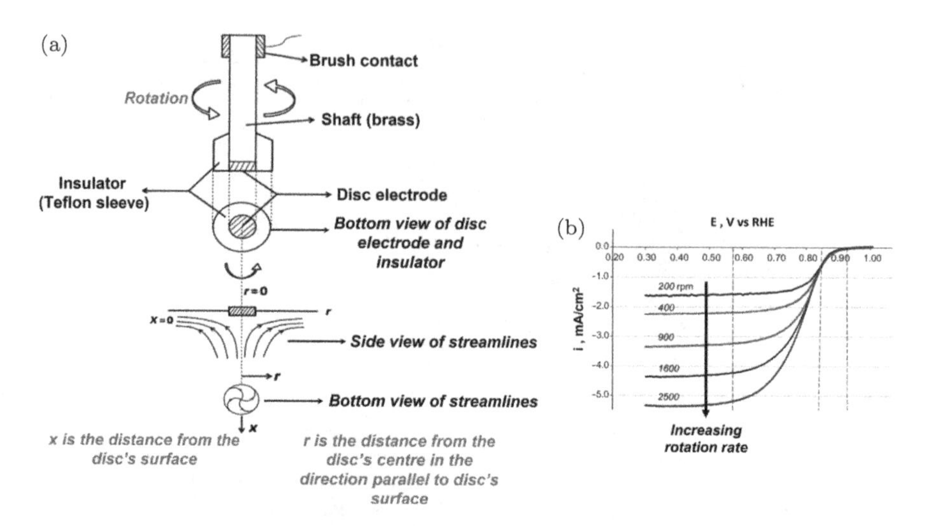

Fig. 10.7 (a) side and bottom views of the rotating disc electrode setup and the corresponding flow pattern, (b) the influence of the rotation speed on the polarisation curves during oxygen reduction on Pt in 0.5 M H_2SO_4. Figures are adapted from [10.18].

When investigating gas-evolving electrochemical reactions (e.g. OER), the rotating disc electrode helps to remove the bubbles that develop at the electrode surface, which diminish the effective surface area of the electrode. However, even at fast rotation speeds of 10,000 rpm or more, this technique does not always remove nano and micro bubbles, and it is necessary to resort to alternatives, such as a floating electrode arrangement [10.19].

A more advanced but also more complex hydrodynamic technique is the rotating ring disc electrode [10.9]. It detects short-lived or stable species formed during an electrochemical reaction. In this configuration the disc electrode is surrounded by an insulating barrier, which in turn is surrounded by a secondary working electrode (ring). The flow pattern under rotation is such that the products generated at the disc electrode can be detected at the ring electrode. A bipotentiostat is used to control the potential at the disc and ring electrodes. The geometry of the setup (width of insulating gap, width of the ring) is of special importance to ensure accurate measurements.

10.3.6 *The Electrochemically Active Surface Area* [10.20]

To evaluate the intrinsic performance of an electrocatalyst, it is necessary to quantify the electrochemically active surface area (ECSA). The most common method for determining the ECSA is to measure its double layer capacitance (C_{dl}) and divide it by the specific capacitance of a $1\,cm^2$ flat surface (C_s):

$$ECSA = C_{dl}/C_s. \tag{10.17}$$

C_s values are typically in the range 0.01–0.1 $mF{\cdot}cm^{-2}$. C_s for metals can be measured and generally accepted values have been published. However, C_s values for many emerging electrocatalysts (such as metal oxides, sulfides or phosphides) are as yet unknown and it is difficult to determine them. Most electrochemical studies consider 0.040 $mF{\cdot}cm^{-2}$ as a universal C_s value. C_{dl} can be experimentally determined by means of CV (Eq. (10.14)) or by EIS by means of the modified Randles circuit approximation (Fig. 10.6). However, both measurements are reliable only for electrocatalysts with high electrical conductivities. ECSA of metal electrocatalysts can also be determined by measuring the coulombic charge of a specific Faradaic reaction, i.e. by CO stripping or hydrogen underpotential deposition.

10.4 Electrocatalysis for the Production of Sustainable Fuels and Chemicals

10.4.1 *Development of Electrocatalysts*

In view of the crucial environmental challenge of minimising the anthropogenic CO_2 emissions, electrocatalysis offers several possibilities. Solar, wind and hydroelectricity can be used to drive uphill reactions such as water electrolysis and CO_2 electroreduction. Likewise, excess renewable electricity can be stored in chemicals, thus offering ways to address issues related to the intermittent nature of renewables. The development of active, stable and selective electrocatalysts is the key to improving the efficiency of these processes. Minimisation of costs is a key marketing prerequisite; thus, electrocatalysts should offer the much-improved utilisation of noble metals or rely fully on cheap materials [10.2]. To improve activity an attempt should be made to improve the intrinsic activity of a catalyst or to increase its active surface area or both [10.21].

Theoretical investigations have played an important role in rationalising the trends of electrocatalytic performance and consequently in the design of new materials. Based on a solid understanding of the key properties that affect performance, several methodologies have been studied for the development of electrocatalysts with increased intrinsic activity. These are based mainly on tuning the binding strength of reactants or reaction intermediates using advanced nanostructuring techniques, which enable the control of shape or crystal structure, core-shell design, catalyst encapsulation (e.g. confinement), alloying and phase stabilisation by adsorption or intercalation of small molecules or ions.

Similar to thermal catalysis, catalyst nanostructuring and dispersion on highly porous supports (e.g. carbon nanofibers/nanotubes, graphene) are common strategies for increasing the active surface area. To improve exposure of active sites, methods for engineering the catalyst morphology have been proposed. These include the development of three-dimensional mesoporous structures, etching by cathodic corrosion and plasma treatment. Reducing the particle size is also an efficient approach to maximising the number of catalytically active species over a given surface area. To this end, atomically dispersed catalysts have demonstrated an intriguing performance as electrocatalysts [10.22]. Moreover, in dispersed catalysts the support can also have a significant effect on electrocatalysis. Since electrochemical reactions involve the transfer of both mass and charge, catalyst supports should have not only a large surface area, but also electronic conductivity. Furthermore, interactions between metal catalysts and the support can enhance activity by affecting the electronic structure of the supported metal and by inducing spillover phenomena [10.23].

As discussed in Section 10.2, apart from the reactant molecules, ions and electrons also take part in electrochemical reactions at the electrode/electrolyte interface. This has consequences for the design of electrodes. For instance, in a cell with a solid electrolyte (PEM or solid oxide ceramic), reactions take place at so-called "**triple phase boundaries** (TPBs)" where three types of reactants co-exist: ions, electrons and gases. Hence, to improve the efficiency of devices, the length of TPB should be maximised. Figure 10.8 shows two state-of-the-art electrode designs for fuel cells. PEM-type electrode assemblies (Fig. 10.8a) include a highly porous substrate with large electronic conductivity (e.g. carbon cloth), i.e. the gas diffusion layer. Owing to its porosity, it offers unobstructed mass transport, while also enabling electrons to reach the TPBs. The catalyst layer is deposited on the gas diffusion layer; it consists of (i) the catalyst nanoparticles (e.g. Pt), which can be either unsupported or supported on an electron conducting and porous support (e.g. Vulcan carbon), and (ii) a binding agent with H^+ conductivity (e.g. Nafion® ionomer solution), which extends the H^+ pathways in the porous electrode and thus allows H^+ to move to/from the TPB.

Figure 10.8b shows the corresponding situation for a solid oxide fuel cell (SOFC). In this case, an YSZ (yttria stabilised zirconia) film is the

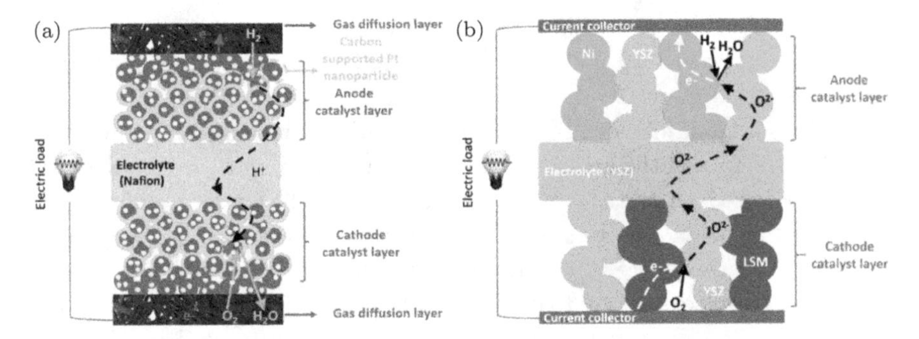

Fig. 10.8 State-of-the-art electrode design of (a) PEM, and (b) solid oxide fuel cells (SOFCs).

state-of-the-art O^{2-} conducting electrolyte. The anode and cathode consist of a porous Ni/YSZ cermet and a $La_{0.8}Sr_{0.2}MnO_3$ (LSM)/YSZ composite respectively, enabling gas transport. Ni and LSM provide the necessary electron conductivity, while YSZ allows O^{2-} to reach the TPB and provides mechanical support for the anode and cathode.

There are benchmarking protocols to measure the activity and stability of catalysts towards the half-reactions related to water electrolysis (HER and OER) [10.24, 10.25]. In general, to assess the electrocatalytic activity of a material, the following parameters are compared:

(i) The Tafel slope, which is an inherent property of the catalyst and indicates the extent of the potential necessary to increase the reaction rate by an order of magnitude. Optimal catalysts should exhibit low Tafel slopes.
(ii) The exchange current density, which is a kinetic parameter and, thus, depends on the actual experimental conditions. An optimal catalyst should exhibit high exchange current density.
(iii) The overpotential required to drive a current density of 10 mA·cm^{-2} (typically noted as η_{10}). The specific current density is chosen because it is the expected current density for an integrated solar water-splitting device under one-sun illumination operating at 10% solar-to-fuels efficiency. The best electrocatalysts should operate at low overpotentials. Reasonable η_{10} values for HER and OER are about 0.1 and 0.35 V, respectively.
(iv) The specific activity (i.e. current density divided by the ECSA), which indicates the intrinsic activity of the catalyst.

(v) The mass activity (i.e. current density divided by the catalyst loading), which is indicative of the intrinsic activity of the catalyst and also quantitatively indicates catalyst utilisation.

(vi) The Faradaic efficiency can be determined experimentally by quantifying the rate of the generated products and dividing it by the expected rate based on the current (i.e. Faraday's law). These measurements are of paramount importance during OER with specific classes of electrocatalysts (e.g. metal phosphides), since the observed current might also correspond to changes in the catalyst oxidation state. In the case of multiproduct reactions (e.g. CO_2 electroreduction) Faradaic efficiency measurements are essential and indicate both the selectivity towards CO_2 electroreduction versus the competing HER and are also linked to the selectivity towards a specific product (e.g. CO, methanol).

(vii) The stability is of paramount importance for practical applications. Ideally, it should be assessed both at a fixed overpotential and under variable power (e.g. $>1{,}000$ CV cycles).

10.4.2 *Hydrogen Evolution Reaction*

The HER is probably the most investigated electrochemical reaction. It is a two-electron reaction, which occurs in two steps. The first, the Volmer reaction, includes the formation of intermediate H_{ads} atoms. The second step may occur either by an electrochemical process (Heyrovsky reaction) or by a chemical process (Tafel reaction). The Tafel step proceeds by a reverse Langmuir–Hinshelwood mechanism, which involves the reaction of two H atoms adsorbed on neighbouring sites. The Heyrovsky step resembles the Eley–Rideal mechanism in heterogeneous catalysis (Section 4.1.2), whereby an adsorbed species reacts with a species that is not accommodated at the surface (here the electrode). HER can proceed via either the Volmer–Tafel or the Volmer–Heyrovsky route. In the former case, electron transfer occurs only in the first step, while in the latter case both steps involve electron transfer. As a result, the potential dependence of the kinetics differs. In practice, Tafel slopes of 120, 40 and 30 $mV{\cdot}dec^{-1}$ indicate that the rate-determining step is the Volmer, Heyrovsky and Tafel reaction respectively.

	Alkaline medium	**Acid medium**
Volmer reaction	$H_2O + e^- \rightarrow H_{ads} + OH^-$	$H^+ + e^- \rightarrow H_{ads}$
Heyrovsky reaction	$H_{ads} + H_2O + e^- \rightarrow H_2 + OH^-$	$H_{ads} + H^+ + e^- \rightarrow H_2$
Tafel reaction	$2\,H_{ads} \rightarrow H_2$	$2\,H_{ads} \rightarrow H_2$

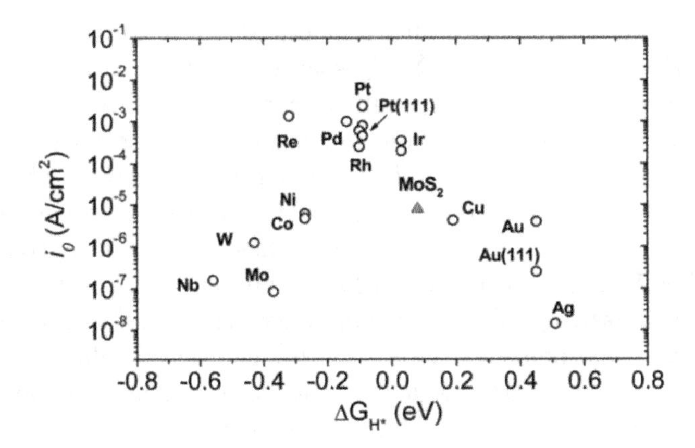

Fig. 10.9 The exchange current density during HER depends on the free energy of hydrogen adsorption at the electrocatalyst and shows a Volcano plot. Note the position of MoS_2 among the metals. Reprinted with permission from [10.26]. Copyright (2007) Science.

As shown in the volcano-type plot of Fig. 10.9, the activity of various metal electrocatalysts in the HER depends on the strength of the metal-hydrogen bond [10.26]. According to Sabatier's principle (cf. Section 6.6), an active HER electrocatalyst should not bind H atoms too strongly or too weakly. Metals on the left of the volcano curve adsorb H atoms too strongly and, thus, the Tafel or Heyrovsky steps are rate-determining. In contrast, metals to the right of the volcano curve bind H atoms but only weakly, in which case the Volmer reaction is the rate-determining step.

HER activity is optimal for the platinum group metals. However, their low availability and concomitant high cost mean that they are unsuitable for large-scale application. Ni is generally employed as HER electrocatalyst for alkaline water electrolysis, mainly because it shows reasonable activity (Fig. 10.9) and it is relatively inexpensive. However, Ni electrocatalysts also suffer from deactivation. Alloying is often employed to eliminate these problems. In acidic environments, as in PEM water electrolysers, the low pH imposes further limitations on the choice of materials, thus constituting Pt as the state-of-the-art HER electrocatalyst.

In the search for alternative HER-active electrocatalysts, which are stable in acidic media, several transition metal sulfides and phosphides seem to be promising [10.2]. Such materials are also well known for their catalytic activity towards hydrodesulfurisation (HDS) (Chapter 8). Indeed, the mechanisms of HER and HDS both relate to the strength of the hydrogen bond with the catalyst. Of the sulfides, MoS_2 is the most promising emerging HER electrocatalyst. MoS_2 is used industrially as an HDS catalyst and its edge sites are the catalytically active sites (Section 8.2.1.1). This is also the case for the HER reaction [10.26]. However, it is a semiconducting material with a marked anisotropic behaviour in conductivity. Thus, certain types of MoS_2 catalysts show limited charge transport properties. In the first decade of the 21st century research has focused on optimising the design of MoS_2-based electrodes, either by improving the exposure of edge sites or by improving electron conductivity. The combination of various electrocatalyst morphologies (nanoparticles, nanosheets, films, composites) and conducting supports has led to the development of catalysts with promising performance. Doping has been explored to increase the intrinsic activity of MoS_2. During doping with Fe, Co or Ni, unsaturated sulfur atoms can form additional S–H bonds and promote HER. Intercalation of lithium ions into the van der Waals gap of MoS_2 can also affect its electronic properties and, thus, alter HER activity. Furthermore, the support can play a key role, and improve the HER activity of MoS_2, as has been demonstrated with doped graphene-supported MoS_2.

Transition metal phosphides (TMPs) of iron, cobalt and nickel form another class of non-precious materials with good HER activity [10.27–10.30]. TMPs have also successfully been used as anodes in Li-ion batteries and as HDS catalysts (Section 8.2.2). Even though none of these phosphides outperforms Pt (Fig. 10.10a), their low cost makes them attractive for practical applications. Composition engineering of TMPs can tailor the electronic structure of TMPs and, thus, affect intrinsic electrocatalytic activity. This can be achieved either by tuning the stoichiometric ratio of M/P or by doping. Improved HER activity has been reported for bimetallic TMPs (e.g. Fe_xCo_yP), TMPs doped with non-metals (N, S), as well as for hybrids (e.g. Ni_2P–CoP). Apart from the composition, catalytic properties are also affected by crystal and electronic structures. Monoclinic NiP_2 is a narrow band gap semiconductor, while cubic NiP_2 is a semimetal. Figure 10.10b shows a substantial difference in HER activity for these two NiP_2 polymorphs and demonstrates that not only the overall TMP composition, but also the bulk arrangement affects electrocatalysis [10.28]. Structural engineering, aimed at increasing the active surface area, has been explored with hollow and nanoarray structures (nanowires, nanosheets, nanorods), which generally show better catalytic performance than planar structures.

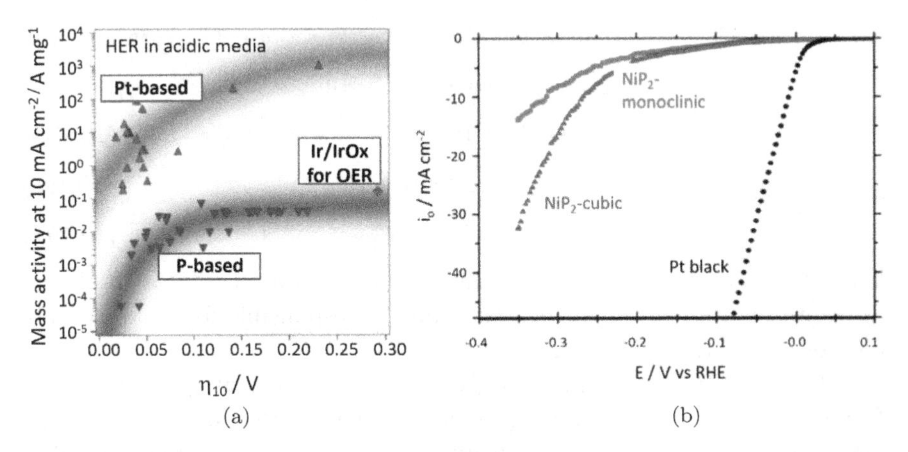

Fig. 10.10 (a) Mass activity of various HER electrocatalysts in acidic media. Reproduced with permission from [10.27]. Copyright (2019) Macmillan Publishers Ltd. (b) HER activity in H_2SO_4 of two different NiP_2 polymorphs and of Pt black. Reproduced with permission from [10.28]. Copyright (2019) the American Chemical Society.

10.4.3 *Oxygen Evolution Reaction*

In the electrolysis of water, oxygen forms at the anode (OER). OER involves the exchange of four electrons and is therefore a complex reaction, which requires a relatively high potential of 1.6–1.8 V. Overall, the OER reaction in acidic and alkaline media, respectively, is represented by the following equations:

$$2\,H_2O \rightarrow 4\,H^+ + 4\,e^- + O_2, \tag{10.18a}$$

$$4\,OH^- \rightarrow 2\,H_2O + O_2 + 4\,e^-. \tag{10.18b}$$

Due to its complexity, the OER is a much more sluggish reaction than the HER. The OER mechanism is not yet fully understood. Various reaction mechanisms have been proposed in recent years [10.31], such as the electrochemical oxide path and the oxide path. In acidic media, most of the mechanisms propose water adsorption on the catalyst as the first step. Depending on the pathway, the following steps can include charge transfer (H^+, e^-), formation or cleavage of chemical bonds and adsorption or desorption of intermediates. The most widespread mechanism consists of at least four elementary steps, all of which involve the transfer of one electron [10.32]. These read as following, where * represents a catalytic site:

$$H_2O + * \rightarrow HO* + H^+ + e^-, \tag{10.19a}$$

$$HO* \rightarrow O* + H^+ + e^-, \tag{10.19b}$$

$$O* + H_2O \rightarrow HOO* + H^+ + e^-, \tag{10.19c}$$

$$HOO* \rightarrow * + O_2 + H^+ + e^-. \tag{10.19d}$$

The elementary step with the highest activation energy is probably associated with O–O bond formation and is responsible for the relatively high overpotential of this reaction.

IrO_2 and RuO_2 are the most active OER electrocatalysts. However, RuO_2 suffers from deactivation (dissolution) in both alkaline and acidic media. Thus, the scarce and expensive IrO_2 is still the state-of-the-art OER catalyst. The development of low-cost and OER-active electrocatalysts, which can withstand the acidic environment of PEM water electrolysis cells, is a great challenge in electrocatalysis.

There is a great deal of interest in oxides with the perovskite structure as OER electrocatalysts; most of the proposed mechanisms indicate that the metal centres of such oxides act as the active sites. However, for perovskites with an increased concentration of oxygen vacancies (i.e. by the incorporation of Sr^{2+}, as in $La_1 - xSr_xCoO_{3-\delta}$), lattice oxygen is also found to participate in the OER and, thus, a second parallel mechanism can operate. In the first decade of the 21st century the theoretical work of Nørskov and co-workers made an important contribution to the understanding of the OER [10.32] by correlating OER activity to the difference in binding energies between the reaction intermediates HOO* and HO* (Fig. 10.11). These findings opened the way for the development of emerging materials with improved intrinsic OER activity (i.e. $LaNiO_3$) as well as for the development of novel electrocatalyst architectures, which enable breaking the scaling relationships (i.e. by controlling the adsorption of HO and HOO on two different sites or by stabilising the HOO intermediates).

Various descriptors have been proposed to govern the dependence of OER activity and electronic structure. The occupancy of the e_g orbitals of 3d transition metals is currently the most recognised descriptor. It depends on the number of d electrons and on the spin state of transition metal ions; OER activity is optimal when the e_g occupancy is close to unity. However, this theory takes into account that the transition metal is the only active site, whereas in highly covalent late-transition-metal perovskites (e.g. $La_1 - xSr_xCoO_{3-\delta}$) both the metal and the oxygen atoms can serve as active

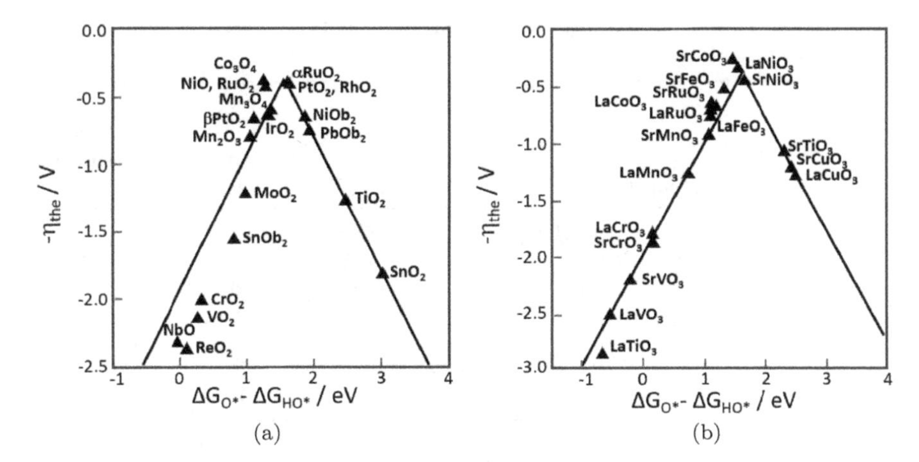

Fig. 10.11 Theoretical dependence of OER activity of (a) metal oxides and (b) perovskites, on the standard free energy of two intermediates. Reproduced with permission from [10.32]. Copyright (2011) Wiley.

sites. Alternative proposed descriptors include the magnetic moment on the conduction plane atoms in perovskites, metal–oxygen covalency, the d-band centre, the p-band centre, the number of valence electrons remaining in the metal atom upon oxidation ("outer" electrons), degree of geometric tilting, charge transfer energy, metal–oxygen hybridisation and unsaturated metal cations. A magnetic model has also been proposed, which correlates the OER activity with the spin polarisation accumulated on the metal/oxygen atoms in the bulk material [10.33].

10.4.4 CO_2 *Electroreduction*

Electrochemical conversion of CO_2 into chemical feedstocks is a promising way to close the carbon loop [10.34–10.36]. Electroreduction of CO_2 can proceed by transfer of 2, 6, 8, 12 or 18 electrons and can lead to a wide variety of gaseous and liquid products of interest in the chemical industry (Table 10.4). As a reduction, CO_2 electroreduction is a cathodic reaction, and in a CO_2 electrolyser it is coupled with the anodic OER, leading to a total cell reaction:

$$x\,\mathrm{CO_2} + y\,\mathrm{H_2O} \rightarrow \mathrm{Product} + z\,\mathrm{O_2}. \tag{10.20}$$

The activity and product selectivities of CO_2 electroreduction depend on many factors, including applied potential, catalyst composition, structural and morphological characteristics of the catalyst, electrolyte composition

Table 10.4 CO_2 electroreduction pathways, their standard potential and relation to the global market.

Reaction	E^o(V)	Number of electrons	Annual global production (MT)
$CO_{2(g)} + 2H^+ + 2e^- \leftrightarrow CO_{(g)} + H_2O_{(l)}$	−0.106	2	150
$CO_{2(g)} + 2H^+ + 2e^- \leftrightarrow HCOOH_{(l)}$	−0.250	2	0.6
$CO_{2(g)} + 6H^+ + 6e^- \leftrightarrow CH_3OH_{(l)} + H_2O_{(l)}$	+0.016	6	110
$CO_{2(g)} + 8H^+ + 8e^- \leftrightarrow CH_{4(g)} + 2H_2O_{(l)}$	+0.169	8	250
$CO_{2(g)} + 12H^+ + 12e^- \leftrightarrow C_2H_{4(g)} + 4H_2O_{(l)}$	+0.064	12	140
$CO_{2(g)} + 12H^+ + 12e^- \leftrightarrow C_2H_5OH_{(g)} + 3H_2O_{(l)}$	+0.084	12	77
$CO_{2(g)} + 18H^+ + 18e^- \leftrightarrow C_3H_7OH_{(g)} + 5H_2O_{(l)}$	+0.095	18	0.2

as well as pH, CO_2 concentration, electrode morphology and design of the electrochemical cell. The latter two parameters are important for ensuring a sufficient supply of CO_2 to the electrode surface. In cells with aqueous electrolytes, CO_2 diffuses through the electrolyte and, thus, such cells are more susceptible to mass transport limitations at high current densities. However, practical applications require operation at high current densities (i.e. > 0.2 A·cm^{-2}). To ensure CO_2 supply at each catalyst site, electrodes based on gas diffusion layers are used.

When CO_2 electroreduction takes place at an electrode, the competing HER might also take place. Detailed product analysis is therefore necessary to determine whether current densities correspond to conversion of CO_2 or to hydrogen production. Figure 10.12 shows the general trends of various catalytic systems to producing specific products. The main conclusion drawn from this figure is that high Faradaic efficiencies can be achieved at the expense of current densities, while Faradaic efficiencies are considerably larger for C1 products.

Beginning with the systematic work of Hori in the 1980s [10.37], it is well-established that Au, Ag and Zn are selective for CO production, Pb and Sn towards formic acid, while Cu is selective for methane and C2 products. Copper is the most studied catalyst for CO_2 electroreduction. The unique ability of Cu to reduce CO_2 by pathways involving more than two electrons has been related to the balance of its CO binding energy and its capacity to protonate adsorbed CO compared to other transition metals [10.35]. A different mechanism has been proposed for formate formation, which occurs on Pb, In, Sn or Bi and is believed to involve the weakly adsorbed $CO_2^{\cdot-}$ radical [10.35].

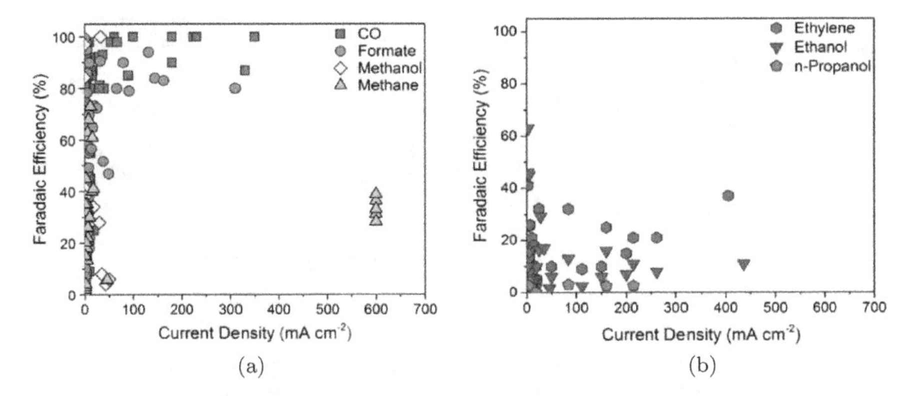

Fig. 10.12 Faradaic efficiency versus current density for CO_2 electroreduction toward (a) C1 and (b) other products. Reproduced with permission from [10.36]. Copyright (2018) the American Chemical Society.

Overall, strategies for catalyst optimisation focus on the modification of surface adsorption properties. As for HER and OER electrocatalysts, nanostructuring is widely applied to preferentially expose specific sites with adsorption properties that differ from the bulk material. In the case of nanodispersed catalysts, metal-support interactions may also play a role, as in thermally driven heterogeneous catalysis. The development of bimetallic catalyst compositions (e.g. Cu-based alloys with d-block metals) is another way to tune adsorption properties. New classes of materials such as MoS_2 and nitrogen-doped carbon materials also function as active catalysts for CO_2 electroreduction.

An alternative technology for the electrochemical reduction of CO_2 is based on CO_2–H_2O co-electrolysis with solid oxide ceramics as the electrolyte [10.2]. The CO produced at the cathode originates not only from CO_2 electroreduction, but also from the catalytic reverse water–gas shift reaction ($CO_2 + H_2 \rightarrow CO + H_2O$), which can also take place at the cathode (Fig. 10.13a). The CO + H_2 mixture (syngas) produced at the cathode can subsequently transform into synthetic fuels by thermal catalysis (e.g. the Fischer–Tropsch process, cf. Section 6.4.1.3).

The majority of studies on high temperature CO_2–H_2O co-electrolysis use Ni–YSZ composite electrodes, similar to those used in solid oxide fuel cells. However, these cathodes have stability issues under the redox conditions that dominate co-electrolysis (Fig. 10.13b). Cathode degradation has been attributed to Ni grain growth and consecutive loss of Ni percolation and loss of the Ni–YSZ contact. Moreover, carbon formation,

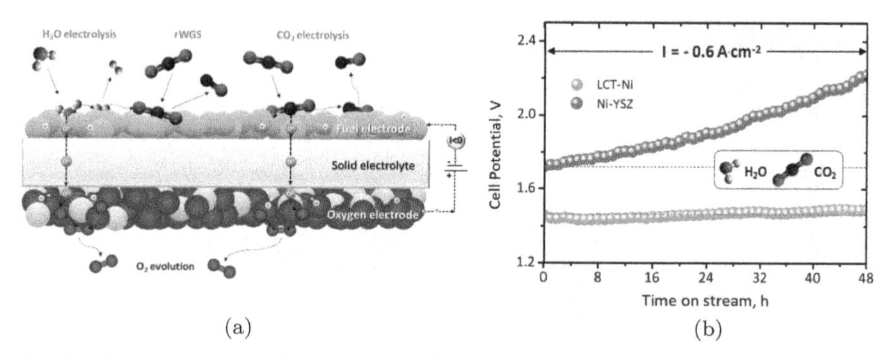

Fig. 10.13 (a) Processes during CO_2–H_2O co-electrolysis in a solid oxide electrochemical cell. (b) Durability study during CO_2–H_2O co-electrolysis at a current density of $0.6\,A{\cdot}cm^{-2}$ on a commercial Ni–YSZ composite and a Ni-exsolved perovskite (Ni-decorated $La_{0.43}Ca_{0.37}Ni_{0.06}Ti_{0.94}O_3$, denoted as LCT–Ni). Reproduced with permission from [10.38]. Copyright (2019) Elsevier.

especially on electrodes with low porosity, is another common source of degradation. Perovskites are the most typical alternative materials employed in this field. Perovskites with mixed ionic-electronic conductivity can accommodate several kinds of defects under redox conditions, allowing them to maintain stability and functionality in redox environments. By controlling the A-site deficiency in ABO_3 perovskites, nanoparticles of transition metals can be exsolved from the backbone of the oxide to the surface in reducing environments. Such perovskites, decorated with transition metal nanoparticles, show significantly improved activity and stability compared to Ni–YSZ composites during CO_2–H_2O co-electrolysis, as shown in Fig. 10.13b [10.38].

10.4.5 *Other Electrochemical Processes*

As explained above, the electrolysis of water as well as of $CO_2 + H_2O$ mixtures is severely limited by the high potential required for the OER and its slow kinetics. Therefore, pairing the cathodic half-reaction with an alternative oxidation reaction of a lower thermodynamic onset potential at the anode can significantly lower the potential demands to drive the overall process.

The concept is schematically presented in Fig. 10.14. In this case, the anodic OER of a conventional water electrolyser is replaced by the methanol electrooxidation reaction. The power required to drive the production of hydrogen is reduced by more than a factor of two, which translates to

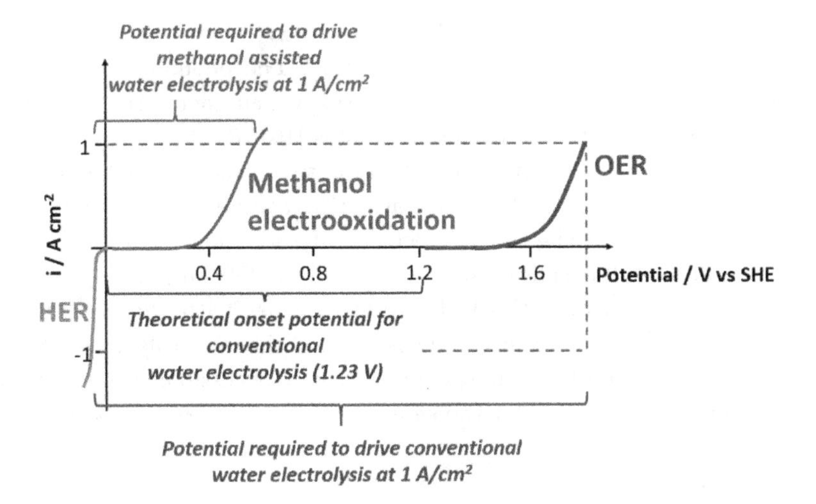

Fig. 10.14 Polarisation curves for HER, OER and methanol electrooxidation, indicating the electrochemical potential windows for conventional and methanol-assisted water electrolysis in acid medium at 25 °C and 1 bar.

significantly reduced operational costs. Moreover, this concept provides flexibility with regard to the choice of the anodic half-reaction, which can now be used for the simultaneous generation of useful chemicals, instead of producing oxygen via the OER. In this process, anodic half-reactions can involve the electrooxidation of various compounds, such as methanol, ethanol, propanol, glycerol, urea, ammonia, methane, carbon and formic acid [10.39–10.42]. From the practical perspective, these organics are widely available from biomass, industrial wastewater (e.g. Fischer–Tropsch wastewater) or natural reserves (e.g. methane which otherwise would be simply flared in the oil industry) or can be the main by-products of the chemical industry (e.g. glycerol is a by-product of industrial biodiesel and soap production). In each of these cases, the development of more active and selective electrocatalysts is the key to improvement.

Solid oxide electrochemical cells can also be used as reactors in gas phase catalytic reactions, where the reaction takes place on the gas-exposed surface of an electrode. The catalytic activity changes dramatically when polarizing the electrode/solid electrolyte interface. Apart from increases in catalytic activity, changes in selectivity have also been observed. Changes are reversible, since after interruption of the polarisation, the rate returns to its initial value. This phenomenon [10.43] is known as the non-Faradaic modification of catalytic activity or electrochemical promotion of catalysis and has been studied in more than 100 catalytic reactions, on a large variety of catalysts, and solid ion conductors. The phenomenon is caused by the formation of an effective double layer, which affects the strength of chemisorption of reactants and intermediates [10.44].

In electrochemical promotion of catalysis, electrical polarisation is employed to affect the rate of a catalytic reaction. In a similar way, polarisation can affect the rate of electrocatalysis by means of the so-called triode operation of fuel cells or batteries [10.45, 10.46]. This concept is based on the introduction of a third electrode together with the conventional anode and cathode. This electrode is run in the electrolytic mode and permits fuel cell operation under previously inaccessible cell potentials. Such designs hold promise for the significant reduction of overpotentials.

References

10.1 A. J. Bard, L. R. Faulkner, *Electrochemical Methods: Fundamentals and Applications*, 2nd Edition, Wiley-VCH, New York, 2000.

10.2 F. M. Sapountzi, J. M. Gracia, C. J. Weststrate, H. O. A. Fredriksson, J. W. Niemantsverdriet, Electrocatalysts for the generation of hydrogen, oxygen and synthesis gas, *Prog. Energy Combust. Sci.* 58, 1–35, 2017.

10.3 J. Tafel, Über die Polarisation bei kathodischer Wasserstoffentwicklung, *Zeit. Physik. Chem.* 50A, 641–712, 1905.

10.4 J. A. V. Butler, Studies in heterogeneous equilibria. Part II. — The kinetic interpretation of the Nernst theory of electromotive force, *Trans. Faraday Soc.* 19, 729–733, 1924; J. A. V. Butler, Studies in heterogeneous equilibria. Part III. A kinetic theory of reversible oxidation potentials at inert electrodes, *Trans. Faraday Soc.* 19, 734–739, 1924.

10.5 T. Erdey-Grúz, M. Volmer, Zur theorie der wasserstoffüberspannung, *Zeit. Phys. Chem.* 150 (A), 203–213, 1930.

10.6 V. G. Mairanovsky, in *Organic Electrochemistry in the USSR in Electrochemistry in a Divided World*, Ed. F. Scholz, 1st Edition, Springer International Publishing AG, Switzerland, 2015.

10.7 W. Grubb, Catalysis, electrocatalysis and hydrocarbon fuels, *Nature* 198, 883–884, 1963.

10.8 G. C. Botte, Electrochemical manufacturing in the chemical industry, *Electrochem. Soc. Interface* 23(3), 49–55, 2014.

10.9 N. Eliaz, E. Gileadi, *Physical Electrochemistry*, 2nd Edition, Wiley-VCH, Weinheim, 2018.

10.10 R. Holze, *Experimental Electrochemistry: A Laboratory Textbook*, 2nd Edition, Wiley-VCH, Weinheim, 2019.

10.11 S. Chen, in *Practical Electrochemical Cells in Handbook of Electrochemistry*, Ed. C. G. Zoski, 1st Edition, Elsevier, Amsterdam, 2007.

10.12 C. Lefrou, P. Fabry, J. C. Poignet, *Electrochemistry: The Basics with Examples*, Springer Verlag, Berlin, 2012.

10.13 A. S. Bandarenka, E. Ventosa, A. Maljusch, J. Masa, W. Schuhmann, Techniques and methodologies in modern electrocatalysis: Evaluation of activity, selectivity and stability of catalytic materials, *Analyst* 13, 1274–1291, 2014.
10.14 N. Elgrishi, K. J. Rountree, B. D. McCarthy, E. S. Rountree, T. T. Eisenhart, J. L. Dempsey, A practical beginner's guide to cyclic voltammetry, *J. Chem. Educ.* 95, 197–206, 2018.
10.15 R. G. Compton, C. E. Banks, *Understanding Voltammetry*, 3rd Edition, World Scientific Publishing, Singapore, 2018.
10.16 E. Barsoukov, J. R. Macdonald, Eds., in *Impedance Spectroscopy: Theory, Experiment and Applications*, 2nd Edition, Wiley-VCH, New Jersey, 2005.
10.17 A. S. Badarenka, Exploring the interfaces between metal electrodes and aqueous electrolytes with electrochemical impedance spectroscopy, *Analyst* 138, 5540–5554, 2013.
10.18 C. Du, Q. Tan, G. Yin, J. Zhang, in *Rotating Disk Electrode Method in Rotating Electrode Methods and Oxygen Reduction Electrocatalysts*, Eds. W. Xing, G. Yin, J. Zhang, Elsevier, Amsterdam, 2014.
10.19 P. Jovanovic, K. Stojanovski, M. Bele, G. Drazic, G. K. Podborsek, L. Suhadolnik, M. Gaberscek, N. Hodnik, Methodology for investigating electrochemical gas evolution reactions: Floating electrode as a means for effective gas bubble removal, *Anal. Chem.* 91, 10353–10356, 2019.
10.20 M. Lukaszewski, M. Soszko, A. Czerwinski, Electrochemical methods of real surface area determination of noble metal electrodes-an overview, *Int. J. Electrochem. Soc.* 11, 4442–4460, 2016.
10.21 Z. W. She, J. Kibsgaard, C. F. Dickens, I. Chorkendorff, J. K. Nørskov, T. F. Jaramillo, Combining theory and experiment in electrocatalysis: Insights into materials design, *Science* 355, 146–158, 2017.
10.22 M. Liu, L. Wang, K. Zhao, S. Shi, Q. Shao, L. Zhang, X. Sun, Y. Zhao, J. Zhang, Atomically dispersed metal catalysts for the oxygen reduction reaction: Synthesis, characterisation, reaction mechanisms and electrochemical energy applications, *Energy Environ. Sci.* 12, 2890–2923, 2019.
10.23 J. Park, D. S. Lee, H. E. Kim, A. Cho, S. Kim, J. W. Han, H. Lee, J. H. Jang, J. Lee, Investigation of the support effect in atomically dispersed Pt on WO_{3-x} for utilisation of Pt in the hydrogen evolution reaction, *Angew. Chem.* 131, 16184–161888, 2019.
10.24 C. C. L. McCrory, S. Jung, J. C. Peters, T. F. Jaramillo, Benchmarking heterogeneous electrocatalysts for the oxygen evolution reaction, *J. Am. Chem. Soc.* 135, 16977–16987, 2013.
10.25 C. C. L. McCrory, S. Jung, I. M. Ferrer, S. M. Chatman, J. C. Peters, T. F. Jaramillo, Benchmarking hydrogen evolving reaction and oxygen evolving reaction electrocatalysts for solar water splitting devices, *J. Am. Chem. Soc.* 137, 4347–4357, 2015.
10.26 T. F. Jaramillo, K. P. Jørgensen, J. Bonde, J. H. Nielsen, S. Horch, I. Chorkendorff, Identification of active edge sites for electrochemical H_2 evolution from MoS_2 nanocatalysts, *Science* 317, 100–102, 2007.

10.27 J. Kibsgaard, I. Chorkendorff, Considerations for the scaling-up of water splitting catalysts, *Nat. Energy* 4, 430–433, 2019.

10.28 B. Owens-Baird, J. Xu, D. Y. Petrovykh, O. Bondarchik, Y. Ziouani, N. Gonzalez-Ballesteros, P. Yox, F. M. Sapountzi, H. Niemantsverdriet, Y. Kolen'ko, K. Kovnir, NiP_2: A story of two divergent polymorphic multifunctional materials, *Chem. Mater.* 31, 3407–3418, 2019.

10.29 Y. Wang, B. Kong, D. Zhao, H. Wang, C. Selomulya, Strategies for developing transition metal phosphides as heterogeneous electrocatalysts for water splitting, *Nano Today* 15, 26–55, 2017.

10.30 F. M. Sapountzi, E. D. Orlova, J. P. S. Sousa, L. M. Salonen, O. I. Lebedev, G. Zafeiropoulos, M. N. Tsampas, J. W. Niemantsverdriet, Y. V. Kolen'ko, *Energy Fuels* 34, 6423–6429, 2020.

10.31 E. Fabbri, T. J. Schmidt, Oxygen evolution reaction-the Enigma, *ACS Catal.* 8, 9765–9774, 2018.

10.32 I. C. Man, H. Y. Su, F. Calle-Vallejo, H. A. Hansen, J. I. Martinez, N. G. Inoglu, J. Kitchin, T. F. Jaramillo, J. K. Nørskov, J. Rossmeisl, Universality in oxygen evolution electrocatalysis on oxide surfaces, *ChemCatChem.* 3, 1159–1165, 2011.

10.33 T. Lim, J. W. Niemantsverdriet, J. Gracia, Layered antiferromagnetic ordering in the most active perovskite catalysts for the oxygen evolution reaction, *ChemCatChem.* 8, 2968–2974, 2016.

10.34 D. Gao, R. M. Aran-Ais, H. Sang Jeon, B. Roldan Cuenya, Rational catalyst and electrolyte design for CO_2 electroreduction towards multicarbon products, *Nat. Catal.* 2, 198–210, 2019.

10.35 G. O. Larrazabal, A. J. Martin, J. Perez-Ramirez, Building blocks for high performance in electrocatalytic CO_2 reduction: Materials, optimisation, strategies and device engineering, *J. Phys. Chem. Lett.* 8, 3933–3944, 2017.

10.36 M. Jouny, W. Luc, F. Jiao, General techno-economic analysis of CO_2 electrolysis systems, *Ind. Eng. Chem. Res.* 57, 2165–2177, 2018.

10.37 Y. Hori, in *Modern aspects of electrochemistry No.42*, Eds. C. G. Vayenas, R. E. White, M. E. Gamboa-Aldeco, Springer Verlag, New York, pp. 89–189, 2008.

10.38 V. Kyriakou, D. Neagu, E. I. Papaioannou, I. S. Metcalfe, M. C. M. van de Sanden, M. N. Tsampas, Co-electrolysis of H_2O and CO_2 on exsolved Ni nanoparticles for efficient syngas generation at controllable H_2/CO ratios, *Appl. Catal. B* 258, 117950–117958, 2019.

10.39 Y. X. Chen, A. Lavacchi, H. A. Miller, M. Bevilacqua, J. Filippi, M. Innocenti, A. Marchionni, W. Oberhauser, L. Wang, F. Vizza, Nanotechnology makes biomass electrolysis more energy efficient than water electrolysis, *Nat. Commun.* 5, 4026, 2014.

10.40 F. M. Sapountzi, M. N. Tsampas, H. O. A. Fredriksson, J. Gracia, J. W. Niemantsverdriet, Hydrogen from electrochemical reforming of C1-C3 alcohols using proton conducting membranes, *Int. J. Hydrogen Energy,* 42, 10762–10774, 2017.

10.41 S. Verma, S. Lu, P. J. A. Kenis, Co-electrolysis of CO_2 and glycerol as a pathway to carbon chemicals with improved technoeconomics due to low electricity consumption, *Nat. Energy* 4, 466–474, 2019.
10.42 H. Ju, S. Badwal, S. Giddey, A comprehensive review of carbon and hydrocarbon assisted water electrolysis for hydrogen production, *Appl. Energy* 231, 502–533, 2018.
10.43 M. Stoukides, C. G. Vayenas, *J. Catal.* 70, 137–146, 1981.
10.44 P. Vernoux, L. Lizarraga, M. N. Tsampas, F. M. Sapountzi, A. De Lucas-Consuegra, J. L. Valverde, S. Souentie, C. G. Vayenas, D. Tsiplakides, S. Balomenou, E. A. Baranova, Ionically conducting ceramics as active catalyst supports, *Chem. Rev.* 113, 8192–8260, 2013.
10.45 S. P. Balomenou, C. G. Vayenas, Triode fuel cells and batteries, *J. Electrochem. Soc.* 151, A1874–A1877, 2004.
10.46 F. M. Sapountzi, S. C. Divane, M. N. Tsampas, C. G. Vayenas, Enhanced performance of CO poisoned proton exchange membrane fuel cells via triode operation, *Electrochim. Acta* 56, 6966–6975, 2011.

Questions

10.1. Electrochemical reactions are either driven by or produce electricity. Which thermodynamic parameter(s) dictate(s) whether electricity will be produced or consumed?

10.2. (a) Describe the operation principles of a PEM fuel cell and a PEM water electrolyser and write the half-reactions that occur at each electrode and the overall cell reaction. (b) Repeat the same for a fuel cell and a water electrolyser with a solid oxide electrolyte of O^{2-} conductivity.

10.3. Explain overpotential and discuss the main overpotential types. How can each type be minimised?

10.4. A PEM fuel cell is supplied with pure hydrogen at the anode and pure oxygen at the cathode. (a) What is the expected cell potential under open-circuit conditions (current is zero)? (b) Give the possible reasons for experimentally measuring lower open-circuit cell potential values than those predicted.

10.5. The current density of a 5 cm^2 PEM water electrolyser at 1.85 V is 60 $mA{\cdot}cm^{-2}$. The electrolyser operates at 95% Faradaic efficiency. (a) What is the rate of hydrogen production, expressed in mol $H_2{\cdot}s^{-1}$? (b) How much is the power input? (c) What is the productivity of the electrolyser, expressed in $kWh{\cdot}Nm^{-3}$ of hydrogen?

10.6. Why should an electrochemical cell with a three-electrode configuration be used for investigating the performance of an electrocatalyst towards a specific half-reaction?

10.7. Two electrocatalysts are investigated for their activity towards HER by means of the three-electrode configuration. Electrocatalyst-1 presented a Tafel slope of $40\,mV{\cdot}dec^{-1}$ and an exchange current density of $0.4\,mA{\cdot}cm^{-2}$, while a current density of $10\,mA{\cdot}cm^{-2}$ was reached at an overpotential of 55 mV. Electrocatalyst-2 presented a Tafel slope of $58\,mV{\cdot}dec^{-1}$, while an exchange current density of $0.2\,mA{\cdot}cm^{-2}$ and a current density of $10\,mA{\cdot}cm^{-2}$ were reached at an overpotential of 112 mV. Which electrocatalyst is more suitable for use as a cathode in a water electrolyser?

10.8. In electrocatalysis, the measured currents are indicative of the reaction rate. To compare different electrocatalytic systems, currents are normalised by the geometric surface area, the mass of the catalyst or the electrochemical active surface area. Discuss the different kind of information that can be obtained using the different ways of normalizing the current.

10.9. What is the difference between the total specific surface area of an electrocatalyst determined by BET and the ECSA?

10.10. Which strategies improve the activity of a given electrocatalyst?

10.11. It is known that MSIs catalysts can greatly affect catalytic performance of nano-dispersed catalysts. MSIs have also been found to play a role in electrocatalysis. Discuss common features and differences between MSIs in catalysis and electrocatalysis.

10.12. Discuss the mechanism of the hydrogen and oxygen evolution reactions. Which are the proposed descriptors of catalytic activity for each of these reactions?

10.13. CO_2 electroreduction is carried out in an electrochemical cell. CO and methanol are the only C-containing products while hydrogen is also produced at the cathode. At a specific current density, the Faradaic efficiency of hydrogen is 10% and that of CO is 45%. What will be the Faradaic efficiency for methanol? What will be the selectivity towards methanol?

10.14. Can an electrochemical system operate with a Faradaic efficiency higher than 100%?

Answers

Chapter 1

1.1 A Mars–van Krevelen mechanism consists of two steps, which can be carried out separately. For instance, in the oxidation of SO_2 to SO_3 it is possible to flow SO_2 over V_2O_5, so that the reaction $SO_2 + V_2O_5 \rightarrow SO_3 + V_2O_4$ takes place. When all the V_2O_5 has converted to V_2O_4, the catalyst is regenerated in a flow of O_2 ($V_2O_4 + 0.5\ O_2 \rightarrow V_2O_5$). Thereafter, the procedure can be repeated. During regeneration SO_3 is not produced, which is inefficient.

One way to ensure continuous production of SO_3 is to use two reactors, one to oxidise SO_2 to SO_3 and the other to regenerate the catalyst. When reactor 1 loses activity and reactor 2 has been regenerated, the gas streams are switched and reactor 2 is used for the oxidation of SO_2 and reactor 1 for regeneration. This is referred to as a swing process.

Another possibility is to feed both SO_2 and O_2 to the same reactor filled with V_2O_5. This works only when the re-oxidation of V_2O_4 is faster than the reduction of V_2O_5, otherwise only some of the catalyst is in the form of V_2O_5 and can produce SO_3.

1.2 In the Mars–van Krevelen mechanism SO_2 takes up an O atom from the catalyst, not from the O_2 molecule. Thus, if unlabelled $S^{16}O_2$ and $V_2{}^{16}O_5$ and labelled $^{18}O_2$ are used, initially the product will be $S^{16}O_3$. Later increasing amounts of $S^{16}O_2{}^{18}O$ form, because the catalyst will be re-oxidised by $^{18}O_2$, and ^{18}O atoms will be deposited on the catalyst surface and can then be taken up by $S^{16}O_2$.

If a Mars–van Krevelen mechanism does not occur and SO_2 reacts with O_2 or with O atoms from dissociated O_2, then $S^{16}O_2^{18}O$ will form at the beginning of the reaction.

1.3 Methanol may react with another methanol molecule by etherification

$$2\ CH_3OH \rightarrow CH_3OCH_3 + H_2O.$$

Methanol may also react with the acetic acid product by esterification

$$CH_3OH + CH_3COOH \rightarrow CH_3COOCH_3 + H_2O.$$

Both reactions take methanol away from the carbonylation reaction ($CH_3OH + CO \rightarrow CH_3COOH$) and, thus lower the efficiency in the use of methanol.

Water leads to a loss of CO in the WGS reaction:

$$H_2O + CO \rightarrow H_2 + CO_2.$$

The H_2 that forms in WGS reaction may lead to loss of methanol and acetic acid:

$$H_2 + CH_3OH \rightarrow CH_4 + H_2O$$

$$H_2 + CH_3COOH \rightarrow CH_3CHO + H_2O.$$

Water can also lead to a loss of the CH_3I co-catalyst:

$$H_2O + CH_3I \rightarrow HI + CH_3OH.$$

A lower CH_3I concentration will give a lower rate of acetic acid formation, while the formation of HI leads to greater corrosion.

1.4 In the reaction $2NCl_3 \rightleftharpoons N_2 + 3Cl_2$, $\Delta n = 2 > 0$. When the number of molecules increases then bonds break. This costs energy and, thus, $\Delta H^0 > 0$.

The equilibrium constant of the reaction is greater than one at room temperature and, thus, $\Delta G^0 = -RT\ln K < 0$. Because $\Delta n > 0$, $\Delta S^0 > 0$ and $-T\Delta S^0 < 0$. Thus, $\Delta H^0 > 0$, $-T\Delta S < 0$ and $\Delta G^0 = \Delta H^0 - T\Delta S^0 < 0$. Then ΔH^0 cannot be very large and positive, because in that case ΔG^0 would become positive.

When $\Delta H^0 > 0$, heat is consumed not created in the reaction; thus, heat formation cannot be the cause of the explosion. The explosion is due to the creation of four gas molecules (with a very large volume) from two solid molecules. This explanation holds for almost all explosives!

1.5 The ratio of the forward and backward reactions does not depend on the presence of a catalyst. Therefore, the Ni/Al_2O_3 methanation catalyst should also work in the reverse steam reforming reaction.

On the other hand, thermodynamics requires different conditions: low T for methanation and high T for steam reforming. The less stable γ-Al_2O_3 support can be used for the Ni methanation catalyst and the more stable α-Al_2O_3 support for the Ni steam reforming catalyst.

1.6 The hydrogenation of N_2 to NH_3 and of CO to CH_3OH are exothermic reactions because the number of molecules decreases in the reactions ($\Delta n = -2$); bonds form and heat is set free.

Another explanation is that N_2 is iso-electronic with CO (same number of valence electrons) and that their thermodynamic properties should be similar. Thus, if the ammonia formation is exothermic, then CO hydrogenation is probably exothermic too.

1.7

$$ZnO + H_2S \rightarrow ZnS + H_2O$$

$$ZnO + 2\,HCl \rightarrow ZnCl_2 + H_2O.$$

1.8 The $Cu/ZnO/Al_2O_3$ catalyst in the second WGS reactor is more active than the Fe_2O_3 catalyst in the first WGS reactor. The WGS reaction is exothermic and lower temperature therefore enables higher conversion and less unreacted CO.

1.9 Fe is a non-noble metal and is easily oxidised by oxygen-containing molecules such as CO and CO_2 ($CO + Fe \rightarrow C + FeO$, $CO_2 + Fe \rightarrow CO + FeO$). Because ammonia synthesis must occur with a metallic catalyst, oxidation of Fe to FeO diminishes the activity of the catalyst.

1.10 The enthalpy change in the methanation of CO is $\Delta H^0 = -206\,kJ \cdot mol^{-1}$ and that of the ammonia synthesis is only $\Delta H^0 = -92.5\,kJ \cdot mol^{-1}$. This means that, at the same temperature, $\Delta G^0 = \Delta H^0 - T\Delta S^0$ will be much more negative in the methanation reaction than in the ammonia synthesis reaction, and the equilibrium constant $K = \exp(-\Delta G^0/RT)$ will be much higher. For instance, because K(methanation)/K(ammonia) $\sim 10^8$ at 450 °C, a much higher pressure (25 MPa) is required to obtain reasonable conversion of $N_2 + 3\,H_2$ than of synthesis gas. The search for a good ammonia synthesis catalyst was performed by means of the reverse ammonia decomposition reaction, while the best catalyst for the methanation reaction can be explored with the forward reaction.

Chapter 2

2.1 1. Bauxite is extracted with a strong base (NaOH) to form $NaAl(OH)_4$. Fe, Si and Ti impurities stay behind in the solid.
2. The $NaAl(OH)_4$ solution is seeded with $Al(OH)_3$ gibbsite crystals and $Al(OH)_3$ precipitates.
3. In a third step the $Al(OH)_3$ is dehydrated to boehmite, AlO(OH), at a pH around 6.
4. Boehmite is heated to γ-Al_2O_3 at 500–600 °C.

2.2 Boehmite and diaspore, with formula AlO(OH), are formed by dehydration of bayerite and gibbsite, with formula $Al(OH)_3$. Dehydration costs energy ($\Delta H^0 > 0$) and the evaporating water makes $\Delta S^0 > 0$. With $\Delta G^0 = \Delta H^0 - T\Delta S^0$, the fully hydroxylated structures are stable at low T because then the effect of $\Delta H^0 > 0$ prevails. On the other hand, at high T the term $T\Delta S^0$ dominates and stabilises the dehydrated structures.

2.3 The oxygen anions in bayerite and boehmite are hexagonally close-packed (ABC stacking), while those of diaspore (which has the same AlO(OH) stoichiometry as boehmite) are hexagonally close-packed (AB stacking). Therefore, it is easier for bayerite to transform into boehmite than into diaspore, even though diaspore is slightly more stable than boehmite. α-Al_2O_3 has the same AB stacking as diaspore and can, therefore, form directly upon heating diaspore. On the other hand, the reaction of γ-Al_2O_3 and η-Al_2O_3 to α-Al_2O_3 occurs by a rearrangement of the oxygen lattice and, thus, at high temperature (Fig. 2.13).

2.4 Fe^{3+} and Al^{3+} cations have the same charge and a similar radius [$r(Al^{3+}) = 0.057$ nm and $r(Fe^{3+}) = 0.067$ nm]. Thus, the lattice energy of iron hydroxides, oxides and oxyhydroxides is about the same as the energy of aluminum hydroxides, oxides and oxyhydroxides, respectively, and the structures of iron and aluminum hydroxides, oxides and oxyhydroxides are usually the same.

2.5 Magnetite (Fe_3O_4) is an inverse spinel, $[Fe^{3+}][Fe^{2+}, Fe^{3+}]O_4$, with one Fe^{3+} cation in tetrahedral position and the other Fe^{3+} cation and the Fe^{2+} cation in octahedral positions. The spinel and the inverse spinel structures have one M^{2+} and two M^{3+} cations. Iron cations can have a charge of 2+ or 3+, but the Al cation must have a 3+ charge. Thus, Al_3O_4 does not exist.

2.6 A pore volume of $0.5\,ml{\cdot}g^{-1}$ means that 0.25 ml solution fills the pores of 0.5 g Al_2O_3. To obtain a loading of 1 wt.% Pt, $0.01 \times 0.5/195 =$

2.56×10^{-5} at Pt must be put on 0.5 g Al_2O_3. The Pt concentration in the 0.25 ml solution thus is $2.56 \times 10^{-5}/0.25 \times 10^{-3} = 0.128$ at Pt·l^{-1}.

2.7 One kg of a 1 wt.% Pt/Al_2O_3 catalyst with 100% Pt dispersion contains $(1000 \times 0.01/195) \times 6 \times 10^{23} = 3.1 \times 10^{22}$ accessible Pt atoms. They have a surface area of $3.1 \times 10^{22} \times 6 \times 10^{-20} = 1{,}850\,\text{m}^2$.

2.8 When zeolites are contacted with D_2 gas the OH groups become OD groups. Assuming that the O–H and O–D force constants are equal, the frequencies of the OH and OD vibrations are inversely proportional to the square root of the reduced mass $1/m = 1/m_1 + 1/m_2$, with m_1 and m_2 being the atomic weights of the vibrating atoms. The ratio of the OH/OD frequencies then becomes $\sqrt{(1/1 + 1/16)/(1/2 + 1/16)} = \sqrt{17/9} = 1.37$, which agrees quite well with the observed ratio of $3750/2760 = 1.36$.

2.9 NMR and IR intensities are proportional to the population difference between the lower and upper energy levels between which the energy transition takes place: Intensity $\sim [1 - \exp(-E/kT)]$. For $\nu(\text{NMR}) = 400\,\text{MHz} = 4 \times 10^8\,\text{s}^{-1}$, $E = h\nu = 6.6 \times 10^{-34} \times 4 \times 10^8 = 26.4 \times 10^{-26}$ J. At room temperature, $kT = 1.4 \times 10^{-23} \times 293 = 4.1{\cdot}10^{-21}$ J. This means that for NMR at 400 MHz, $E/kT = 26.4 \times 10^{-26}/4.1 \times 10^{-21} = 6.4 \times 10^{-5} \ll 1$ and that the intensity $\sim$ [1 – exp(–E/kT)] $= E/kT = 6.4 \times 10^{-5}$.

IR radiation of $3{,}000\,\text{cm}^{-1}$ wavenumber has a wavelength of 1/3,000 cm, a frequency $\nu = c/\lambda = 3 \times 10^{10}/3{,}000^{-1} = 9 \times 10^{13}\,\text{s}^{-1}$ and an energy $E = h\nu = 6.6 \times 10^{-34} \times 9 \times 10^{13} = 59.4 \times 10^{-21}$ J. Thus, $E/kT = 59.4 \times 10^{-21}/4.1 \times 10^{-21} = 14.5$ and the IR intensity $\sim [1 - \exp(-E/kT)] \sim 1$.

Therefore, the intensity of IR radiation of $3{,}000\,\text{cm}^{-1}$ is about $1/(6.4{\cdot}10^{-5}) \sim 15{,}600$ times stronger than the intensity of 400 MHz NMR. In general, IR radiation is much more sensitive than NMR radiation.

2.10 The particle size can be calculated from the width of an XRD peak by means of the Scherrer equation $D_{\text{hkl}} = K{\cdot}\lambda/[\cos\theta_{\text{hkl}}{\cdot}(\delta\theta)_{\text{hkl}}]$, in which D_{hkl} is the diameter of the particle perpendicular to the crystal plane (*hkl*), $K = 0.9$, $\lambda = 0.15$ nm for Cu $K\alpha$ radiation, θ is the angle at which the X-ray is reflected from the crystal plane *hkl* and $\delta\theta$ is the width at half height of the XRD peak in radians. Pattern 4 in Fig. 2.8 shows a width at half height of $4° = 4 \times 2\pi/360 = 0.07$ rad for the XRD line of pseudoboehmite at $2\theta = 28°$. Thus, the diameter of the particles is $0{\cdot}9 \times 0.15/[(\cos 14) \times 0.07] = 0.135/[0.97 \times 0.07] = 2.0$ nm.

Chapter 3

3.1 Branched alkanes are smaller than normal alkanes and have less contact with neighbouring molecules in the liquid phase. Because van der Waals forces are additive, they are lower for branched alkanes and the boiling points are also lower.

3.2 The coverage of a surface decreases with decreasing P and strongly decreases with increasing T, because $\sigma \sim P/\sqrt{T} \cdot \exp(Q/RT)$.

3.3 HHe is a stable molecule because it has three electrons, two in the bonding molecular orbital and one in the antibonding molecular orbital between the H 1s and He 1s orbitals. Thus, there is 100% energy gain and 50% energy loss, resulting in a net gain.

3.4 The minimum in the adsorption-energy diagram (Fig. 3.16b) of H_2 on Au is above zero energy. As a consequence, H_2 will not dissociate on Au. A hydrogen atom can form a bond with an Au atom and become located in the metastable energy well shown in Fig. 3.16b. However, if the hydrogen atom has a too high kinetic energy, it escapes from the energy well and combines with another H atom to H_2.

3.5 Three bonding MOs are filled in N_2, three bonding MOs and one anti bonding MO in O_2 and one bonding MO in H_2. Therefore, bonds in these molecules are threefold, double and single, respectively, and the bond energy of N_2, O_2 and H_2 is 945, 497 and 436 $kJ{\cdot}mol^{-1}$, respectively.

3.6 The bond between an alkene and a metal atom is visible in the infrared spectrum as a low frequency M–C vibration. A decrease in the frequency of the C=C vibration in the infrared or Raman corresponds to weakening of the C=C bond.

3.7 One g Pt/Al_2O_3 catalyst with 0.3 wt.% Pt contains with $0.003/195 = 1.54 \times 10^{-5}$ mol Pt. Adsorption of 6 μmol H_2 per g catalyst means that 12×10^{-6} atoms of H adsorb. With a H/Pt = 1 stoichiometry, the Pt dispersion is thus $12 \times 10^{-6}/1.54 \times 10^{-5} = 0.78$ (78%).

Chapter 4

4.1 The rate equation for a bimolecular reaction according to the Eley–Rideal mechanism is

$$r = k\mathrm{N}\theta_\mathrm{A}\mathrm{P}_\mathrm{B} = \mathrm{NkP_B}\frac{\mathrm{K_A P_A}}{1 + \mathrm{K_A P_A}}.$$

The order of the reaction is always 1 in B, between 1 and 0 in A and between 2 and 1 in total. The rate has a maximum for $\theta_A = 1.0$ and $K_A P_A \gg 1$.

The rate equation for a bimolecular reaction according to the Langmuir–Hinshelwood mechanism is

$$r = kN\theta_A N\theta_B = kN^2 \frac{K_A P_A K_B P_B}{(1 + K_A P_A + K_B P_B + K_P P_P)^2}.$$

The order of the reaction is between 1 and –1 in A and in B and between 2 and 0 in total. The rate has a maximum for $\theta_A = \theta_B = 0.5$.

When adsorption of A and B is weak ($K_A P_A \ll 1$ and $K_B P_B \ll 1$), the order in A and in B is 1 in Eley–Rideal as well as in Langmuir–Hinshelwood kinetics. In this case, it is impossible to distinguish between the two mechanisms on the basis of kinetic results alone.

4.2 When product inhibition occurs in a monomolecular reaction, the adsorption of the product B is strong and the adsorption of the reactant A is relatively weak. Thus, $Q_A \ll Q_B$ and Q_A can be neglected in Eq. (4.9):

$$E_{\text{app}} = E_0 - Q_A + Q_B \approx E_0 + Q_B.$$

4.3 In a mixture of acetylene and ethene, acetylene adsorbs much more strongly than ethene and hinders the hydrogenation of ethene. In the hydrogenation of pure ethene, there is no inhibition.

4.4 In first order reactions the molecules react independently of each other. Only their own diffusivity may play a role. In higher order reactions, several molecules react together and the diffusivity of several molecules plays a role and diffusion sooner becomes a problem.

4.5

$$r = \frac{c_0}{\frac{1}{k_{LG} a_{LG}} + \frac{1}{k_{LS} a_{LS}} + \frac{1}{k\eta c_A^n a}} \quad \text{and, thus,}$$

$$\frac{c_0}{r} = \frac{1}{k_{LG} a_{LG}} + \frac{1}{k_{LS} a_{LS}} + \frac{1}{k\eta c_A^n a},$$

where a is the specific surface of the internal catalyst surface and is proportional to the amount of catalyst m. Therefore, $1/r = A + 1/m$ when a_{LG} and a_{LS} are constant. With faster stirring, the gas bubbles decrease in size and their specific surface area a_{LG} increases. As a consequence,

$$A = \frac{1}{c_0 k_{LG} a_{LG}} + \frac{1}{c_0 k_{LS} a_{LS}}$$

decreases and $1/r$ decreases.

4.6 When HCOOD is admitted to Au, it decomposes to adsorbed CO_2 and H and D atoms: HCOOD $\rightarrow$ CO_2+H+D. The H and D atoms combine to H_2, D_2 and HD at a statistical ratio of 1:1:2, and all molecules desorb to the gas phase. Thus, the gas product of HCOOD decomposition is CO_2, H_2, D_2 and HD at a ratio of 4:1:1:2.

Chapter 5

5.1 In XPS and Auger spectroscopy, electrons are ejected from the sample and the kinetic energy is measured. If the electrons were to travel through the gas or liquid phase, then they would collide with molecules, lose energy and the kinetic energy would be lower. For these reasons, XPS and Auger spectroscopy samples are usually measured under vacuum.

5.2 In Auger spectroscopy, an electron is ejected from an inner shell. For MgO, the 1s electron of an O^{2-} ion (with electron configuration $1s^22s^22p^6$) is ejected, giving the $1s^12s^22p^6$ configuration. Thereafter, an electron from a higher shell fills the hole in the 1s shell and another electron from a higher shell is ejected and detected. Thus, in the final state, there are two missing electrons, leaving two holes in the 2s and/or 2p shell. There are then three final states: one state with two holes in the 2s shell, one state with one hole in the 2s shell and one in the 2p shell and one state with two holes in the 2p shell. The corresponding Auger spectral lines have the kinetic energy $2E_{2s} - E_{1s}$, $E_{2s} + E_{2p} - E_{1s}$ and $2E_{2p} - E_{1s}$, respectively.

5.3 In a laboratory XPS instrument, the source of the excitation energy is an X-ray tube with a fixed energy, e.g. 1486 eV for an Al anode and 1256 eV for a Mg anode. Because the binding energy of electrons in orbitals is fixed, the kinetic energy of the ejected electrons is also fixed, $E_{kin} = h\nu - E_{B}$. Only a few elements have electrons in orbitals with a binding energy that gives a kinetic energy of about 50 eV.

To measure all elements at maximum surface sensitivity, it is necessary to vary the energy of the excitation, so that $h\nu - E_B$ becomes about 50 eV for every element. Variable X-ray energy cannot be produced with a standard X-ray tube but can be produced in a synchrotron radiation source, as found, for example, in the USA, Europe, China and Japan.

5.4 Electrons which are ejected by X-rays from atoms in the bulk of a material and interact with the electrons of other atoms in the material, lose energy. As a consequence, they have a lower kinetic energy and, because $h\nu = E_{B} + E_{kin}$, have an apparent higher binding energy E_B

than electrons which do not interact with other electrons. Therefore, XPS peaks of bulk materials have a tail to higher binding energy (Fig. 5.6). Electrons ejected from Au atoms at the surface reach the detector without interaction with atoms and, therefore, a monolayer of Au will have an XPS spectrum with sharp 4f and 4d peaks and no tails to high energy.

5.5 The formal oxidation state of the N atoms in NH_4NO_3 is +5 in the nitrate group and −3 in the ammonium group. In both cases, the actual charge may be closer to zero. Nevertheless, the charge difference between the two N atoms is substantial and, with a linewidth of about 1 eV, there are two distinct XPS peaks.

5.6 The enrichment factor χ is determined by $-kT \ln \chi = \mathrm{m}(E_{\mathrm{BB}} - E_{\mathrm{AA}})/4 + \alpha[(2p+m)x_{\mathrm{b}} - 2px_s - m/2]$. The first term is usually much larger than the second term, so that $\chi = \exp[-m(E_{\mathrm{BB}} - E_{\mathrm{AA}})/4\mathrm{kT}]$, where m is the number of missing neighbours. Surface enrichment increases as the number of missing neighbours (as at kinks and steps) increases.

5.7 Figure 5.14 shows that the sublimation energy of Pd is higher than that of Ag. Thermodynamics, therefore, predicts that the surface of a PdAg alloy is enriched in Ag. A surface layer of Pd on top of the Ag is thermodynamically unstable (Eq. (5.2)) and can exist only when the diffusion of the atoms is slow, as is the case at low temperature.

5.8 The sublimation energy of In is lower than that of Sn (Fig. 5.14). At low temperature, when enthalpy is more important than entropy, the surface will be enriched in In. At high temperature, entropy is important and the surface will have the average (1:1) composition.

5.9 Zn atoms have 10 3d electrons. If there were only 3d orbitals, the 3d band in solid Zn would be filled and no bonding would occur between the atoms in solid Zn. However, there are 4s orbitals in Zn atoms, and two bands form when solid Zn forms: one 3d band and one 4s band. The bands overlap and the electron configuration of a Zn atom in solid Zn is $(3\mathrm{d})^{10-\mathrm{x}}(4\mathrm{s})^{\mathrm{x}}$, not $3\mathrm{d}^{10}$. Both bands are partially filled, and both contribute to bonding and to metallic behaviour. The 3d band is filled to a large extent and the 4s band is filled to a small extent. Therefore, the bonding in Zn is weak and the melting point is low.

Chapter 6

6.1 There is less steric hindrance in the hydrogenation of short alkyl chains than of long alkyl chains at the metal surface. Therefore, the chance for chain growth (α) is lower for short chains and a relatively large number of short alkanes forms. Ethane is an exception because ethene can be inserted into the growing alkyl chain: $C_i + C_2 \rightarrow C_{i+2}$.

6.2 Metal-to-ethene bonds form when ethene adsorbs on a metal surface, at the expense of the ethene bonds and the bonds of the metal surface. Thus, the C—C bond in adsorbed ethene elongates, and the M—M distance also becomes longer.

6.3 Reactions are either monomolecular or bimolecular. For a trimolecular reaction, three molecules or species would have to be at the same location at the same time, which is highly unlikely. Because H_2 molecules are unreactive, H atoms are necessary in hydrogenation reactions. Since a trimolecular reaction (A + H + H) is unlikely, only $A + H \rightarrow AH$ followed by $AH + H \rightarrow AH_2$ is possible.

6.4 As discussed in Section 4.1.1.1, when adsorption is weak, the apparent activation energy is $E_{\text{app}} = E_0 - Q_{\text{A}}$, where E_0 is the activation energy of the reaction at the catalyst surface and Q_{A} is the heat of adsorption. In the reaction of ethene to adsorbed ethyl, $E_0 = 88\,\text{kJ}\cdot\text{mol}^{-1}$ and $Q_A = 62\,\text{kJ}\cdot\text{mol}^{-1}$, giving $E_{\text{app}} = 26\,\text{kJ}\cdot\text{mol}^{-1}$.

6.5 According to Section 4.1.1.1, at high coverage of ethene, the apparent activation energy of the reaction of ethene to adsorbed ethyl is $E_{\text{app}} = E_0$, equivalent to the actual intrinsic barrier.

6.6 The ratio of the forward and backward reactions of a reversible reaction is a thermodynamic constant and does not depend on a catalyst (Section 1.1). If the rate of the forward reaction were to change from one metal surface to another, then the rate of the reverse reaction would change accordingly. Both reactions would have to proceed in the same way: either structure-sensitive or structure-insensitive. That one reaction is structure-insensitive and its reverse reaction is structure-sensitive, thus, is a contradiction.

6.7 When methanol forms by non-dissociative chemisorption, $^{12}C^{16}O$ forms $^{12}CH_3{}^{16}OH$, and $^{13}C^{18}O$ leads to the formation of $^{13}CH_3{}^{18}OH$. In dissociative chemisorption, $^{12}C^{16}O$ splits into ^{12}C + ^{16}O and $^{13}C^{18}O$ splits into ^{13}C + ^{18}O, and ^{12}C might combine with ^{18}O and ^{13}C with ^{16}O during the formation of methanol. In the latter case, four methanol molecules form: $^{12}CH_3{}^{16}OH$, $^{12}CH_3{}^{18}OH$, $^{13}CH_3{}^{16}OH$ and $^{13}CH_3{}^{18}OH$. A double labelling experiment distinguishes between dissociative and non-dissociative chemisorption.

6.8 An experiment, in which a mixture of CO and CO_2 is hydrogenated, shows whether CO or CO_2 is the source of methanol when one of the two molecules is labelled, for instance ^{13}CO and $^{12}CO_2$. Double labelling is not necessary.

6.9 To produce C_nH_{2n+2} alkanes by the FT reaction, a H_2/CO ratio slightly larger than 2 is necessary: $n\,CO + (2n+1)\,H_2 \rightarrow C_nH_{2n+2} + n\,H_2O$. However, Fe is not only a good FT catalyst but also a reasonable WGS catalyst. The H_2O that is produced as by-product in the FT reaction can react with CO to CO_2 and the overall reaction becomes $2n\,CO + (n+1)\,H_2 \rightarrow C_nH_{2n+2} + n\,CO_2$. In that case, a H_2/CO ratio slightly larger than 0.5 is sufficient.

6.10 The FT product is the result of a polymerisation-type mechanism. As a consequence, all alkane, alkene and alcohol molecules are linear and have a low octane number and a high cetane number. The FT product is unsuitable for gasoline, but advantageous in diesel.

6.11 According to Section 5.3, the element with the lower sublimation energy will take up the energetically unfavourable kink and step positions. Figure 5.14 shows that Au has a much lower sublimation energy than Ru; thus, Au atoms will cover the defect sites.

6.12 It is believed that the mechanism of the shift reaction is $Cu + H_2O \rightarrow CuO + H_2$ followed by $CuO + CO \rightarrow Cu + CO_2$ and that it is an example of the Mars–van Krevelen mechanism.

6.13 The argon in the purge gas in NH_3 synthesis comes from the air (air contains 0.9% argon) that is used in the second autothermic steam-reforming reactor.

6.14 Urea is produced from CO_2 and NH_3 in two steps:

$$2\,NH_3 + CO_2 \rightarrow NH_4{-}O{-}C(O){-}NH_2 \rightarrow H_2N{-}C(O){-}NH_2 + H_2O.$$

Urea is one of the most important fertilisers.

Chapter 7

7.1 1. First NaY is made from sodium aluminate and sodium silicate at a Si/Al ratio of 2 to 3 by hydrothermal synthesis in an autoclave above 100 °C: $NaAlO_2 + Na_2SiO_3 + NaOH + H_2O \rightarrow Na_x(AlO_2)_x(SiO_2)_y(H_2O)_z$. An organic structure-directing agent is not required in the synthesis of NaY.

2. Then the Na^+ cations must be exchanged for protons. Because of the lability of NaY in acid solution, this cannot be done by treatment with an acid (e.g. HCl). The exchange has to be done in two steps. First, the Na^+ cations are replaced by NH_4^+ cations

by treatment of NaY with a solution of NH_4Cl (NaY + $NH_4^+ \rightarrow NH_4Y + Na^+$). To remove all Na^+ cations this exchange is repeated two times.

3. The product is filtered and dried and then ammonia is removed by calcination, giving HY ($NH_4Y \rightarrow HY + NH_3$).

7.2 Scrambling of the carbon atoms of pentane as well as isomerisation of pentane to isopentane can take place on an acid catalyst at the same fast rate, because both reactions go through a cyclopropane intermediate.

7.3 The activation energy of the isomerisation of the tertiary carbenium ion of 2,3-dimethylbutane to the secondary carbenium ion of 2,2-dimethylbutane is 56 kJ·mol^{-1}, 42 kJ·mol^{-1} for the difference between the tertiary and secondary cation plus 14 kJ·mol^{-1} for the 1,2-methyl shift.

7.4 The isomerisation of 1-^{13}C-but-1-ene to cis and trans 1-^{13}C-but-2-ene proceeds by protonation of 1-^{13}C-but-1-ene to the $^{13}CH_3$–C^+H—CH_2—CH_3 carbocation and deprotonation to cis or trans 1-^{13}C-but-2-ene. It appears that this process requires an activation energy of 41 kJ·mol^{-1}. The shift of the ^{13}C label from the carbon atom C1 to the carbon atom C2 proceeds by the same $^{13}CH_3$–C^+H—CH_2—CH_3 carbocation and then, by means of a cyclopropyl mechanism, from a secondary to a secondary cation. The activation energy therefore should be 32 + 14 = 46 kJ·mol^{-1} in addition to the 41 kJ·mol^{-1}, thus being 46 + 41 = 87 kJ·mol^{-1}.

7.5 When isobutene forms in a bimolecular reaction, two butene molecules react to an octane carbenium ion, which isomerises to the iso-octane cation, which cracks to two isobutene molecules. During these reactions fast scrambling reactions also take place, with an "exchange" of carbon atoms. As a result, the ^{13}C atoms are distributed all over the C_8 molecule, thus, all over the isobutene molecules, and a 1:2:1 mixture of isobutene molecules with two, one and zero ^{13}C atoms will form. On the other hand, all the isobutene molecules will have one ^{13}C atom when the reaction of butene to isobutene is monomolecular.

7.6 The CH_3—$^{13}CH_2$—CH_2—C^+H—CH_3 carbocation can undergo a 1,2 H shift to the CH_3—$^{13}CH_2$—C^+H—CH_2—CH_3 carbocation, which reacts to the $^{13}CH_3$—CH_2—C^+H—CH_2—CH_3 carbocation by a cyclopropane mechanism. A 1,2 shift results in the $^{13}CH_3$—C^+H—CH_2—CH_2—CH_3 carbocation, which isomerises to the $^{13}CH_3$—$C^+(CH_3)$vCH_2—CH_3

and $^{13}CH_3-CH_2-C^+(CH_3)-CH_3$ tertiary carbenium ions by a cyclopropane mechanism followed by a 1,2 H shift.

7.7 In industry, pentane and hexane, but not heptane, are isomerised. The cracking side reaction is unlikely to occur for pentane and hexane but is much more likely for heptane. Cracking of pentane and hexane is unlikely because the cracking of C_5 cations and of most C_6 cations leads to very unstable primary carbocations. Only the secondary $CH_3-CH(CH_3)-CH_2-C^+H-CH_3$ cation would crack to a secondary cation, but the equilibrium between $CH_3-CH(CH_3)-CH_2-C^+H-CH_3$ and $CH_3-C^+(CH_3)-CH_2-CH_2-CH_3$ is strongly on the side of the latter. Cracking of several C_7 cations, on the other hand, is more likely to occur, because C_7 cations can go from secondary to tertiary or from tertiary to secondary cations, which leads to a loss of isomerisation selectivity.

7.8 Ethene converts to ethanol over an acid catalyst. Propene forms isopropanol (2-propanol) and butene forms 2-butanol; in both cases the alkene reacts with a proton to a secondary carbenium ion, which reacts with water to the secondary and not the primary alcohol.

7.9 Ethylbenzene cannot isomerise to xylene, but it can be hydrogenated to ethylcyclohexene on a metal catalyst and then isomerised on an acid catalyst to dimethylcyclohexene by a six-ring contraction and ring expansion (as does cyclohexane to methylcyclopentane). Dehydrogenation of dimethylcyclohexene gives xylene.

7.10 Acid-catalysed isomerisation of xylene occurs by a methyl shift. For toluene the methyl shift is detectable only when one of the aromatic carbon atoms is labelled, revealing a change in the relative position of the methyl C and labelled aromatic C atom.

7.11 When an alkylcarbenium ion breaks by β scission an alkene and another alkylcarbenium ion are formed:

$$CH_3-{}^+CH-CH_2-CH(CH_3)_2 \rightarrow CH_3-CH{=}CH_2 + CH_3-C^+H-CH_3.$$

If an alkylcarbenium ion breaks by α scission a carbene and another alkylcarbenium ion are formed:

$$CH_3-C^+H-CH_2-CH(CH_3)_2 \rightarrow CH_3-CH(\cdot\cdot) + {}^+CH_2-CH(CH_3)_2.$$

The carbene $CH_3—CH(\cdot\cdot)$ is neutral, because it carries two electrons on the CH carbon atom. A carbene is a very highly energetic species and that makes its formation very unlikely. This can also be seen from the breaking of a β or α bond in the ethyl carbenium ion:

$$CH_3{-}^{+}CH_2 \rightarrow H^+ + CH_2{=}CH_2,$$

and

$$CH_3{-}^{+}CH_2 \rightarrow CH_2{-}CH(\cdot\cdot) + H^+.$$

7.12 The reverse β scission reaction between an alkylcarbenium ion and an alkene produces a secondary carbenium ion:

$$R^+ + CH_2{=}CH{-}CH_3 \rightarrow R{-}CH_2{-}^{+}CH{-}CH_3.$$

However, with ethene a primary carbenium ion would form and this has a high energy

$$R^+ + CH_2{=}CH_2 \rightarrow R{-}CH_2{-}^{+}CH_2.$$

Therefore, cationic polymerisation of propene is possible, but not of ethene.

7.13 ZSM-5 has a stronger acidity than zeolite Y and, thus, leads to more cracking of the hydrocarbons. As shown in Section 7.2.2 (Fig. 7.24), this leads to more isobutene and propene.

Chapter 8

8.1 When a MoS_2 crystal is cleaved perpendicular to the basal plane through the S atoms of the $(10\bar{1}0)$ planes and all S atoms in the cleavage plane go to one of the two resulting crystal fragments, two types of MoS_2 edges are created (Fig. 8.2). In one type of edge the Mo atoms are sitting on top of four S atoms in a square planar fashion (Fig. 8.2, left, Mo edge, 0% S); this is the Mo edge. In the other edge the Mo atoms are coordinated by six S atoms in trigonal pyramidal fashion (Fig. 8.2, left, S edge, 100% S); that edge is called the S edge.

The Co atoms in Co-promoted MoS_2 are located on the S edge, tetrahedrally coordinated by S atoms with 50% sulfur coverage.

8.2 As explained in [8.10], an STM instrument measures the current that flows between a sharp tip and the surface below it when a voltage is applied between the tip and the surface. Electrons tunnel through the vacuum between tip and surface (Fig. 8.13). When the tip is moved across the sample in the x-y plane, changes in the surface height and in the density of states change the current. These changes can be mapped in images. STM can routinely image individual atoms at surfaces in ultra-high vacuum as well as in air, gas, water and other liquids at ambient pressure and at temperatures from near 0 to 500–600 K.

8.3 Mo—M and Ni—M compounds with M = P are good hydrotreating catalysts because they have metallic properties, which makes them good hydrogenation and hydrogenolysis catalysts. Other metallic Mo–M and Ni–M compounds might also be good hydrotreating catalysts. If they are too strong in hydrogenolysis they may not only remove heteroatoms such as S and N but also break too many C–C bonds. In that case less valuable gaseous hydrocarbons would be formed.

8.4 In a hydrogenolysis reaction ($R{-}CH_2{-}CH_2{-}SH + H_2 \rightarrow R{-}CH_2{-}CH_3 + H_2S$) hydrogen is consumed, but not in an elimination reaction ($R{-}CH_2{-}CH_2{-}SH \rightarrow R{-}CH{=}CH_2 + H_2S$).

8.5 Benzothiophene is similar to indole (Section 8.3.2) with a six-ring and a five-ring, which share two C atoms, and with the S atom in the five-ring. The main product of the HDS of benzothiophene is ethylbenzene. It is formed by hydrogenation of the double bond in the five-ring of benzothiophene, followed by hydrogenolysis or elimination to 2-ethylthiophenol. To remove the S atom from this molecule, the benzene ring does not have to be hydrogenated, because thiophenol gives mainly benzene.

The first step in the HDN of indole is also hydrogenation of the double bond in the five-ring. Then hydrogenolysis or elimination gives 2-ethylaniline. The main reaction to remove the N atom from aniline is hydrogenation to cyclohexylamine, followed by elimination of NH_3. Therefore, the main product of the HDN of indole is ethyl-cyclohexane.

8.6 HDN requires the hydrogenation of six rings, while HDS does not need that, as seen in the Answer to Question 8.4. Therefore, a higher H_2 pressure is needed in HDN than in HDS.

8.7 Organic esters and acids can undergo HDO by CO or CO_2 removal. It is difficult to distinguish between these mechanisms because even if CO is formed, CO_2 can be quickly formed as well by the WGS reaction ($CO + H_2O \rightarrow CO_2 + H_2$).

8.8 In the HDS of naphtha one wants to remove S atoms by hydrodesulfurisation but avoid hydrogenation of alkenes, because their high octane numbers make them valuable components in gasoline. The loss in octane number can be minimalised by utilising the fact that the alkenes are mainly concentrated in the low-boiling fraction of the

FCC naphtha, whereas most of the sulfur compounds are present in the high-boiling fraction. The FCC naphtha is distilled before HDS and each fraction is desulfurised under different conditions. The mild catalyst that is used to treat the light fraction preserves the alkenes. The heavy fraction contains a relatively low concentration of alkenes but most of the sulfur impurities, including benzothiophene and alkylbenzothiophenes. These compounds require more severe conditions but, because of the small amount of alkenes in this heavy fraction, the overall loss in alkenes is small.

8.9 Figure 8.5 shows that Ni_2P contains two types of Ni atoms. One Ni atom is coordinated by four P atoms and the other by five P atoms.

8.10 The feed in the FT process is synthesis gas, which must be sulfur-free, because Fe and Co catalysts are sulfur-sensitive. The feed in the MTH process is methanol. Methanol is produced over a Cu catalyst, which is also sulfur-sensitive. Because the feeds in the FT and MTH processes are sulfur-free, the products, too, will be sulfur-free and does not have to be desulfurised.

Chapter 9

9.1

$$O_2 + 2* \rightarrow 2O*$$

$$CO + O* \rightarrow CO_2 + *.$$

CO reacts directly upon collision with chemisorbed oxygen atoms. Thus, the surface is not covered by CO.

In an ER mechanism only one of the two reactants is chemisorbed, and a switch from one chemisorbed molecule to another, therefore, is impossible. Thus, surface oscillations are impossible.

9.2 A Mars–van Krevelen reaction can take place in one as well in two reactors. In one reactor, the feed contains the reactant as well as the oxidant, while a swing process takes place in a two-reactor system. The reactant is fed to reactor one and is oxidised by the catalyst, while oxygen or air is fed to the second reactor to re-oxidise the reduced catalyst. With time, the process is reversed and oxygen or air is fed to reactor one and reactant to reactor two.

9.3 By labelling the oxygen gas or the oxygen atoms of the metal oxide catalyst, it is possible to determine whether the oxygen atom from the oxygen gas or the oxygen atom from the catalyst is taken up by the reactant. If a Mars–van Krevelen mechanism applies, then the product takes up the oxygen atom from the catalyst. However, the labelling experiment is valid only under the initial conditions, because, after some time, the surface of the catalyst is covered by labelled as well as unlabelled oxygen atoms.

9.4 (a) According to Le Chatelier, the conversion of SO_2 to SO_3 increases when the concentration of one of the two reactants is higher than demanded by stoichiometry (SO_2:$O_2 = 2$). When one of the reactants is in high excess, the yield decreases.

(b) The exothermic oxidation of SO_2 is negatively influenced by an increase in temperature.

(c) The number of molecules decreases in the reaction $2\,SO_2 + O_2 \rightarrow 2\,SO_3$, indicating that pressure has a positive influence on conversion.

(d) A catalyst can never have an effect on the equilibrium conversion, but it does improve the rate of the reaction.

9.5 The reaction between a surface amide species and gas-phase NO is highly unlikely, because it would be an Eley–Rideal reaction. The statistical chance for a collision between a gas-phase molecule and a molecule on the surface is very low.

9.6 WO_3 increases the acidity of the V_2O_5/TiO_2 SCR catalyst because the Brønsted acidity of a W^{6+}–OH group is higher than that of V^{5+}–OH and Ti^{4+}–OH groups. As the charge on the metal cation increases, the repulsion of the H^+ becomes stronger and the strength of the acid increases.

9.7 When oxygen gas is labelled in the Wacker reaction of ethene with oxygen, unlabelled acetaldehyde is produced. This indicates that the oxygen atom in acetaldehyde does not originate from the oxygen gas.

9.8 At one time, butanol was produced by a $C_2 + C_2$ strategy. Aldol condensation of acetaldehyde gave crotonaldehyde, which was hydrogenated to butanol:

$$CH_3CHO + CH_3CHO \rightarrow CH_3\text{—}CHOH\text{—}CH_2\text{—}CHO$$

$$\rightarrow CH_3\text{—}CH\text{=}CH\text{—}CHO + H_2O$$

$$CH_3\text{—}CH\text{=}CH\text{—}CHO + 2\,H_2 \rightarrow CH_3\text{—}CH_2\text{—}CH_2\text{—}CH_2OH.$$

New processes for the production of butanol are based on the $C_3 + C_1$ method and make use of inexpensive propene and CO in a hydroformylation reaction:

$$CH_3{-}CH{=}CH_2 + CO + H_2 \rightarrow CH_3{-}CH_2{-}CH_2{-}CHO$$

$$CH_3{-}CH_2{-}CH_2{-}CHO + H_2 \rightarrow CH_3{-}CH_2{-}CH_2{-}CH_2OH.$$

9.9 The EO catalyst consists of an α-Al_2O_3 support, a silver catalyst and Cs and Cl promoters. The support is neutral to avoid isomerisation of EO to acetaldehyde and has wide pores, which enable the EO to separate immediately from the catalyst, before it undergoes combustion to CO_2. Silver is the best metal for the formation of EO, with the fewest side reactions. The Cs and Cl promoters increase EO selectivity by decreasing the rate of combustion.

9.10 The EO process runs with excess ethene so as to stay outside the explosion limits and to maintain low conversion and low production of heat. Furthermore, the feed gas contains about 50% methane as diluent gas (which has a high thermal conductivity) to avoid overheating and runaway combustion.

9.11 HOCl is produced *in situ* from Cl_2 and water and Cl_2 chlorinates propene:

$$CH_3{-}CH{=}CH_2 + Cl_2 \rightarrow CH_3{-}CHCl{-}CH_2Cl.$$

Due to the HCl, chloropropanol forms an ether:

$$2\,CH_3{-}CH(OH){-}CH_2Cl \rightarrow (CH_3)(CH_2Cl)CH{-}O{-}HC(CH_2Cl)(CH_3) + H_2O.$$

The Cl atom in chloropropanol is hydrolysed by water:

$$CH_3{-}CH(OH){-}CH_2Cl + H_2O \rightarrow CH_3{-}CH(OH){-}CH_2OH + HCl.$$

9.12 Both benzene and butane can be oxidised to maleic anhydride, but in the oxidation of benzene two carbon atoms are lost to CO_2. For the same reason, the exothermicity in the oxidation of benzene is higher than in the oxidation of butane. Benzene is more expensive than butane.

9.13 The oxidation of butane to maleic anhydride is highly exothermic, and it is very important to achieve efficient removal of heat. A multitubular reactor or, even better, a fluid-bed reactor is suitable for the control of heat production.

Chapter 10

10.1. The Gibbs free energy of the reaction.

10.2. (a) Electrolyte conducts H^+. Fuel cell, Anode: $H_2 \rightarrow 2\,H^+ + 2\,e^-$, Cathode: $^1/_2\,O_2 + 2\,H^+ + 2\,e^- \rightarrow H_2O$. Water electrolyzer, Anode: $H_2O \rightarrow {}^1/_2\,O_2 + 2\,H^+ + 2\,e^-$, Cathode: $2\,H^+ + 2\,e^- \rightarrow H_2$. (b) Electrolyte conducts O^{2-}. Fuel cell, Anode: $H_2 + O^{2-} \rightarrow H_2O + 2\,e^-$, Cathode: $^1/_2\,O_2 + 2\,e^- \rightarrow O^{2-}$. Water electrolyzer, Anode: $O^{2-} \rightarrow {}^1/_2\,O_2 + 2\,e^-$, Cathode: $H_2O + 2\,e^- \rightarrow H_2 + O^{2-}$.

10.3. Overpotential is the difference between the theoretical and experimentally observed potential to drive an electrochemical reaction at a certain rate. Concentration overpotential is due to slow mass transport and can be minimised by correct electrode design and by tuning operation conditions. Ohmic overpotential is due to low ionic/electronic conductivity in the cell and depends mainly on the electrolyte. Activation overpotential is due to slow kinetics and can be optimised by appropriate electrocatalyst selection and design of electrode/electrolyte interface.

10.4. (a) 1.23 V at standard conditions according to thermodynamics. (b) Lower open-circuit potentials can be due to hydrogen crossover.

10.5. (a) Total current at 1.85 V is 300 mA. Theoretical hydrogen production rate (100% Faradaic efficiency) is 1.55×10^{-6} mol $H_2 \cdot s^{-1}$. Faradaic efficiency is 95%, so the actual rate is 1.48×10^{-6} mol $H_2 \cdot s^{-1}$. (b) Power input $= 1.85\,V \times 0.3\,A = 0.55\,W$ (or $0.11\,W \cdot cm^{-2}$). (c) Rate of H_2 production: 1.48×10^{-6} mol $H_2 \cdot s^{-1} = 0.128\,L \cdot h^{-1} = 1.28 \times 10^{-4}\,Nm^3 \cdot h^{-1}$. Productivity $= 0.55\,W/1.28 \times 10^{-4}\,Nm^3 \cdot h^{-1} = 4.29\,kWh \cdot Nm^{-3}$.

10.6. The three-electrode configuration contains two electrical circuits. The first circuit drives the reaction and the second circuit senses potential changes. Thus, this configuration allows for better control and measurement of the current-potential relationship.

10.7. Electrocatalyst-1 is more active and exhibits a lower Tafel slope and higher exchange current, while lower overpotential is required to drive a current density of 10 $mA \cdot cm^{-2}$.

10.8. Normalisation by the geometric surface area (current density) is meaningful only for the comparison of similar catalytic systems (e.g. gas diffusion electrodes with similar catalyst loading); it cannot determine whether activity differences are due to differences in intrinsic activity or in catalyst surface area. Normalisation per mass of catalyst is of practical interest because it indicates the degree of effective catalyst utilisation and, thus, relates activity to catalyst cost. Normalisation per ECSA indicates the rate per number of available active sites and is, therefore, the most appropriate parameter for comparing the performance of various electrocatalysts.

10.9. The total specific surface area is the catalyst area that is accessible to (gas or liquid) reactants, while ECSA is the area that is accessible to reactants, ions and electrons.

10.10. Use of advanced nanostructuring techniques to preferentially expose active sites (e.g. exposure of edge sites of MoS_2) or to improve exposure of active sites (e.g. atomically dispersed catalysts), dispersion onto highly porous supports to increase active surface area and possibly to enable metal-support interactions.

10.11 In both thermal catalysis and electrocatalysis, the support can enhance the catalytic activity of nanodispersed metal nanoparticles either by affecting the electronic structure of the metal or by inducing spillover phenomena. In thermal catalysis, the support should have large surface area, whereas in electrocatalysis the support should also exhibit electronic conductivity.

10.12. HER can proceed either by the Volmer–Tafel (electron transfer only in the first step) or by the Volmer–Heyrovsky route (electron transfer in both steps). HER activity depends on the strength of the metal catalyst-hydrogen bond. The OER mechanism is not yet fully understood; the most widespread mechanism consists of four elementary steps, all of which involve transfer of one electron. The step with the highest activation energy is assumed to be the one associated with the O–O bond formation. Various descriptors have been proposed to govern OER activity; the e_g occupancy of the transition metal 3d orbitals is currently the most recognised descriptor.

10.13. The observed cell current should correspond to the production of H_2, CO and CH_3OH. Therefore, the Faradaic efficiency of CH_3OH should be $100\% - 10\% - 45\% = 45\%$. During CO_2 electroreduction, two and six electrons are involved in the production of CO and CH_3OH, respectively. Since the Faradaic efficiency of CO is 45%, $r_{CO}/(i/2F) = 0.45$. Similarly, $r_{CH3OH}/(i/6F) = 0.45$. Thus, it seems that $r_{CO} = 3\ r_{CH3OH}$. Taking into account that selectivity towards

CO and CH_3OH are $r_{CO}/(r_{CO}+r_{CH3OH}) = 3/4$ and $r_{CH3OH}/(r_{CO} + r_{CH3OH}) = 1/4$, respectively, and CO selectivity is 75% and CH_3OH selectivity 25%.

10.14. The Faradaic efficiency of pure electrochemical reactions can never exceed 100%. However, when electrocatalysis is used as external activation knob for altering the rate of a gas-phase catalytic reaction (within the frame of the electrochemical promotion of catalysis), the rate is not subjected to Faraday's law and Faradaic efficiency greater than unity is obtained.

Index

B

C

D

I

P

Q

R

S

U

V

W

X

Y

Z

CPSIA information can be obtained
at www.ICGtesting.com
Printed in the USA
JSHW031714080722
27837JS00004B/16